ESSAYS

ON THE

LANGUAGES, LITERATURE, AND RELIGION OF NEPAL AND TIBET

ESSAYS

ON THE

LANGUAGES, LITERATURE, AND RELIGION OF NEPAL AND TIBET:

TOGETHER WITH FURTHER PAPERS ON THE

GEOGRAPHY, ETHNOLOGY, AND COMMERCE OF THOSE COUNTRIES.

B. H. HODGSON, ESQ.

GYAN

Published by

Gyan Publishing House
5, Ansari Road
Daryaganj, New Delhi-110002
Phone: 011-47034999, 9811692060
E-mail: books@gyanbooks.com

Distribution Network
gyanbooks.com
India, USA, Canada, UK, Australia, France

ISBN: 978-81-212-9290-0 (PB)
First Published, 1874

2nd Impression 2023

Printed at: Gyan Press, Delhi.

ESSAYS ON THE LANGUAGES, LITERATURE, AND RELIGION OF NEPAL AND TIBET
Author: B. H. HODGSON

ESSAYS

ON THE

LANGUAGES, LITERATURE, AND RELIGION OF NEPÁL AND TIBET

ESSAYS

ON THE

LANGUAGES, LITERATURE, AND RELIGION OF NEPÁL AND TIBET:

TOGETHER WITH FURTHER PAPERS ON THE

GEOGRAPHY, ETHNOLOGY, AND COMMERCE OF THOSE COUNTRIES.

BY

B. H. HODGSON, ESQ.

HONORARY MEMBER OF THE GERMAN ORIENTAL SOCIETY;
CORRESPONDING MEMBER OF THE FRENCH INSTITUTE; CHEVALIER OF THE LEGION OF HONOUR;
MEMBER OF THE ASIATIC SOCIETIES OF CALCUTTA, LONDON, AND PARIS; OF THE
ETHNOLOGICAL AND ZOOLOGICAL SOCIETIES OF LONDON; AND
LATE BRITISH MINISTER AT THE COURT OF NEPÁL.

Reprinted, with Corrections and Additions, from "Illustrations of the Literature and Religion of the Buddhists," Serampore, 1841; and "Selections from the Records of the Government of Bengal, No. XXVII.," Calcutta, 1857.

LONDON:
TRÜBNER & CO., 57 & 59 LUDGATE HILL.
1874.

NOTICE.

WHEN Professor J. Summers was about to start the *Phœnix,* a monthly magazine for China, Japan, and Eastern Asia, the first number of which appeared in July 1870, he solicited and obtained permission of Mr B. H. Hodgson to reprint in it those contributions of his to the "Journal of the Bengal Asiatic Society," which bear on the ethnology, languages, and religion of Tibet and Nepál. The plan Professor Summers had in view is sketched out in the following editorial note with which the series of reprints is prefaced :—

"The present and following papers (to be given in successive numbers of the *Phœnix*) are from the pen of Mr Brian H. Hodgson, and originally appeared in the Bengal Asiatic Society's Journal, between the years 1828 and 1838. Upon the subject of ethnology, Mr Hodgson's views have since that time been improved and extended, and we purpose, when we have completed the present series of papers, chiefly devoted to Buddhism, to reproduce in the *Phœnix* those improved and extended views of Tibetan and Nepaulese races and languages, from No. 27 of 'Selections from the Records of the Government of Bengal,' wherein they were published in the year 1857. But those 'Selections' form a work even more inaccessible to men of letters in Europe than the 'Journal of the Bengal Society;' and we believe, therefore, that we shall be doing a service to the learned of Europe by making Mr Hodgson's researches into northern Buddhism and ethnology more generally and easily accessible."—*Phœnix,* vol. i. p. 43.

Mr Hodgson's "improved and extended views," so far as Buddhism is concerned, were found embodied in numerous marginal notes in his own copy of the "Illustrations of the Literature and Religion of the Buddhists" (Serampore, 1841). In the same way many manuscript additions were made by him in his own copy of the "Selections." All these corrections and additions have been introduced into the text of the present reprint, though they represent, as is only just to Mr Hodgson to state, various phases of his views, ranging over a period of nearly thirty years.

Professor Summers further proposed to Mr Hodgson to issue these reprints in a collected form as a separate publication, to which proposition the latter gave his ready consent.

At p. 96 of vol. ii. of the *Phœnix* the reprints from the "Selections" commence, and proceed *pari passu* with those from the "Illustrations" to p. 26. of vol. iii., where the last article of the latter (on the Pravrajyá Vrata) terminates. In consequence of this arrangement, the Editor of the present work found it necessary to begin a fresh pagination with the Second Part. References to this part have, therefore, in the index been marked by a II. prefixed to the Arabic figure, showing the page.

Eight pages of the papers on the Commerce of Nepál were remaining to be set up when Professor Summers' acceptance of an appointment in Japan put a stop to the publication of the *Phœnix*, and to the completion of the separate re-issue in accordance with his original design. Under these circumstances, it was thought best to place the materials, as left by Mr Summers on his departure, in the hands of Messrs Trübner & Co., with a view to their eventual publication. Only the above-mentioned article has subsequently been completed.

On comparison with the two former collective publications, the present one will be found to have excluded three short articles contained in the "Illustrations" (IX. Remarks on an Inscription in the Rancha and Tibetan characters; X. Account of a Visit to the Ruins of Simroun; XII. Extract of Proceedings of the Royal Asiatic Society), which were considered as of a sufficiently ephemeral nature to be omitted, and articles IV., V., and XI. 1. 2. of the "Selections" (Route from Káthmándú to Darjeeling; Route of Nepalese Mission to Pekin; Some account of the systems of Law and Police as recognised in the State of Nepál; and on the Law and Legal Practice of Nepál, as regards familiar intercourse between a Hindu and an Outcast). These last-mentioned would in due course have appeared in the *Phœnix*, and have been incorporated in the separate reprint, but for the sudden discontinuance of that magazine. This is more especially to be regretted in the case of the papers on Nepalese Law, which still remain the only trustworthy source of information on that subject. The same may, in fact, be said of most other papers by Mr Hodgson, especially those on the Tribes and Languages of the Northern Non-Aryans adjacent to India, which are scattered over periodicals now scarce and little accessible, and would be well worth preserving in a collected form, inasmuch as on all these questions, both those treated of in the present volume and those bearing on the ethnology and glossology of the Himalayan tribes, he has almost exclusively remained master of a field of research in which he had been the first to break ground.

The foregoing statement will explain the somewhat ungainly form of the present publication, without, however, it is hoped, detracting from its substantial usefulness, as placing within the reach of scholars matter which few of them have means or opportunity to consult in the "Journal of the Bengal Asiatic Society," or in the "Selections from the Records of the Government of Bengal."

Should the present volume be favourably received, the remaining papers of Mr Hodgson will probably be given in another volume or two.

CONTENTS.

PART I.

ON THE LANGUAGES, LITERATURE, AND RELIGION OF NEPÁL AND TIBET.

PART II.

LIST OF ADDITIONS AND CORRECTIONS.

PART I.

Page 3, line 14, at languages, *add* foot note, "see on to pp. 29–36."
,, 8, note "*," misplaced, belongs to the word "Buddha," four lines lower.
,, 15, *for* "Najra" *read* "Vajra."
,, 19, note. *For* "see No. 15" *read* "see on to the Pravrajyá Vrata," p. 139, *et cæt.*
,, 21, *for* "list of Bhotiya books" *read* "list (that of Bhotiya books.")
,, 20, *for* "emigration" *read* "immigration."
,, 22–32, heading of all, *for* "religion of Bhot" *read* "religion of Nepál."
,, 33, heading, *erase* "List of Buddhist works."
,, 34, for the same heading *read* "List of Jathágatas."
,, 23, note, *for* "ought" *read* "sought."
,, 24, line 19, *for* "and" *read* "an."
,, 25, 8 lines from bottom, *for* "meditation" *read* "mediation."
,, 26, 6 lines from bottom, *for* "articular" *read* "particular."
,, 30, line 14, *for* "Dharma" *read* "Dhyáni."
,, 39, 1 line from bottom, *for* "were sent" *read* "sent by me to Royal Asiatic Society."
,, 49, line 12 from bottom, at the word "them," insert the footnote "‖": "This is probably an error. Sakya taught orally; but his immediate disciples (Kasyapa, Ánanda, and Upáli) reduced his doctrines to writing."
,, 52, line 10 from top, *for* "bhikshari" *read* "Khikshari."
,, 60, line 14 from top, *for* "are" *read* "is."
,, 60, line 18, *after* "reduced" *read* "them."
,, 89, line 9 from top, *for* "mortals" *read* "morals."
,, 93, *erase* the whole of the Dwiamnaya and Triamnaya, and *substitute* as follows:—

Dwiamnaya.

Upáya.	Prajná.
Prajná.	Upáya.

The first is theistic; the second, atheistic.

Triamnáya.

Buddha.	Dharma.	Buddha.
Dharma.	Buddha.	Sangha.
Sangha.	Sangha.	Dharma.

The first and third of this series are theistic (diverse); the second is atheistic, Buddha=Upáya, Dharma=Prajná.

,, 98, in note, 4 lines from bottom, *for* "pp. 137–9 of vol. i." *read* "for full list of Sanskrit works, see pp. 36–39 aforegone."
,, 101, note "*." *Add* to note, "The identity in question has since been upheld by Cunningham, Wilson (of Bombay), Chapman (of Madras), and Colonel Yule."

Page 102, at word "published" in last line, *add* footnote "‡," "These drawings have since been presented to the French Institute."
,, 116, in note, line 10 from bottom, *for* "above" *read* "about."
,, 126, at title, *add* as footnote "‡," "From Royal Asiatic Society's Transactions, vol. ii., dated July 11, 1829."
,, 133, at title, *add* footnote "‡," "From 'Oriental Quarterly Magazine,' No. III. A.D. 1827."
,, 139, at title, *add* footnote "§," "From volume on *Buddhism*, printed at Serampore A.D. 1841."
,, 140, line 5 from bottom, *for* "Pravra" *read* "Pravrajyá."
,, 141, note, *for* "Gardhar" *read* "Gandhar."
,, 142, *add* to the note, "See enumeration of all the principal objects of Buddhist worship above given," pp. 93–96.

PART II.

Page 12, line 9, *for* "reach" *read* "reaches."
,, 13, lines 19 and 23, complete the brackets after 4000 and after *et cæt.*
,, 14, line 1, *for* "Lescha" *read* "Lepcha;" line 5, *for* "Kaya" *read* "Vayu;" line 16, *for* "Leschu" *read* "Lepcha;" line 19, *after* "craftsmen," *add*, "of which the names are as follows:—In the mountains. In the valley."
,, 14, line 5 from bottom in note, *for* "Tháruh" *read* "Thárú," and bracket the words, "not own name," and *also* the word "Sallyan." *Add* to note, "Many of the Awalias will be found spoken of in the paper on Náyakot, herein given."
,, 15, at the words "Nepál, J.A.S.B., May 1833," *add* in note "†," given herein, at p. 39.
,, 17, line 13, *for* "viverrula" *read* "viverricula." Last line, *for* "Galophasis" *read* "Gallophasis."
,, 19, line 11, *for* "to" *read* "too."
,, 21, let the words at bottom of diagram run all through.
,, 25, line 14, *for* "plateau" *read* "plateaux."
,, 29, line 6 from bottom, at word "omitted," *add* footnote, "In the 'Bengal Asiatic Journal' for June 1848, may be seen a sample of the Khas tongue."
,, 29, line 3 from bottom, at words "broken tribes," *add* footnote, "See a paper thereon expressly, in the sequel of this work."
,, 30, line 8 from top, *add* footnote "§," "For the tribes EAST of Bhutan, round Assam, and thence down the Indo-Chinese frontier, see papers in the sequel."
,, 31, in note, *for* "4500" *read* "4000."
,, 32, line 9, *after* "Dravidian," *add*, "Mundarian or Hó-Sontal."
,, 32, line 11, at word "dialects," *add*, "See them, as hereto annexed."
,, 33, line 3, at word "weavers," *add* footnote, "See list of them aforegone, at p. 14."
,, 34, *for* "4500–4700" *read* "4000."
,, 39, line 17, *for* "caste" *read* "cast."
,, 40, line 4 from top, *for* "some" *read* "about 100."
,, 46, line 1, *for* "already" *read* "always."
,, 46, line 5 from top, at word "Kusunda," *add* footnote as follows, "+":—"Since accomplished, and the result given hereinafter in the paper on the broken tribes."
,, 46, line 8 from bottom, at word "Haiyu," *erase* note "§," and *substitute* "Haiyu, Hayu, vel Vayu." For more on this tribe, see Treatise hereinafter given on the Vayu and Bahing.

Page 53, the headings, *for* "Tibetan" *read* "Chepang;" and *for* "Shopa" *read* "English."

,, 57, line 4 from bottom, at word "Denwar," *add* in footnote "+," "See paper on broken tribes, before referred to."

,, 60, line 14 from bottom, *for* "dialect" *read* "dialects;" and *add* footnote: "See paper on broken tribes, complete vocabulary of these tongues, and compare 13, 14 supra, Part II."

,, 61, line 14 from top, *for* "overhang" *read* "overhanging."

,, 65, line 7 from top, at word "tongues" *add* footnote "†," "See the former instance here alluded to, in the paper on the Caucasian affinities of the Tibetans as given in the sequel."

,, 65, line 7 from bottom, *for* "Trochu" *read* "Thochu," and last line, *for* "Khor" *read* "Hor."

,, 66, line 15 from top, at word "Kuenlun" *add* footnote, "Is not the Karakorum the western prolongation of the Nyenchhen, and distinct from the Kuenlun, though curving up to it on nearing the Pamer?"

,, 67, line 12 from bottom, at word "Pekin," *add* as footnote, "See this itinerary hereinafter given."

,, 69, line 1, at word "Indochinese," *add* footnote, "The paper on the Indo-Chinese borderers herein."

,, 69, line 20, at word "Caucasus" *add* footnote, "See paper on these affinities in the sequel."

,, 72, in note, *for* "tribunal" *read* "tribe."

,, 76, *add* to second note, "They are given as corrected in the sequel."

,, 85, line 9 from bottom, *erase* the repeated "no end." Line 7, *for* "drawback" *read* "drawbacks."

,, 87, *for* "weed" *read* "weeds." In note, *for* "4500" *read* "4000."

,, 88, three lines from bottom, *for* "an" *read* "any."

,, 89, *before* "timber" *insert* "tea," and *add* the following footnote "†:"—"The growth of tea in the lower region, and its sale in Tibet as well as in the plains, are now affording great and increasing means of profitable employment to settlers."

,, 89, note "‖." *For* "1832" *read* "1831," and *add* at the end of this note: "The trade papers in question are given in the sequel; and observe that the tea trade with Tibet is now adding greatly to our means of successful competition with Russia."

,, 90, note, last line but one, *for* "whp" *read* "why."

,, 92, 4 lines from bottom, at the word "rupees," *add* in footnote: "See note '†,' in next page."

,, 97, line 22 from top, *for* "or Takyeul" *read* "and Takyeul;" and *for* "line of transit" *read* "lines of transit."

,, 98, line 13, *after* "Kothees," *add* "or houses of business firms."

,, 100, line 14 from bottom, *for* "th" *read* "the."

,, 113, line 3 from top, at the word "assertion," *add* note as follows:—"To judge from the statements lately made (1872) by a member of the British Embassy in Nepál, it would seem that the present condition of Nepál's commerce with us, as well as that of ours with her, calls loudly for the attention of our Government."—*Note* of 1873.

PART I.

ON THE LANGUAGES,* LITERATURE, AND RELIGION OF NEPAUL AND TIBET.

WITHIN the mountainous parts of the limits of the modern kingdom of Nepaul, there are thirteen distinct and strongly-marked dialects spoken. These are the Khas or Parbattia, the Magar, the Gurung, the Sunwar, the Kachari, the Haiyu, the Chepang, the Kasunda, the Múrmi, the Newari, the Kiranti, the Limbuan, and the Lapachan. With the exception of the first (which will be presently reverted to) these several tongues are all of Trans-Himalayan stock, and are closely affiliated. They are all extremely rude, owing to the people who speak them having crossed the snows before learning had dawned upon Tibet, and to the physical features of their new home (huge mountain barriers on every hand) having tended to break up and enfeeble the common speech they brought with them.

At present the several tribes or clans to which these dialects are appropriated, can hardly speak intelligibly to each other, and not one of the dialects, save the Newari or language of Nepaul Proper (and the Lapcha, which with the Limbu belongs now to Sikim), can boast a single book, or even a system of letters, original or borrowed. The Newari has, indeed, three systems of letters, of which more will be said in the sequel; and it has also a small stock of books in the shape of translations and comments from and upon the sacred and exotic literature of the Newars. But the Newari tongue has no dictionary or grammar; nor is its cultivation ever thought of by those, numerous as they are, who devote their lives to the sacred literature of Buddhism. It may be remarked, by the way, that the general and enduring effects of this addiction to an exotic medium, in preference to the vernacular, have been, to cut off the bridge leading from speculation to practice, to divorce learning from utility, and to throw a veil of craftful mystery over the originally popular and generous practical Institutes of the religion this people profess.

Before proceeding to a brief comparison of Newari and of the language of Tibet, with a view to indicate the Northern stock of the former tongue, it will be better to notice the Khas or Parbattia Bhasha, since the subject may be dismissed in a few words, and will not need revertence to.

The only language of Southern origin spoken in these Hills is the Khas or Parbattia—an Indian Prakrit, brought into them by colonies from below (twelfth to

---

* For these languages, see on to the Paper at p. 29 of Part II., "On the Aborigines of the Himálaya," with its annexed "Comparative Vocabulary."

fifteenth century of Christ) and now so generally diffused, that, in the provinces West of the Kali river, it has nearly eradicated the vernacular tongues, and, though less prevalent in the provinces East of that river, it has, even in them, as far as the Trisul Ganga, divided the empire of speech almost equally with the local mother tongues. The Parbattia language is terse, simple, sufficiently copious in words, and very characteristic of the unlettered but energetic race of soldiers and statesmen who made it what it is. At present it is almost wholly in its structure, and in eight-tenths of its vocables, substantially Hindee. Yet several of its radical words still indicate an ancient barbarous stock. And I have no doubt that the people who more especially speak it (the Khas) were originally what Menu calls them, viz., barbarous mountaineers of a race essentially the same with the several other races of Nepaulese Highlanders. Few persons except Brahmans and professional scribes or Khardars are regularly taught the Parbattia language ; but most gentlemen speak, and many read and write it with ease and correctness ; the court where all so often assemble, being the nucleus of unity and refinement. This language, however, has no literature properly so called, and very few and trivial books. It is always written in the Devanagari characters, and, as a language of business, is extremely concise and clear.

The Gorkhalis speak the Parbattia Bhasha, and to their ascendency is its prevalence, in later times, to be mainly ascribed.

Considering that Nepaul Proper, or the country of the Newars, has long been the metropolis of Gorkhali power, it is rather remarkable that the fashionable and facile Parbattia has not made any material impression on the Newari language. The causes of this (not wholly referable to modern times) are probably, that the fertility and facility of communication characterising the level country of the Newars, soon gave consistency and body to their speech, whilst their religion (Buddhism) made them look with jealousy, as well on the more ancient Hindoo immigrants, as on the more modern Hindoo conquerors. In the mountainous districts, strictly so called, the case was different ; and, besides, from whatever reason, the tide of immigration into these regions from the South set chiefly on the provinces west of the Trisul Ganga. There too, to this day, Brahmanical Hindúism principally flourishes, its great supporters being the Khas, and, next to them, the Magars and Gurungs. Those southern immigrants were refugees from Moslem bigotry ; and were so numerous as to be able to give the impress of their own speech and religion to the rude and scattered highlanders. The prior establishment of Buddhism in Nepaul Proper prevented these Brahmanical southerns from penetrating there, where, however, ages before, some southerns had found a refuge. These latter were Buddhists, fleeing from Brahmanical bigotry. They came to Nepaul Proper about two centuries after Christ. Buddhism had previously been established therein, and these immigrants were too few to make a sensible impression on the speech or physiognomy of the prior settlers, already a dense and cultivated population. It is difficult to chronologize these events. But apparently the Sakavans came into Nepaul when Kapila was destroyed by the King of Kosala.

For the rest, the population of the kingdom of Nepaul is principally Bauddha; preferring for the most part the Tibetan model of that faith: the Newars are the chief exception, and the vast majority of them are Buddhists, but not Lamaites. Between the Buddhism of Tibet and that of Nepaul Proper, (or of the Newars) the differences are,

1st. That the former still adheres to, whilst the latter has rejected, the old monastic institutes of Buddhism; 2nd. that the former is still, as of old, wholly unperplexed with caste; the latter, a good deal hampered by it; and that, lastly, the Tibetan Buddhism has no concealments, whilst the Nepaulese is sadly vexed with a proneness to withhold many higher matters of the law from all but chosen vessels.

CONNEXION OF THE LANGUAGE OF NEPAUL PROPER WITH THAT OF TIBET.

I proceed now to indicate that affinity of the language of the Newars to the language of the Tibetans which I have already adverted to. I had extended this vocabulary (in an amplified form) to the whole of the languages above-mentioned: but the results were, for several reasons, liable to question in detail, so that I prefer holding them back for the present, though there can be no doubt of the general facts, that these dialects are of northern origin, and are closely connected.

The language of Nepaul Proper or the Newari, has, as already intimated, much in common with that of Bhot or Tibet. It is however, a poorer dialect than that of Lassa and Digarchi; and it has, consequently, been obliged to borrow more extensive aid from Sanskrit, whilst the early adoption of Sanskrit as the sole language of literature has facilitated this infusion. The following is a comparison of a few terms:—

English.		*Newari.*	*Bhotiya.*
The World.	*(S)	Sansar.	Jambu Ling.
God.	(S)	Bhagawan.	Lhá.
Man.	(S)	Manno, or Mijan.	Khiyoga.
Woman.		Misá. †	Pemi, Kemi.
Quadruped.	(S)	Pasu, Pepanchu.	Tendú.
Bird.		Jhongo.	Djia and Chabi, Byu pron. Chu.
Insect.	(S)	Kicha.	Bú.
A Worm.		Dalambi.	Dalabú.
Fire.		Mih.	Mha and Mih.
Air.	(S)	Phoy.	Lha-phú and Lhawa.
Earth.		Chá.	Sha.
Water.		P. Lo. C. Luk. B. Gná.	Chú.
The Sun.	(S)	Suraj.	Nima.
The Moon.	(S)	Chandrama.	Dawa.
The Stars.	(S)	Nagú.	Kerma.
A Mountain.	(S)	Parba.	Rajhi and Lumba.
A River.		Khussi.	Changbo
Father		Boba and Opju	Ava and Aba
Mother		Má	Amma

* The (S) indicates a Sanskrit origin.

† *Mi-sa* woman, *mi-jan* man, from the Tibetan root *mi* 'man.'

English.	*Newari.*	*Bhotiya.*
Grandfather	Adjhu	Adjhu
Grandmother	Adjhama	Adzhi
A Child	Mocha	Namú ? Piza. Bú
A Boy	Kay Mocha and Bhajú	Phú
A Girl	Miah Mochu and Mejú	Pamú
Uncle	Kakka	Aghu
Aunt	Mamma	Ibi, Asa
Summer	(S) Tápullá	Chapaha
Winter	Chilla	Gun ? Khyabu
Grain	(S) Ann	Soh ? Du
Rice	Jaki, Wá	Brá
Wheat	Cho	Tho
Barley	Tacho	———
Marriage	(S) Biah	Páma
Birth.	Macha-Bullo.	Kesin.
Death.	Sito.	Lhesin.
A House.	Chén.	Khim.
A Stone.	Lohu.	Ghara ? To. Do.
A Brick.	Appa.	Arpa.
Temple.	(S) Dewa.	Lha-Kang.
An Image.	Kata Malli, Patima.	Toto, Thu.
A Bridge.	Ta and Taphu.	Sambá.
A Tree.	Sima.	Ston-bba or Tongbá.
A Leaf.	Sihau and Hau.	Loma or Lapti.
A Flower.	Swang.	Meto, or Mendo.
A Fruit.	Si.	Brebú.
A Horse.	Sallo.	Tapu or Tábá.
A Bull.	Doho.	Sandhí.
A Cow.	Mása and Sá.	Pago.
A Buffalo.	Mia.	Mye.
A Dog.	Khicha.	Khigo or Khibo.
A Cat.	Bhow.	Guré.
A Jackal.	Dhong.	Kipchang.
A Sister.	Kihin.	Chamu ? Numu.
A Brother.	Kinja.	Chou ? Gnú.
Own Family.	Thajho and Tha Mannu.	Pin.
Kinsfolk.	Phuki.	Phebin.
Strangefolk.	Kato & Miah-Ping.	Chomi.
The Head.	Chong.	Wu or Go.
The Hair.	Song.	Tar or Ta.
The Face.	Qua.	Tongba.
The Eye.	Mikha.	Mhi.
The Nose.	Gniá.	Gná.
The Mouth.	Mhutu.	Kha.

English.	*Newari.*	*Bhotiya.*
The Chin.	Mano.	Koma.
The Ear.	Nhiapo.	Nhamjo.
The Forehead.	Kopa.	Prála.
The Body.	Mho.	Zhúbú.
The Arm.	Laha, Lappa.*	Lakpa, Lá-g-pa.
The Leg.	Túti.	Kangba.
Right.	Jou.	Yeba.
Left.	Khou.	Yémá.
A Month.	Lá.	Lá-wa.
A Year.	Dat'chí.	Lochik.
Day.	Gni or Nhi.	Nain, Nyi-n-mo.
Night.	Chá.	Chan.

With regard to the Newari words, I can venture to say they may be relied on, though they differ somewhat from Kirkpatrick's, whose vocabulary, made in a hurry, exhibits some errors, especially that of giving Sanskrit words instead of the vernacular. It is remarkable that the Newars, (those that pretend to education, and those who are wholly illiterate), are apt to give a stranger, a Sanskrit, instead of their own Newari, name for any object to which their attention is called for the purpose of naming it. This habit owes its origin to the wish to be intelligible, which the Newars know they cannot be in speaking their own tongue. The real poverty of the Newari is, also, no doubt, another cause, and its want of words expressive of general ideas: thus, Creation, God, have no Newari names, and the Sanskrit ones have therefore been borrowed of necessity: the like is true of the word Mankind, for which, as well as for the two former words, I have not been able, after great pains, to obtain any vernaculars. When a Newar would express the idea of God, without resorting to Sanskrit, he is driven to periphrasis, and says Adjhi Deo, which word is compounded of Adjhu, a Grandfather, and Deo; and thus, by reverence for ancestors, he comes to reverence his maker, whom he calls, literally, the father of his father, or the first father. I am quite aware the foregone and following meagre examples of Newari will not go far to establish the affinity of this language. The subject must be reserved for the future; but, in the mean time, I may observe that the northern stock, and intimate affinity of Newari and of the other dialects before enumerated, (excepting the Khas or Parbattia), are written as palpably upon the face of these languages as upon the physiognomy and form of the races who speak them.

As for the Bhotiya words, I cannot wholly vouch for them, few as they are, having obtained them from a Lama, who was but little acquainted with Newari or Parbattia. The majority are, I believe, sufficiently accordant with the Lhassa model, but some may be dialectically corrupted. Still, however, they will equally serve, (as far as they go), to illustrate my assertion that the root and stock of Newari are Trans-himalayan and northern; for there are many dialects on *both*

* Lappa, (almost identical with the Bhotiya Lakpa) means the true arm, or upper half of the limb. Laha means the whole.

sides of the snows, and some of the inferior Tibetan dialects may, very probably, come nearer to Newari than the best, or that of Lhassa.

The twelfth word in the Newari column, Water, is given according to the sub-dialects of the Valley. Water is Lo at Patan, Luk at Katmandu, and Gnā at Bhatgong; these places being the capitals of as many kingdoms before the Gorkha conquest, though situated in very close vicinity to each other.

With respect to the numerals of the decimal scale, the resemblance is strikingly close.

NUMERALS.

	Bhotiya.	*Newari.*
1.	Chi.	Chi.
2.	Gni.	Na Shi.
3.	Sum.	Swong.
4.	Zhi.	Pih.
5.	Gnah.	Gniah.
6.	Tukh.	Khu.
7.	Tun.	Nha or Nhasso.
8.	Ghiah.	Chiah.
9.	Gun.	Gun.
10.	Chu (Thampa, an expletive merely.)	Sánho.
11.	Chú-chi.	Saran-chi.
12.	Chú-gni.	Saran Nassi.
13.	Chu (P.) súm, (the letter (P) written but scarcely audibly uttered.)	
14.	Chu (P.) Zhi.	Saran Pih.
15.	Cheánga.	Saran Gniah.
16.	Churu.	Saran Khu.
17.	Chuptin.	Saran Nha.
18.	Chopkia.	Saran Chiah.
19.	Churko.	Saran Gun.
20.	Né shu (thampa.)	Saran Sanho.
21.	” ”	Ni Chi.
22.	” ”	Ni Nassi.
23.	” ”	Ni Swong.
24.	” ”	Ni Pih.
25.	” ”	Ni Gniah.
26.	” ”	Ni Khu.
27.	” ”	Ni Nhi.
28.	” ”	Ni Chiah.
29.	” ”	Ni Gun.
30.	Sum chu (thampa.)	Ni Shao.
31.	” ”	Swi Chi.
32.	” ”	Swi Nassi.
33.	” ”	Swi Swong.
34.	” ”	Swi Pih.

	Bhotiya.	*Newari.*
35.	Sum chu (thampa.)	Swi Gniah.
36.	,, ,,	Swi Khu.
37.	,, ,,	Swi Nha.
38.	,, ,,	Swi Chiah.
39.	,, ,,	Swi Gun.
40.	Zhe-chu (thampa.)	Swi Sánho.
41.	,, ,,	Pi Chi.
42.	,, ,,	Pi Nassi.
43.	,, ,,	Pi Swong.
50.	Gna-chu (thampa.)	Gniayú or Pi-Sanho, or merely by pausing on the last letter of Gniah or :5 and thus also 60, 70, &c. are formed out of 6, 7, &c.
60.	Tukh-chu (thampa.)	Qui.
70.	Tun ,,	Nhaiyu.
80.	Gheah ,,	Chaiye.
90.	Gu (P.) ,,	Guye.
100.	Gheah ,,	Saché.
1,000.	Tong-tha-che.	Do-ché.
10,000.	Thea.	Zhí-dot.
100,000.	Bum.	Lak-chi.

Nor is the variation, after passing the ten, of any importance, the principle of both being still the same; that is, repetition and compounding of the ordinals; thus, ten and one, ten and two, are the forms of expression in both, and so, twice, &c. The Bhotiya word thampa, postfixed to the decimally increasing series, is a mere expletive, and often omitted in speech. The Newari names of the figures from one to ten, as given by Kirkpatrick, are not correct, and hence the difference between the Newari and Bhotiya names has been made to appear greater than it is: in fact, it seems to me, that even the little difference that remains in the present specimens may be resolved into mere modes of utterance. Although the following offer no verbal resemblances, the principle on which they are formed presents several analogies.

Bhotiya and Newari names of the twelve months.

	Newari.	*Bhotiya.*
February.	—— ——	Dagava (or Láwa) Tangbu.
March.	Chongchola or Chilla.	(Láwa) Gnipa.
April.	Bachola or Néla.	,, ,, Sumba.
May.	Túchola ,, Swola.	,, ,, Zhiba.
June.	Dil'la ,, Péla.	,, ,, Gnappa.
July.	Gung'la ,, Gniāla.	,, ,, Tuakpu.
August.	Yung'la ,, Khola.	,, ,, Tumba.
September.	Koula ,, Nhúla.	,, ,, Gnappa.
October.	Kozla ,, Chála.	,, ,, Guabba.
November.	Thingla ,, Gungla.	,, ,, Chuba.
December.	Puéla ,, Séla.	,, ,, Chu-chikpa.
January.	Sel'la ,, Zhin'chala.	,, ,, Chu-gnipa.
February.	Chil'la ,, Zhin'nala.	—— ——

The second set of Newari names is formed merely by compounding the word La, a month, with the names of the cardinals, one, two, etc. As for the first set of names, there too we have the final La; and the prefixes are mere characteristic epithets of the seasons; thus, February is called Chilla; but Chilla means also the cold month, or winter.

The Bhotiyas, like the Newars, have no simple names for the months, but call them periphrastically the first, etc., month. Dawa and Lawa both mean a month; but in speech this word is never prefixed, save in speaking of the first Bhotiya month or February, for from February their year begins. What Tangbu means, I know not, unless it be the same with Thampa, the word that always closes the series of numbers, 10, 20, 30, etc. The names of all the others are easily explained, they being compounds of the numbers 2, 3, etc., with the syllable pá or bá—evidently the Lá of the Newars—postfixed.

		Newari names of the seven days of the week.		*Bhotiya names of the seven days.*	
Sunday,	(S)	Adhwina,	or	Chanhu,	Nima.
Monday,	(S)	Swomwa,	,,	Nenhu,	Dawa.
Tuesday,	(S)	Ongwa,	,,	Swonhu,	Mimer.
Wednesday,	(S)	Budhwa,	,,	Penhu,	Lhákpa.
Thursday,	(S)	Bussowa,	,,	Gnianhu,	Phōorboo.
Friday,	(S)	Sukrawa,	,,	Khonhu,	Pasang.
Saturday,	(S)	Sonchowa,	,,	Nhainhu,	Pemba.

The first of the Newari series are wholly corrupt Sanskrit, and the second formed by compounding the word Nhi or Gni, a day, with the cardinals: the Newars have no simple words of their own, expressive of the seven days.

A variety of characters is met with in the Nepaulese and Bhotiya books, some of which are now obsolete. A manuscript, of which a copy is forwarded, contains a collection of these alphabets, each bearing a separate designation. Of the Newari, three kinds of letters are most familiarly known, and four of the Bhotiya.

WRITTEN CHARACTERS OF NEPAUL PROPER.

The three Newari alphabets (so to speak) are denominated Bhanjin Mola, Ranja, and Newari. Whether these three sorts of letters were formerly used by the Siva Márgi Newars, I cannot say; but old Bauddha* works exhibit them all, especially the two former. Newari alone is now used by both sects of Newars for profane purposes; and for sacred, both often employ the Devanágarí, oftener the Newari. If the Siva Márgi Newars ever used (which I doubt) Bhanjin Mola, or Ranja, at least they do so no longer; and the Newars of the Buddha faith having long ceased ordinarily to employ those letters in making copies of their Scriptures, few can now *write* them, and the learned only (who are accustomed to refer to their old works) can *read* them with facility.

In regard to the origin of these letters, we may at once refer the Newari to Nágari; but the other two present at first sight more difficulties. Dr Carey was, some time back, of opinion that they are mere fanciful specimens of caligraphy

* For Buddha read Bauddha, et sic passim, where the word is used adjectively.

This notion is refuted by the fact of their extensive practical application, of which Dr. Carey was not aware when he gave that opinion. By comparing one of them (the Ranja) with the fourth alphabet of the Bhotiyas, it will be seen, that the general forms of the letters have a striking resemblance. And as this Lanja or Ranja is deemed exotic by the Bhotiyas, I have no doubt it will prove the same with the Newari letters so called: for the words Lanja, Lantza, and Ranja are one and the same. Of the Bhanjin Mola, it may be observed that it has a very ornate appearance, and, if the ornamental parts were stripped from the letters, they (as well as the Ranja) might be traced to a Devanágarí origin, from the forms of which alphabet the Bauddhas might possibly alter them, in order to use them as a cover to the mysteries of their faith. The Bauddha literature is, originally, Indian. Now, though probability may warrant our supposing that those who originated it, together with its religion, might alter existing alphabetical forms for the purpose above hinted at, it will not warrant our conjecturing, that they would undergo the toil of inventing entirely new characters. All these systems of letters follow the Devanágarí arrangement, nor should I hesitate to assign them all a Devanágarí origin. Indeed it is well known to the learned, that there were anciently in the plains of India many sorts of written characters, since become extinct: and I have no doubt that the letters adverted to were part of these.

WRITTEN CHARACTERS OF TIBET.

Of the Bhotiya characters, four kinds are distinguishable; but only two of them are known by name to the Newars: they are called (in Tibet as well as here) Uchhen and Umen. The first are capitals: the second, small letters: the third, running hand; and the fourth, as already observed, equivalent to the Nepaulese Ranja. There is also a character in use in and near Tibet which is ascribed to the Sokpa, who, with the Hor or Horpa, constitute the nomad population of Tibet, of Túrki, and Mongol etymon respectively.

LITERATURE OF BHOT OR TIBET.

The term Bhot is the Sanskrit, Tibet the Persian name, Bod the native one, but probably only a corruption of the first term, and, if so, the Tibetans had not any general name for themselves (Bod-pa) or their country when their Indian teachers first came among them in the 7th centuary, A.D.

The great bulk of the literature of Bhot (as of Nepaul) relates to the Bauddha religion. In Bhot the principal works are only to be found at the larger monasteries: but numerous Bhotiya books of inferior pretensions, are to be obtained at Katmandu from the poor traffickers and monks who annually visit Nepaul on account of religion and trade.

The character of the great part of these latter, or the Bhotiya books *procured in Nepaul*, is that of popular tracts, suited to the capacity and wants of the humbler classes of society, among whom it is a subject of surprise, that literature of any kind should be so common in such a region as Bhot, and, more remarkably so, that it should be so widely diffused as to reach persons covered with filth, and destitute of every one of those thousand luxuries which (at least in our ideas) precede the great luxury of books.

Printing is, no doubt, the main cause of this great diffusion of books. Yet the very circumstance of printing being in such general use, is no less striking than this supposed effect of it; nor can I account for the one or other effect, unless by presuming that the hordes of religionists, with which that country [Tibet] swarms, have been driven by the *tedium vitæ*, to these admirable uses of their time.

The invention of printing, the Bhotiyas got from China; but the universal use they make of it is a merit of their own. The poorest individual who visits this valley from the north is seldom without his Pothi [book], and from every part of his dress dangle charms [Jantras,] made up in slight cases, the interior of which exhibits the neatest workmanship in print.

Some allowance, however, should also be made for the very familiar power and habit of *writing*, possessed by the people at large: another feature in the moral picture of Bhot, hardly less striking than the prevalence of printing or the diffusion of books, and which I should not venture to point out, had I not had sufficient opportunities of satisfying myself of its truth among the annual sojourners in Nepaul who come here in hundreds to pay their devotions at the temple of the self-existent Supreme Buddha [Swayambhu Adi Buddha].

In the collections forwarded to the Society will be found a vast number of manuscripts—great and small—fragments, and entire little treatises—all which were obtained [as well as the small printed tracts] from the humblest individuals. Their number and variety will, perhaps, be allowed to furnish sufficient evidence of what I have said regarding the appliances of education in Tibet, if due reference be had, when the estimate is made to the scanty and entirely casual source whence the books were obtained in such plenty.

The many different kinds of writing which the MSS. exhibit will, perhaps, be admitted yet further to corroborate the general power of writing possessed by almost all classes of the people. Or, at all events, these various kinds and infinite degrees of penmanship, present a curious and ample specimen of Bhotiya proficiency in writing, let this proficiency belong to what class or classes it may.

Something of this familiar possession of the elements of education, which I have just noticed as characterising Bhot, may be found also in India; but more, I fear, in the theory of its institutions than in the practice of its present society, because of the successive floods of open violence which have, for ages, ravaged that, till lately, devoted land. The repose of Bhot, on the other hand, has allowed its pacific institutions full room to produce their natural effect; and hence we see a great part of the people of Bhot able to write and read.

In whatever I have said regarding the Press, the general power and habit of writing, or the diffusion of books, in Bhot, I desire to be understood by my European readers with many grains of allowance. These words are names importing the most different things in the world in the favoured part of Europe, and in Asia. The intelligent resident in Hindoostan will have no difficulty in apprehending the exact force which I desire should be attached to such comprehensive phrases, especially if he will recollect for a moment that the press, writing and books, though most mighty engines, are but engines; and that the example of China proves to us indisputably, they may continue in daily use for ages in a vast society, without once falling into the hands of the strong man of Milton; and consequently, without waking one of those many sublime energies, the full developement of which in Europe has shed such a glorious lustre around the path of manin this world.

The printing of Bhot is performed in the stereotype manner by wooden planks; which are often beautifully graved: nor are the limited powers of such an instrument felt as an inconvenience by a people, the entire body of whose literature is of an unchanging character.

The Bhotiya or Tibetan writing, again, often exbihits specimens of ready and graceful penmanship. But then it is never employed on any thing more useful than a note of business, or more informing than the dreams of blind mythology; and thus, too, the general diffusion of books (that most potent of spurs to improvement in our ideas) becomes, in Bhot, from the general worthlessness of the books diffused, at least but a comparatively innocent and agreeable means of filling up the tedious hours of the twilight of civilization.

SANSKRIT BAUDDHA LITERATURE OF NEPAUL.

With respect to the authorities of the Buddhist religion or their sacred scriptures, the universal tradition of the followers of this creed (supported by sundry notices in their existing works) asserts, that the original body of their scriptures amounted, when complete, to eighty-four thousand volumes—probably sútras or aphorisms, and not volumes in our sense.

The most authoritative of the books of the Buddhists now extant in Nepaul in the sacred language of India, as subsequently to be enumerated, are known, collectively, and individually, by the names of Sútra and Dharma.

In a work called the Pújá Khand there is the following passage:—

"All that the Buddhas have said, as contained in the Mahá Yána Sútra, and the rest of the Sútras, is Dharma Ratna," or precious science. Hence the Scriptures are also frequently called "Buddha Vachana," the words of Buddha. Sákya Sinha first gave definite form and systematic force to these words, if indeed he did not wholly originate them; and, in this important respect Sákya is to Buddhism what Vyása is to Brahmanism.

The old books of these religionists universally assert this; the modern Bauddhas admit it in the face of that host of ascetics whom the easiness of latter superstition has exalted to the rank of an inspired teacher. The sacred chronology of the sect is content with assigning Sákya to the Kali Yuga, and profane chronology is a science which the Buddhists seem never to have cultivated. But the best opinion seems to be that Sákya died about four and a half centuries before our era. In the subsequent enumeration of the chief Sanskrit authorities of the Buddhists it will be seen that Sákya is the "Speaker" in all the great works. This word answers to "hearer," and refers to the form of the works, which is, for the most part, that of a report of a series of lectures or lessons delivered verbally by Sákya to his favourite disciples, but sometimes diverging into dialogue between them. That Sákya Sinha was substantially the originator of this creed, such as it has come down to our times, is thus I think demonstrable from the uniform tenour of the language of the great scriptural authorities of the sect, wherein, either before or after the enunciation of every cardinal text, stand the words, 'thus said Sákya Sinha,' or, 'so commanded Sákya Sinha.' Sákya Sinha therefore must be concluded to be the founder of this creed, which took its existing written form from the hands of his earliest disciples, or Kásyapa, Ananda, and Upáli.

Adverting now to the technical arrangement, or classification of these works, I may observe that they are primarily divided into Esoteric and Exoteric, and that these classes are ordinarily termed Tantras and Puránas by the Buddhists as well as by the Brahmanists, though the former would likewise seem to convey this distinction by the words Upadesa and Vyákarana. Vyákarana is also employed in the sense of narration as opposed to speculation. Gáthá, Játaka, Avadána, etc., seem to be subdivisions.

The word Sútra as explained, "Múla Grantha," "Buddha Vachana," (chief book, words of Buddha,) has been held to be equivalent to the Sruti of the Brahmans, as has their Smriti to the Bauddha Vyákarana. But, apt as Buddhism is to forget the distinction of divine and human nature, this analogy must be allowed to be somewhat defective; and, in fact, the Sútra of the Buddhists often comprehends not only their own proper "Buddha Vachana," but also "Bodhisatwa and Bhikshu Vachana," (words of Bodhisatwa and of Bhikshu); which latter the Brahmans would denominate "Rishi Vachana," and of course, assign to the Smriti, or comments by holy men upon the eternal truth of the Sruti.

The Newars assert, that of the original body of their sacred literature but a small portion now exists. A legend, familiar to this people, assigns the destruction to Sankara Achárya; and 'the incomparable Sankara' of Sir W. Jones, is execrated by the Nepaulese Bauddhas as a blood-stained bigot.*

Of the existing Bauddha writings of Nepaul (originally of Indian growth and still found unchanged in the Sanskrit language) by far the most important, of the *speculative* kind, are the five Khandas or parts of the Prajná Páramitá or Rakshá Bhágavatí, each of which contains 25,000 distiches. Of the *narrative* kind, the chief are eight of the nine works called the 'Nava Dharma;' the ninth being the Ashta Sáhasrika Prajná Páramitá. It is a valuable summary of the great work first mentioned, to which, therefore, rather than to the *narrative* class, the Ashta Sáhasrika bears essential affinity. In the sequel will be found a list of all the Sanskrit Bauddha works known to me by name.†

* Sankara is placed in the ninth century of Christ (1,000 years ago), and Sákya, the founder of Buddhism, (for we have nothing authentic before him) certainly was not born sooner than about the middle of the sixth century, B.C. The interval of fifteen centuries may vaguely indicate the period during which Buddhism most flourished in India. The decline of this creed in the plains we must date from Sankara's era, but not its fall, for it is now certain that the expulsion was not complete till the fourteenth or fifteenth century of our era. From the ninth century onwards is comprised the worst period of the persecution.

† See the next paper for this list.

The five Rakshás or Páramitás * are enumerated in order in the immediately subsequent detail. They are of highly speculative character, belonging rather to philosophy than religion. The cast of thought is sceptical in the extreme: endless doubts are started, and few solutions of them attempted. Sákya appears surrounded by his disciples, by whom the arguments on each topic are chiefly maintained, Sákya acting generally as moderator, but sometimes as sole speaker. The topics discussed are the great first principles of Buddhism; † the tenets of the four schools of Bauddha Philosophy are mentioned, but those of the Swábhávika alone largely discussed. The object of the whole work seems rather to be proof of the proposition, that doubt is the end as well as beginning of wisdom, than the establishment of any particular dogmas of philosophy or religion: and from the evidence of this great work it would appear that the old Bauddha philosophers were rather sceptics than atheists.

The nine Dharmas are as follows:

1. Ashta Sáhasrika. 2. Ganda Vyúha. 3. Dasa Bhúmeswara. 4. Samádhi Rája. 5. Lankávatára. 6. Sad Dharma Pundaríka. 7. Tathágata Guhyaka. 8. Lalita Vistara. 9. Suvarna Prabhása.

Divine worship is constantly offered to these nine works, as the 'Nava Dharma,' by the Bauddhas of Nepaul. The aggregation of the nine is now subservient to ritual fancies, but it was originally dictated by a just respect for the pre-eminent authority and importance of these works, which embrace, in the first, an abstract of the philosophy of Buddhism; in the seventh, a treatise on the esoteric doctrines; and in the seven remaining ones, a full illustration of every point of the ordinary doctrine and discipline, taught in the easy and effective way of example and anecdote, interspersed with occasional instances of dogmatic instruction. With the exception of the first, these works are therefore of a narrative kind; but interwoven with much occasional speculative matter. One of them (the Lalita Vistara) is the original authority for all those versions of the history of Sákya Sinha, which have crept, through various channels, into the notice of Europeans.

I esteem myself fortunate in having been first to discover and procure copies of these important works. To meditate and digest them is not for me; but I venture to hint that by so doing only can a knowledge of genuine Buddhism be acquired. Buddhism is not simple, but a vast and complicate structure erected, during ages of leisure, by a literary people. It has its various schools divided by various Doctors; nor is the Buddhism of one age less different from that of another, than the Brahmanism of the Vedas, of the Puránas, and of the Bhágavat. Buddhism prevailed in India sixteen to seventeen centuries, and, as its genius was free, so it had even before its founder's death many sects. And soon after his death, schisms multiplied infinitely despite the three great convocations called to stay them. These councils took place respectively, B.C. 465, B.C. 365, B.C. 231. Let it not be supposed, because these works I have cited were procured in Nepaul, that they are therefore of a local character or mountain origin.

* On the Prajnâ Pâramitâ see Wassiljew's "*Der Buddhismus*," p. 157.
† See the sequel at "Religion of Nepaul and Bhot."

Such a notion is, in every view, utterly absurd; for the works bear intrinsic evidence of the contrary in almost every page; and their language (Sanskrit,) always wholly exotic in Nepaul, most assuredly was *never* cultivated there with a zeal or ability such as the composition of these works must have demanded.

These works were composed by the Sages of Magadha,* Kosala,† and Rájagriha,‡ whence they were transferred to Nepaul by Bauddha Missionaries soon after they had assumed their existing shape.

The Sámbhu Purána is the only local work of importance in the large collection which I have made. Perhaps it may be surmised, that if (as is stated) the fire of Sankara's wrath consumed all but some fragments of the sacred writings of the Buddhists, the ample works now produced must be spurious. But, in the first place, the legend is but a legend; and in the next, exaggeration may reasonably be suspected, both as to number of books then extant and destroyed

The Bauddhas never had eighty-four thousand principal scriptures; || nor *could* Sankara destroy more than a few of those which they really possessed when he came (if he ever came) to Nepaul. The proof of the latter statement is—that Buddhism was, long after Sankara's time, the prevalent and national faith of the Nepaulese Princes and subjects; and that it is so still in regard to the people, notwithstanding the Gorkhali conquest. Sankara (or some other famous Brahmanical controversist) may have converted one of the Princes of the Valley; but the others remained Buddhists; and, no doubt, took care of the faith and property of their subjects. All *old* Bauddha works are written in one of the three sorts of letters now peculiar to Nepaul Proper, usually in Ranja and Bhanjin Mola, and on Palmira leaves. Copies of the Rakshá Bhágavatí or Prajná Páramitá are very scarce. I am of opinion, after five years of enquiry, that there were but four copies of it in the Valley, prior to my obtaining one copy and a half: one copy more I got transcribed from an old one.§ No one had, for some time, been able fully to understand its contents; no new copy had been made for ages; and those few persons, who possessed one or more khands or sections of it, as heir-looms, were content to offer to sealed volumes the silent homage of their pújá (worship). Time and growing ignorance have been the chief enemies of Sanskrit Bauddha literature in Nepaul.

The Bauddha Scriptures are with reference chiefly to their form and style, frequently stated to be of twelve kinds,** known by the following twelve names; 1. Sútras; 2. Geya; 3. Vyákarana; 4. Gáthá; 5. Udána; 6. Nidána; 7. Ityukta:

* The modern Bihar. † Berar. ‡ Rajgir.

|| We should doubtless read aphorism or text (Sútra or bana), not book, with reference to the 84,000 in question. The universality of the notion proves that this definite number has truth, in some sense, attached to it.

The primitive meaning of Sûtra [aphorism, or thread of discourse,] implies that Sákya taught verbally; and if this be so, Sútra only took its present sense of principal scripture after his death. These sayings of Sakya may still be found all over the sacred works of the sect in their original aphoristic form. The destruction of Bauddha books adverted to in the text, has, I fancy, reference to the *plains* of *India*. *There* it was complete eventually; but in the mean while the most valuable works had been saved in Nepaul.

§ These I sent to the Library of the College of Fort William A.D. 1825.

** Twelve kinds of Scriptures, see Wassiljew, p. 118.

8. Játaka; 9. Vaipulya; 10. Adbhuta Dharma; 11. Avadána; 12. Upadesa.

Sûtras are the principal scriptures, (Múla Grantha) as the Rakshá Bhágavatí or Prajná Páramitá; they are equivalent to the Vedas of the Brahmanists. The aphorisms of Sákya are the basis of them, hence the name.

Geyas are works of praise, thanksgiving and pious fervour, in modulated language. The Gíta Govinda of the Brahmanists is equivalent to the Buddhist Gíta Pustaka, which belongs to the Geya.

Vyákarana are narrative works, such as those containing histories of the several births of Sákya prior to his attaining Nirvána; and sundry actions of others who by their lives and opinions have illustrated this religion, with various forms of prayer and of praise. Vyákarana, in the sense of narration, is opposed generally to works of philosophy or speculation, such as the Prajná Páramitá. It also characterises works of an exoteric kind, as opposed to the Upadesa or Tantras.

Gáthás are narrative works, in verse and prose, containing moral and religious tales, (Aneka Dharmakathá) relative to the Buddhas, or elucidative of the discipline and doctrine of the sect. The Lalita Vistara is a Vyákarana of the sort called Gáthá.

Udána treat of the nature and attributes of the Buddhas, in the form of a dialogue between a Buddhist adept and novice.

Nidána are treatises, in which the causes of events are shewn; as for example, how did Sákya become a Buddha? the reason or cause; he fulfilled the Dán, and other Páramitás.*

Ityukta, whatever is spoken with reference to, and in conclusion: the explanation of some prior discourse, is Ityukta.

Játaka treat of the subject of transmigration or metempsychosis, the illustrations being drawn from the 550 births of Sákya.

Vaipulya treat of several sorts of Dharma and Artha, that is, of the several means of acquiring the goods of this world (Artha) and of the world to come (Dharma).

Adbhuta Dharma, on preternatural events.

Avadána, of the fruits of actions or moral law of Mundane existence.

Upadesa treat of the esoteric doctrines, and are equivalent to Tantra, the rites and ceremonies being almost identical with those of the Hindoo Tantras, but the chief objects of worship, different, though very many of the inferior ones are the same. According to the Upadesa, the Buddhas are styled Yogámbara and Digámbara. Tántrika works are very numerous. They are in general disgraced by obscenity and by all sorts of magic and dæmonology. But they are frequently redeemed by unsually explicit assertions of a supreme Godhead. Najra Satwa Buddha is the magnus Apollo of the Tántrikas.

The following is an enumeration of some of the most important individual specimens of the preceding classes.

* Pâramitâ here means virtue, the moral merit by which our escape (passage) from mortality is obtained. Dána, or charity, is the first of the ten cardinal virtues of the Bauddhas; "and other" refers to the remaining nine. Appendix A. of paper III. *Páram* beyond and *itá* gone.

First khand, or section, of the Rakshá Bhágavatí or Prajná Páramitá. It is a Mahá Yána Sútra Sástra. It begins with a relation (by himself) of how Sákya became Bhagaván (deified); and how he exhorted his disciples to study and meditate his principles; and how he explained the doctrine of Avidyá, that is, as long as Avidyá* lasts, the world lasts, when Avidyá ceases, (Nirodha) the world ceases; aliter, Pravritti ends, and Nirvritti* begins. Such are the general contents of the former part of this khand; and the latter part of it is occupied with explanations of Súnyatá and Mahá Súnyatá.* Sákya is the speaker, the hearers are Subhúti, and other Bhikshukas: the style is prose (Gadya).

Second and third khands of the Rakshá Bhágavatí. Contents the same as above.

The fourth khand of the Rakshá Bhágavatí relates how any one becomes Sarvakarmajna, or skilled in the knowledge of all things on earth and in heaven; in a word, omniscient; besides which, the subjects of the former khands are treated of, in continuation, in this.

The fifth khand of the Rakshá Bhágavatí. It is a sort of abstract of the other four which form one work. Besides Avidyá, Súnyatá, and all the other great topics of the prior khands, this khand contains the names of the Buddhas, and Bodhisatwas.

These five khands or divisions are each called Pancha, Vinsati, Sáhasrika, Prajná Páramitá; the three first words indicating the extent of each division, and the two last, the nature of the subject or transcendental wisdom. Sata Sáhasrika is a collective name of the four first khands, to which the fifth is not necessarily adjunct; and indeed it is one of several abstracts of the Sata Sáhasrika, as already stated. Arya Bhágavatí and Rakshá Bhágavatí, or holy Goddess and Goddess of Deliverance, are used, indifferently with Prajná Páramitá, as titles of each or all of these five khands. The five khands are all in prose, and comprise the philosophy of Buddhism.

Ashtasáhasrika Prajná Páramitá, a Mahá Yána Sútra. Another and smaller epitome of the transcendental topics discoursed of at large in the Sata Sáhasrika. It is prose. Sákya is the speaker; and Subhúti and other Bhikshukas,† the hearers.

ASHTA ṢAHASRIKA VYAKHYA.

This is a comment on the last work by Hara Bhadra, in verse and prose.

Ganda Vyúha, a Vyákarana Sástra, contains forms of supplication and of thanksgiving, also how to obtain Bodhijnána, or the wisdom of Buddhism. Prose: speaker, Sákya; hearer, Sudhana Kumára. The Ganda Vyúha is a treatise on transcendentalism by Arya Sanga the teacher of the Yogácharya.

Dasa Bhúmeswara, a Vyákarana, containing an account of the ten Bhúmis.‡ Prose: speaker, Sákya; hearer, Ananda Bhikshuka.

* See the explanation of these terms in the sequel. They form the basis of the philosophy of Buddhism.

† Bhikshu, name of a Buddhist mendicant. See on to section on Religion.

‡ Ten heavens, or ten stages of perfectibility: sometimes thirteen are enumerated and the thirteen grades of the spire of the Chaitya are typical of them. See Laidlay's Fahian, p. 91, and J.R.A.S. xi. 1, 21.

Samádhi Rája, a Vyákarana; an account of the actions by which the wisdom of Buddhism is acquired, and of the duties of Bodhisatwas. Prose: speaker, Sákya; hearers, Rávana and others.

Sad Dharma Pundaríka, a Vyákarana, an account of the Mahá and other Dípa Dánas, or of the lights to be maintained in honour of the Buddhas, and Bodhisatwas; with narrations of the lives of several former Buddhas by Sákya, as well as prophetic indications of the future eminence of some of his disciples. Speakers and hearers, Sákya, Maitreya, Manjusrí, etc.

Lalita Vistara. This is a Vyákarana of the sort called Gáthá. It contains a history of the several births of Sákya, and how, in his last birth, he acquired perfect wisdom, and became Buddha. Verse and prose: speaker, Sákya; hearers, Maitreya and others.

Guhya Samagha, otherwise called Tathágata Guhyaka; an Upadésa or Tantra; contains numerous mantras, with explanations of the manner of performing esoteric rites. Prose and verse: speaker, Bhagaván (*i.e.* Sákya); hearers, Vajra Páni* Bodhisatwa and others.

Suvarna Prabhása, a Vyákarana Sástra; discourses by Sákya for the benefit of Lakshmí, Saraswatí and others; also an account of the Bhágavata Dhátu, or mansions of the deities. Prose and verse: speaker, Sákya; hearers, Litsavi † Kumára, the above named Goddesses and others.

Swayambhu Purána, the greater; a Vyákarana of the sort called Gáthá: an account of the manifestation of Swayambhu or Adi Buddha ‡ in Nepaul, and the early history of Nepaul. Verse: speaker, Sákya; hearer, Ananda Bhikshuka.

Swayambhu Purána, the less, a Gáthá, summary of the above; an account of Swayambhu Chaitya, (or temple). Verse and prose: speaker and hearer, as above.

Karanda Vyúha, an account of Lokeswara Padma Páni. Prose: speaker and hearer, as above.

Guna Karanda Vyúha, a Gáthá; an amplification of the above in verse. Speaker and hearer, as above.

Mahávastu, an Avadána Sástra; an account of the fruits of actions, like the Karma Vipáka of the Brahmans. Prose: speaker and hearer, as before.

Asoka Avadána; an account of the Triad, or Buddha, Dharma, Sangha; also of the Chaityas, with the fruits of worshipping them. Verse: speaker, Upagupta Bhikshuka; hearer, Asoka Rájá.§

Bhadra Kalpika, an Avadána Sástra; a detailed account of the Buddhas of this Kalpa.** Verse and prose; speaker, Sákya; hearers, Upagupta Bhikshuka, with a host of immortals and mortals.

Játaka Málá; an account of the meritorius actions of Sákya in his 565 births,

* Vajra Páni is the æon of Vajra Satwa Buddha, already alluded to as the magnus Apollo of the Tántrikas. See Fahian, p. 135.

† Litsavis are the so called Scyths. Litsabyis in Tibetan. For Sakas, see J.R.A.S. xii. 2, 460.

‡ Swayambhu means self-existent. Adi, first, and Buddha, wise.

§ This is the celebrated friend of Antiochus and builder of the Lâts.

** It is styled the Golden because four Buddhas belong to it, viz., Karkut, Kanaka, Kásyapa, and Sàkya.

prior to his becoming a *Tathágata.* Verse and prose: speaker, Sákya; hearer, Ananda Bhikshu.

Manichura, an Avadána; an account of Manichur Rájá, also of the first birth of Sákya, and of the fruits of his actions. Prose: speaker and hearer as above.

Dwávinsati Avadána, an Avadána Sástra; an account of the fruits of building, worshipping and circumambulating* Chaityas. Verse and prose: speaker, Sákya; hearer, Maitreya.

Nandi Mukha Swaghosha, an Avadána; an account of the great fast called Vasundhará, and of the fruit of observing it. Prose: speaker, Sákya; hearer, Ananda.

Bodhi-charyá, an Avadána Sástra, of the sort called Kávya; contains a highly laudatory account of the virtue of charity and of the Bodhi-Charyá, or Buddhist duties. Verse: speaker, Maitreya; hearer, Sudhana Kumára.

Karuna Pundaríka, an Avadána; an account of Arinemi Rájá; of Samudra Renu, Purohita; of Ratna Garbha, Tathágata; and of Avalokiteswara, (i. e., Padma Páni Bodhisatwa) interspersed with sundry philosophical topics which are discussed by Sákya in a broken manner. Sákya, then, in anticipation of his demise, gives directions as to the mode in which his system is to be taught. Prose: speaker, Sákya; hearers, Maitreya, &c.

Chandomrita Málá, a treatise of prosody; the measures illustrated by verses laudatory of Sákya Sinha. Verse and prose: the author Amrita Bhikshu.

Lokeswara Sataka, a hundred verses in praise of Padma Páni. Verse: author, Vajra Datta Bhikshu.

Saraka Dhárá, with a comment; a Kávya in praise of Arya Tárá, Buddha Sakti. Verse: author, Sarvajna Mitrapáda Bhikshu.

Aparámita Dháraní, an Upadesa;† contains many Dháranís addressed to the Buddhas, who are immortal (Aparámitáyusha Tathágata). Prose: speaker, Sákya; hearer, Ananda Bhikshu.

Dháraní Sangraha, a collection of Dháranís, as Mahá Vairochana's D. Mahá Manjusrí's D. and those of many other Buddhas and Bodhisatwas. Verse: speaker, Sákya; hearer, Vajra Páni.

Pancha Rakshá, an Upadesa Dháraní; an account of the five Buddha Saktis, called Pratisárá, &c.‡ Prose: speaker, Sákya; hearer, Ananda.§

Pratyangirá Dháraní, an Upadesa Dháraní; an account of Pratyangirá Buddha Sakti. Prose: speaker, Sákya; hearer, Ananda Bhikshu.

* This circumambulation is one of the commonest and most pious actions of Buddhist devotion. Mental prayers are repeated all the while, and a small cylinder fixed upon the upper end of a short staff or handle, is held in the right hand and kept in perpetual revolution. This cylinder is called Mani; some leaves of the sacred books are usually enclosed in it. Its use is more common to Tibetans than to Nepaulese. Both people use beads to count their repetitions of holy words.

† Dháranís, though derived from the Upadesa, are exoteric. They are short significant forms of prayer, similar to the Panchánga of the Brahmans. Whoever constantly repeats or wears [made up in little lockets] a dharini, possesses a charmed life.

‡ See classified enumeration of the principal objects of Buddhist worship. But Pratisára is not therein named. These are Tántrika goddesses.

§ The Pancha Rakshá is now used in Courts of Justice to swear Buddhists upon.

Tárá Satnáma, an Upadesa Dháraní, contains an account of Arya Tárá, of her hundred names, her Víja mantras, &c. Verse: speaker, Padma Páni; hearer, Vajra Páni.

Sugatávadána, an Avadána Sástra, contains an account of the feast kept in honour of Sanghas or Bodhisatwas. Verse: speaker, Vasundhará Bodhisatwa; hearer, Pushpaketu Rájakumára.

Sukhavatí Loka, account of the so called heaven of Amitábha Buddha.** Verse: speaker, Sákya; hearers, Ananda and others.

Saptavara Dháraní, an Upadesa of the sort termed Dháraní; an account of the seven Devís (Buddha Saktis) called Vasundhará, Vajra Vidáriní, Ganapati Hridayá, Ushnisha Vijayá, Parna Savarí, Maríchi, Graha Mátriká, together with their Víja mantras. Prose: speaker, Sákya; hearers, Ananda and others.

Kriyá Sangraha, an Upadesa; an account of the Tántrika ritual. Prose: speaker, Sákya; hearers, Vajra Páni, &c., resembles the Mahodadhi of the Brahmans.

Sumaghávadána, an Avadána Sastra; on account of the heaven (Bhuvana) of the Bhikshukas; near the close is a story of the merchant Sumagha and his wife, whence the name of the work. Prose: speaker, Sákya; hearer, Ananda.

Chaitya Pungava, an Avadána on the worship of the Chaityas. Prose: speaker, Sákya; hearer, Suchetana Bhikshuka.

Kathinávadána, an Avadána Sástra; containing an account of the merit and reward of giving the Pindapátra,* Khikshari, Chívara and Nivása to Bhikshukas. Prose: speaker, Sákya; hearer, Kásyapa Bhikshu.

Pindapatrávadána, an account of the begging platter of the Bhikshus, and of the merit of bestowing it to them. Prose: speaker and hearer, as above.

Dhwajágra Keyuri, an Upadesa, or Tántrika Dháraní; an account of Dhwajágra Keyuri, Buddha Sakti. Prose: speaker, Sákya; hearer, Indra Deva (the god).

Graha Mátrika, a Tántrika Dháraní; account of Graha Mátriká, Buddha Sakti. Speaker, Sákya; hearer, Ananda Bhikshu.

Nágapújá, a manual of worship to the Nágas for rain. It is extracted from the Sádhana Málá. It is of the same character as the Vrata Paddhati of the Brahmans.†

Mahákála Tantra, an Upadesa; account of the worship to be paid to Mahákála. Prose: Vajra Satwa Bhagaván (i. e. Buddha); speaker and hearer, his Sakti, named Vajra Sattwátmakí.

Abhidhánottarottara, an Upadesa; account of the esoteric rites. Prose: speaker, Vajra Satwa Bhagaván; hearer, Vajra Páni. The rites prescribed by this book

** Dasabhuvana affords no place for Adi Buddha, or the five Dhyánis.

* The begging platter, staff, and slender habiliments of the Bauddha mendicant are called by the names in the text. The Chívara is the upper, the Nivása the lower garb; see on to No. 15 for dress and discipline of all the four orders. They require also for dress a pair of wooden sandals, an umbrella, and a gandhas or ewer for holding water.

† The high honour paid to the Nágas and Indra in Nepaul carries us beyond the Pauranic era to that older time represented in India by the Vedic gods and ritual.

resemble in character the Tántrika ritual of Brahmanism, and differ from it only in being addressed to different objects.

Vinaya Sútra, Treatise on Discipline. Author, Chandra Kírti Achárya. It is equivalent to the Vyása Sútra of the Brahmans.

Kalpalatávadána, an Avadána, a highly ornate account of the first birth of Sákya, and of the fruits of his actions in that birth. Verse: author, Kshemendra Bhikshu.

Gítá Pustaka, a Geya; a collection of songs on Tántrika topics, by various hands.

Stotra Sangraha, the praises of Buddha, Dharma, and Sangha. In verse of various measures and by various authors.

Divyávadána, an Avadána Sástra, containing various legends of the first birth of Sákya. Verse and prose: speaker, Sákya, hearers, Ananda Bhikshu and others.‡

BHOT LITERATURE IN THE LANGUAGE OF TIBET.

The following list of a more miscellaneous description.||

BHOTIYA WORKS.

Sumáchik; by Thula Lama, written at Khanam in Bhot, on Jurisprudence.

Chama Dam; by Aguchu Lama, at Tíja Nowa; subject similar to the Sagun Pothi of the Hindus.

Charúg; by Thiya Lama, at Gejaketha, on the Jnána Pothi of the Hindus, or divine wisdom.

Chúrúge Chapah; by Yepah Regreh Maha Lama, at Pargreh ah chu, on cure of all diseases.

Tuchurakh; by Suka Lama, at Jab-la Denuk; read by mendicant monks to prosper their petition for alms.

Mani Pothi; by Chufil Lama; at Gumewan; on the use and virtue of the mani or praying cylinder.

Chú Dam; by Gevighup Lama, at Yeparkas, on medicine.

Napache Pothi; by Aberak Lama, at Jatu Lam, on physical science, or the winds, rain, weather.

Kichak; by Kilúah Lama, at Botehi, on witchcraft, demonology, &c.

Tui takh lu; by Rakachandah Lama, at Kubakh, on science of war.

Dutakh-a-si; by Bajachik Lama, at Gnama, read by survivors on the death of a relation, that they may not be haunted by his ghost.

Serua-takh; by Takachik Lama, at Yipurki. To be read by travellers during their wanderings, for the sake of a safe return.

Sata-tu-mah: by Yisahsekar Lama, at Sebhala, read previous to sitting on a panchaet for a prosperous issue thereof.

Kerikh; by Amadatakh Lama, at Asi; to be read for increase of temporal goods.

‡ Since the above was composed, I have added greatly to my stock of Sanskrit works. For their names, see the list appended to next paper—Note of .1837.

|| This list represents merely the odds and ends first got at. Soon after I procured the catalogue of the Kahgyur and ascertained that the great Tibetan Cyclopœdia consisted of translations from those Sanskrit originals whereof a part only had been preserved in Nepaul. I learnt this, and sent the catalogue to Calcutta *before De Körös' appearance there.*

Numbeh; by Titakh Lama; at Bere-ga-hakh; to be read at times of gathering flowers for worship.

Dekmujah; by Múntake-tan Lama, at Múnká; to be read previous to laying the foundation of a house.

Thaka-pah; by Gagamatakh Lama, at Ma-chaclekoh; to be read whilst feeding the sacred fishes at the temple; a very holy act.

Kusa; by Nemachala Lama, at Yeparenesah; to be read at the time of bathing.

Lahassa-ki-pothi; by Uma Lama, at Lassa; to be read before eating, while dinner is serving up, to keep off wicked spirits.

Chandapu; by Grahah Lama, at Jubu-nasah; to be read previous to making purchases.

Sachah; by Urjanh Lama, at Jadún; to be repeated whilst exonerating themselves, that no evil spirit may come up.

Báchah; by Jahadegh Lama, at Maharah; to be read by lone travellers, in forests and bye-ways, for protection.

Kajaw; by Olachavah Lama, at Káráh; to be read by a dead man's relatives to free his soul from purgatory.

Yidaram; by Machal Lama, at Saduri; to facilitate interviews, and make them happy in their issues.

Ditakh; by Chopallah Lama, at Urasikh; to interpret the ominous croaking of crows, and other inauspicious birds.

Káráchakh; by Khuchak Lama, at Pheragiah.

Chala; by Gidu Lama, at Bídakh; to be read at the time of drinking, that no ill may come of the draught.

Kegú; by Tupathwo Lama, at Kabajeh; for increase of years, and a long life.

Chabeh; by Akabeh Lama, at Arí Kalaguh; to be read for removing the inclemencies of the season.

Kaghatukh; by Sugnah Lama, at Bole Káchar; to be read by horsemen, at seasons of journeys that they may come to no harm.

Lúchú; by Nowlah Lama, at Chagúra Kahah; to be read for increase of eloquence and knowledge of languages.

Ghikatenah: by Sujanah Lama, at Seakuhah; to be read by archers for success of their craft.

Baudh Pothi; or history of the founding of the Temple of Kasachit in Nepaul, with other matters appertaining to Buddhism in Nepaul.*

Siri Pothi; by Bistakow Lama; at Jamatakh; a general form of prayer for rich and poor, sick and healthy, man and woman.

The latter of these lists of Bhotiya books is a mere thing of shreds and patches, and, in fact, I have no means of enumerating the standard works of Tibetan literature. But I have no doubt that Tibet is indebted for its literature to Bauddha Missionaries, and Refugees from Hindustan. These individuals carried with them,

* The temples of Kasachit and of Swayambhu Nátha though situated in the Valley of Nepaul, are almost exclusively in the keeping of the Tibetans, and Lamas are the permanent ministering functionaries.

and subsequently procured from India, many of the sacred and profane works of their sect, and, as was their wont, they immediately began to instruct the people of Bhot in their own, that is, in the Sanskrit, letters and language. They had, no doubt, some success in this measure in the first period of their emigration into Bhot; but, in the end, the difficulties of Sanskrit, and the succession of Native teachers to the chairs of the original Indian emigrants, led to the preference of he Bhotiya language, and, consequently, to a translation of all the Sanskrit works they had, and could obtain from India, into the vernacular tongue of the country. This resort to translation took place very early; a circumstance which, aided by the lapse of time, and the further decline of the original literary ardour, inspired by the Indian Refugees, produced, at no distant period from the decease of the first Indian teachers, the oblivion of Sanskrit, and the entire supercession of original Sanskrit versions by translations into Tibetan. The Bhotiyas,* however, although they thus soon lost the Sanskrit language, retained the Devanagari letters. The result of the whole is, that the body of Bhotiya literature now is, and long has been, a mass of translations from Sanskrit; its language, native; its letters,(like its ideas) Indian. To support this view of the case, I have to observe, that even the Nepaulese, much nearer as they are to India, and much more cultivated in some respects as they are, have resorted extensively to vernacular comments, and even translations of their books, which also are Sanskrit; and that, although the Newars have a good language of their own, they have no letters, but such as are clearly of Devanagari origin, and declared by themselves to be so; that all the Bhotiyas, with whom I have conversed, assure me that they got all their knowledge from India; that their books are translations; that the originals, here and there, still exist in Bhot, but that now no one can read them; lastly, that most of the great Bhotiya classics proclaim, by their very names, the fact.† These remarks are applied, of course, to the classics of Bhot, for, in regard to works of less esteem there, I believe such to be not translations, but originals; chiefly legends of the Lamas, and in the vernacular tongue, (the best dialect of which is that spoken about Lassa and Digarchi,) but still, like the translated classics, written in letters essentially Indian.

THE RELIGION OF NEPAUL AND OF BHOT.

An accurate and complete view of the Bauddha system of belief would involve the severe study of a number of the voluminous Sanskrit works above specified,

* Bhot is the Sanscrit, and Tibet the Persian, name of the country. The native name is Bod, a mere corruption of the Sanskrit appellation, proving that the Tibetans had not reached a general designation for their country when the Indian teachers came among them.

† Note of 1837. It is needless now to say, how fully these views have been confirmed by the researches of De Körös. It is but justice to myself to add that the real nature of the Kahgyur and Stangyur was expressly stated and proved by me to the Secretary of the Asiatic Society some time before Mr. De Körös' ample revelations were made. Complete copies of both collections have been presented by me to the Hon. East India Company, and others procured for the Asiatic Society, Calcutta; upon the latter Mr. De Körös worked.

and would demand more time than could be bestowed upon the task by any person, not otherwise wholly unemployed. A few observations must, therefore, suffice in this place on the religious notions of the Bauddhas of this part of India, and in making them I shall keep chiefly in view the facilitation of the study of a new subject on the part of those who may find time and courage to explore the great and new mine of Sanskrit literature which it has been my fortune to discover in Nepaul.

Speculative Buddhism embraces four very distinct systems of opinion respecting the origin of the world, the nature of a first cause, and the nature and destiny of the soul.

These systems are denominated,† from the diognostic tenet of each, Swábhávika, Aiswarika, Yátnika, and Kármika; and each of these, again, admits of several sub-divisions, comprising divers reconciling theories of the later Bauddha teachers, who, living in quieter times than those of the first Doctors, and instructed by the taunts of their adversaries, and by adversity, have attempted to explain away what was most objectionable, as well as contradictory, in the original system.

The Swábhávikas deny the existence of immateriality; they assert that matter is the sole substance, and they give it two modes, called Pravritti, and Nirvritti, or action and rest, concretion and abstraction. Matter itself, they say, is eternal, (however infinitesimally attenuated in Nirvritti); and so are the *powers* of matter which powers possess not only activity, but intelligence.

The proper state of existence of these powers is that of rest, and of abstraction from everything palpable and visible, (Nirvritti), in which state they are so attenuated on the one hand, and so invested with infinite attributes of power and skill on the other, that they want only consciousness and moral perfections to become gods. When these powers pass from their proper and enduring state of rest into their casual and transitory state of activity, then all the beautiful forms of nature or of the world come into existence, not by a divine creation, nor by chance, but spontaneously; and all these beautiful forms of nature cease to exist, when the same powers repass again from this state of Pravritti, or activity, into the state of Nirvritti, or repose.

The revolution of the states of Pravritti‡ and Nirvritti|| is eternal, and with them revolve the existence and destruction of nature or of palpable forms. The Swábhávikas are so far from ascribing the order and beauty of the world to blind chance, that they are peculiarly fond of quoting the beauty of visible form as a proof of the intelligence of the formative powers; and they infer their eternity from the eternal succession of new forms. But they insist that these powers

† My Bauddha pandit assigned these titles to the Extract made from his Sástras, and always used them in his discussions with me. Hence I erroneously presumed them to be derived from the Sástras, and preferable to Mádyámika, &c., which he did not use, and which, though the scriptural denominations, were postponed to those here used on his authority as being less diagnostic. In making the extracts we ought to reach the leading doctrines, and therein I think we succeded.

‡ Pra, an intensive prefix; and Vritti, action, avocation, from *vrit* to turn, move, exist. See on these terms Burnouf, introduction, p.p. 441, 515.

|| Nir, a privitive prefix, and Vritti as before.

are inherent in matter, and not impressed on it by the finger of God, that is, of an absolutely immaterial being. Inanimate forms are held to belong exclusively to Pravritti, and therefore to be perishable; but animate forms, among which man is not distinguished sufficiently, are deemed capable of becoming by their own efforts associated to the eternal state of Nirvritti; their bliss in which state consists of repose or release from an otherwise endlessly recurring migration through the visible forms of Pravritti. Men are endowed with consciousness, as well, I believe of the eternal bliss* of the rest of Nirvritti, as of the ceaseless pain of the activity of Pravritti. But those men who have won the eternity of Nirvritti, are not regarded as rulers of the universe, which rules itself; nor as mediators or judges of mankind still left in Pravritti; for the notions of mediation and judgment are not admitted by the Swábhávikas who hold every man to be the arbiter of his own fate—good and evil in Pravritti being, by the constitution of nature indissolubly linked to weal and woe; and the acquistion of Nirvritti being, by the same inherent law, the inevitable consequence of such an enlargement of his faculties, by habitual abstraction, as will enable a man to know what Nirvritti is. To know this, is to become omniscient, a Buddha; to be divinely worshipped as such, while yet lingering in Pravritti; and to become, beyond the grave, or in Nirvritti, all at least that man can become, and all respecting which some of the Swábhávikas have expressed much doubt, while others of them have insisted that it is eternal repose, and not eternal annihilation§ (Súnyatá); though, adds this more dogmatical school, were it even Súnyatá, it would still be good; man being otherwise doomed to an eternal migration through all the forms of nature; the more desirable of which are little to be wished; and the less so, at any price to be shunned.

From the foregoing sketch it will be seen, that the most diognostic tenets of the Swábhávikas are, the denial of immateriality, and the assertion that man is capable of enlarging his faculties to infinity. The end of this enlargement of human faculties is association to the eternal rest of Nirvritti, respecting the value of which there is some dispute; and the means of it are, Tapas and Dhyána; by the former of which terms, the Swábhávikas understand, not penance, or self-inflicted bodily pain, but a perfect rejection of all outward (Právrittika) things; and, by the latter, pure mental abstraction. In regard to physics, the Swábhávikas do not reject design or skill, but a designer, that is, a single, immaterial, self-conscious being, who gave existence and order to matter by volition. They admit what we call the laws of matter, but insist that those laws are primary

* The doctrine is, that they are; some doctors, however, say no; the question turns on the prior acceptation of Sùnyatà, for which see on.

§ This interpretation of the Swábhávika Súnyatá is *not* the general one, though the opponents of Buddhism have attempted to make it so; for the prevalent sense of the word among the Buddhas, see on. Plotinus contended that the most perfect worship of the Deity consisted in a certain mysterious self-annihilation or total extinction of all our faculties. See M. Laurien's account of Newton's discoveries p. 387. This explains the Sangata doctrine of Dhyána, and partially that of Súnyatá also.

causes, not secondary; are inherent eternally in matter, not impressed on it by an immaterial creator. They consider creation a spontaneity, resulting from powers which matter has had from all eternity, and will have to all eternity. So with respect to man, they admit intellectual and moral powers, but deny that immaterial essence or being, to which we ascribe those powers. Animate and inanimate causation, they alike attribute to the proper vigour of nature, or Swabháva. I believe the Swábhávika to be the oldest school of Buddhist philosophy; but that school has, from the earliest times, been divided into two parties, one called the Swábhávikas simply, whose tenets I have endeavoured to state above, the other termed the Prájnika Swábhávikas, from Prajná,|| the supreme wisdom; viz. of nature.

The Prájnikas* agree with the Swábhávikas, in considering matter as the sole entity, in investing it with intelligence as well as activity, and in giving it two modes, or that of action and that of rest. But the Prájnikas incline to unitize the powers of matter in the state of Nirvritti; to make that unit, deity; and to consider man's *summum bonum*, not as a vague and doubtful association to the state of Nirvritti; but as a specific and certain absorption into Prajná, the *sum* of all the powers, active and intellectual, of the universe. The Aiswarikas admit of immaterial essence, and of a supreme, infinite, and self-existent Deity (Adi Buddha) whom some of them consider as the sole deity and cause of all things, while others associate with him a coequal and eternal material principle; believing that all things proceeded from the joint operation of these *two* principles. The Aiswarikas accept the two modes of the Swábhávikas and Prájnikas, or Pravritti and Nirvritti. But, though the Aiswarikas admit immaterial essence, and a God, they deny his providence and dominion; and though they believe Moksha to be an absorption into his essence, and vaguely appeal to him as the giver of the good things of Pravritti, they deem the connection of virtue and felicity in Pravritti to be independent of him, and the bliss of Nirvritti to be capable of being won only by their own efforts of Tapas and Dhyána, efforts which they too are confident will enlarge their faculties to infinity, will make them worthy of being worshipped as Buddhas on earth, and will raise them in heaven to an equal and selfearned participation of the attributes and bliss of the Supreme Adi Buddha; for such is their idea of Moksha, or absorption into him, or, I should rather say, of union with him. All the Bauddhas agree in referring the use and value of meditation, (earthly and heavenly,) of the rights and duties of morality, and of the ceremo ies of religion, solely to Pravritti, a state which they are all alike taught to contemn; and to seek, by their own efforts of abstraction, that infinite extension of their faculties, the accomplishment of which realizes, in their own persons, a godhead as complete as any of them, and the only one which some of them will acknowledge. The Kármikas and Yátnikas derive their names, respectively, from Karma, by which I understand 'conscious moral agency,' and Yatna, which I interpret

|| Prajná, from pra, an intensitive prefix, and Jnyaná, wisdom, or perhaps, the simple jna.

* See the sequal for a good summary glance at the philosophy of the Prájnikas.

'conscious intellectual agency.' I believe these schools to be more recent than the others, and attribute their origin to an attempt to rectify that extravagant quietism, which, in the other schools, stripped the powers above, (whether considered as of material or immaterial natures,) of all personality, providence and dominion; and man, of all his active energies and duties. Assuming as just, the more general principles of their predecessors, they seem to have directed their chief attention to the phænomena of human nature, to have been struck with its free will, and the distinction between its cogitative and sensitive powers, and to have sought to prove, notwithstanding the necessary moral law of their first teachers, that the felicity of man must be secured, either by the proper culture of his moral sense,* which was the sentiment of the Kármikas, or, by the just conduct of his understanding, a conclusion which the Yátnikas preferred: and this, I believe to be the ground of distinction between these two schools as compared with one another. As compared with their predecessors, they held a closer affinity with the Aiswarikas than with the other schools, inclined to admit the existence of immaterial entities, and endeavoured to correct the absolute impersonality and quiescence of the Causa Causarum, (whether material or immaterial,) by feigning Karma or Yatna, conscious moral, or conscious intellectual, agency, to have been with causation from the beginning. The Kármika texts often hold such a language as this, "Sákya Sinha, who, according to some (the Swabhávikas), sprang from Swabhάva, and, according to others, (the Aiswarikas,) from Adi Buddha, performed such and such Karmas, and reaped such and such fruits from them."

In regard to the destiny of the soul, I can find no essential difference of opinion between the Bauddha and the Brahmanical sages. By all, metempsychosis and absorption are accepted. But absorbed into what? into Brahma, say the Brahmans, into Súnyatá, or Swabháva, or Prajná, or Adi Buddha, say the various sects of the Buddhists. And I should add, that by their doubtful Súnyatá, I do not, in general, understand, annihilation, nothingness, but rather that extreme and almost infinite attenuation which they ascribe to their material powers or forces in the state of Nirvritti, or of abstraction from all particular palpable forms, such as compose the sensible world of Pravritti. By tracing the connextion of Súnyatá with Akása, and through it, with the palpable elements, in the evolution and revolution of Pravritti,† it may be plainly seen, that Súnyatá is the *ubi* and the *modus* of primal entity in the last and highest state of abstraction from all articular modifications such as our senses and understanding are cognizant of.

How far, and in what exact sense, the followers of these diverse and opposite systems of speculation adopted the innumerable deities of the existent Buddhist Pantheon, it must rest with future research accurately to determine. For my part, I have no stomach for the marshalling of such an immense, and for the most

* Notwithstanding these sentiments, which are princpially referable to the state of Pravritti, the Kármikas and Yátnikas still held preferentially to the Tapas and Dhyána, the severe meditative asceticism of the older schools.

† See the Dasákára or ten forms, where the evolution and revolution of each element constitutes a phrase of divine energy.

part useless, host.* But some of the principal objects of worship, with their relation and connexion, may be noticed. The leading, and most fundamental association of these objects is, that of the triad, or three persons named Buddha, Dharma, and Sangha. In the transcendental and philosophic sense, Buddha means 'mind,' Dharma, 'matter,' and Sangha, the concretion of the two former in the sensible or phænomenal world. In a practical and religious sense, Buddha means the mortal author of this religion (Sákya), Dharma, his law, and Sangha, the congregation of the faithful.

The triad is liable to a theistic or atheistic interpretation in the higher or philosophic sense, according as Buddha is preferred or postponed to Dharma.

The next, and a very marked distinction of persons, is established in this creed between those avowed mortals who win the rank and powers of a Buddha by their own efforts, and the Buddhas of a celestial nature and origin.

The most notorious of the former of these are seven† who are all characterised as "Mánushi" or human; of the latter are five or six who are contradistinguished as "Anupapádaká," without parents, and also as "Dhyáni," or divine.

This second appellation of the Celestial Buddhas is derived from the Sanskrit name for that abstracted musing which has found more or less favour with almost all the Asiatic religionists, but which is peculiarly and pre-eminently characteristic of Buddhism.

The Dhyáni Buddhas, with Adi Buddha, their chief, are usually and justly referred to the Theistic school.

The epithet Dhyáni, however, as applied to a class of Buddhas, is obviously *capable* of an atheistic interpretation. It is nevertheless certain, that, in whatever sense other schools may admit this term, or the class of divinities which it characterises, the Aiswarikas (beyond the bounds of Nepaul too)‡ ascribe this creative Dhyána to a *self-existent*, *infinite*, and *omniscient* "Adi Buddha," one of whose attributes is the possession of five sorts of wisdom. Hence he is called "Pánchajnána Atmika;" and it was by virtue of these five sorts of wisdom, that he, by five successive acts of Dhyána, created, from the beginning and for the duration of the present system of worlds, the "Pancha Buddha Dhyáni."

The names and graduation of these Jnánas, Dhyánas, and Buddhas are thus :—

Jnánas.	*Buddhas.*
1. Suvisuddha Dharma Dhátu.	1. Vairochana.
2. Adarsana.	2. Akshobhya.
3. Prativekshana.	3. Ratnasambhava.
4. Sánta.	4. Amitábha.†
5. Krityánushthána.	5. Amoghasiddha.

* See further on for a goodly array.

† Called Vipasyi, Sikhi, Viswabhu, Kakutsanda, Kanakamuni, Kasyapa, and Sákya Sinha. Two others are frequently associated with these to form a series of nine mortal Buddhas, the extra two being Dipankara and Ratnagarbha. But they are much less notorious than the seven, and even of them I find nothing distinct recorded, with the single exception of Sákya, whom I am therefore inclined to regard as the founder of this creed, such at least as it has come down to us in the existing books and existing practical religion of the Buddhists.

‡ For example, in the Ratna Kúta Amitábha and Akshobhya are spoken of, and in the Sarva dharma Mahásánti as well as in the Swayambhú purána and Guna

Dhyánas:—The Dhyána of creation is called by one generic name Loka-Sansarjana; and by five repetitions of this, the five Buddhas were created.

It might be expected, that the supreme Buddha, having created these five celestials, would have devolved on them the active cares of the creation and government of the world. Not so, however; the genius of genuine Buddhism is eminently quiescent, and hence these most exalted æons are relieved from the degradation of action. Each of them receives, together with his existence, the virtues of that Jnána and Dhyána, to the exertion of which, by Adi Buddha, he owed his existence; and by a similar exertion of both, he again produces a Dhyáni Bodhisatwa. The Dhyáni Bodhisatwas are, one by one, in succession, the tertiary and active authors of creation. These creations are but perishable; and, since the beginning of time, three of them have passed away. The present world is, therefore, the work of the fourth Bodhisatwa, who is now Lord of the ascendant, and his worshippers in Nepaul are wont to invest him with all the powers of a supreme and sole God, the "Præsens Divus" being, as usual, everything.‡ When the existing system of worlds shall have run its course, the offices of creator and governor of the next will be assumed by the fifth Bodhisatwa.

The names and lineage of these Dhyáni Bodhisatwas are as follows:—

1. Vairochana.	1. Samantabhadra.
2. Akshobhya.	2. Vajra Páni.
3. Ratnasambhava.	3. Ratna Páni.
4. Amitábha.	4. Padma Páni.
5. Amoghasiddha.	5. Viswa Páni.

The Dhyáni Buddhas and Bodhisatwas are considered to stand in the relation of fathers and sons to each other; and as there are Dhyáni Bodhisatwas, so are there Mánushi Bodhisatwas,§ who again bear to their respective Mánushi Buddhas the connexion of pupil to teacher, of graduate to adept, of the aspirant after the wisdom of Buddhism to him who possesses that wisdom. I should add, that it is competent for a mortal man to become a Buddha,** whilst he yet lingers in the flesh, albeit, the entire fulfilment of the rewards, if not of the prerogatives, of that transcendent character is assigned to a more unearthly state, viz., the state of Nir-

Karanda Vyúha, all Puranic or exoteric works, of which the first is not even obtainable in Nepaul, nor is there any evidence that any of the other works were composed there. See Csoma de Körös in *Bengal Asiatic Society's Journal.*

† Original of the Chinese O-mi-to, a word as utterly without meaning as their Bonze, of which latter the Sanskrit Bandya is the real and significant form. Amitábha is the immeasurably splendid. Bandya is a person entitled to reverence, and the collective or general appellation of all professed or ascetical followers of Buddha. See Crawford's *Archipelago* for a fine representation of Akshobhya, the second Dhyáni Buddha. All the five are represented in the Cave at Bāg.

‡ Hence the celebrity and popularity of his mantra or invocation (Om mani padmē hum), while those of the two other members of that triad to which Padmapani is thus associated as the Sangha, are hardly ever heard of. There is a fine image of Padma Páni at Karnagarh on the Ganges, the old capital of Champa, now Bhagalpur.

§ The nine mortal Bodhisatwas are variously and vaguely set down; see further on. Ananda, Manju Ghosha, and Avalokiteswara, are the only ones of whom anything is known.

** Hence the Divine Lamas of Bhot; though the original idea has been perverted somewhat. They are rather Arhantas.

vritti. In the above remarks I have inserted only the quinary series of Dhyáni Buddhas and Bodhisatwas. But there is, also, a series of six, the Buddha Vajra Satwa, and the Bodhisatwa Vajra Páni, being added to the series of five, to perfect the larger series. Further, as the five material elements,[1] the five senses,[2] and the five respective (outward) seats of sense,[3] are referred to the series of five Buddhas, so intellect,[4] with apprehension[5] and the objects of such apprehension, or the whole phenomena of the universe,[6]* are referred to Vajra Satwa Buddha†. And it should not escape remark, that the above associations give somewhat of the dignity of useful knowledge to what must otherwise have been mere *voces et prœterea nihil.*

Nor is there any want of sufficing original authority for the series of six Celestial Buddhas,‡ any more than for the series of five, though the latter may be, and perhaps is, the older. Wherefore I will take leave in this place to caution the reader against exclusive and confined opinions, founded upon any *one* enumeration he may find; as for instance, that of the Pancha Buddha Dhyáni. Any particular enumeration may have a definite object. But that does not imply that any other and larger enumeration, also with an express object, is *inconsistent* with the other series. It must at the same time be admitted that the ritualists appear to have multiplied these Deities upon very frail and shadowy grounds; and in this way I find the series of six Celestial Buddhas (which as identified with the elements, senses, and mind, I consider valid) augmented to nine by the addition of Vajrakáya, Vajradharma, and Vajrakarma. The next material distinction of persons or divinities in this religion is into Exoteric or Pauránika Buddhas and Esoteric or Tántrika. The first are those ordinarily so called and alone heretofore known to us. The second are more specially styled Yogámbara and Digambara: they form the link of connexion between Jainism and Buddhism; and their statues or images are distinguished either by nudity or by a multiplicity of members: they are wholly unknown to Europeans. I have already adverted to the general character of the Tántrika ritual. It is a strange and unintelligible adjunct of Buddhism, though vouched by numerous scriptural authorities.

The images of the 5 Dhyáni Buddhas, which were sent to the Asiatic Society of Bengal, occupy (and exclusively so of all lower Buddhas) the base of every Mahá chaitya,§ or highest order of temples in Nepaul; and those images are invariably distinguished by the respective differences exhibited in the specimens transmitted, *viz.*, the position of the hands; the nature of the supporters and the particular

(1) Five Bhútas. (2) Five Indriyas. (3) Five Ayatanas. (4) Manas. (5) Dhárana. (6) Dharma.

* The senses are assumed to be inert without Manas; not even sensation, far less perception, or mental realization of sensation, can exist without Manas.

† Vajra Satwa, or the sixth Dhyáni with his appendages, belongs to the Vámácháryas, whose doctrine as to things in general, or the origin, nature, and connexion of material and immaterial phenomena, can hardly be reconciled with the views of the older Dakshinácháryas on these topics.

‡ E grege the Sarva Dharma Mahásánti, said by Mr. Csoma to be the bible of the '*oldest*-Buddhist sect in Tibet.' For authorities for Adi Buddha and the six Celestial Buddhas, see Quotations in Proof, 1837.

§ Temple and monastery are the respective equivalents of Chaitya and of Vihára.

cognizance of each, which is placed between the supporters. Vairochana is seldom figured: the other four celestial Buddhas occupy shallow niches at the base of the hemisphere of the Chaitya, one opposite each cardinal point: Akshobhya to the east, Ratna Sambhava to the south, Amitábha to the west, and Amogha Siddha to the north. Vajra Satwa is seldom represented in statuary form, and never placed in the Chaityas. But pictorial representations of him are frequent in the illuminated Sástras, and I have met with his image or sculptured figure in Viháras.

The Chaitya would appear to be the only *exclusively* Buddhist form of temple. It consists of a solid hemisphere, commonly surmounted by a graduated cone or tetragonal pyramid, the grades of which (the cone or pyramid) are thirteen, and are typical of the thirteen Bodhisatwa heavens of Buddhist cosmography. The cone or pyramid terminates in a *palus* very like a *lingam*, and which is usually surmounted by an umbrella. This part of the structure represents Akanishtha Bhuvana, or the highest heaven, or that of Adi Buddha. The five spokes of the umbrella represent the abodes of the five Dharma Buddhas. Between the hemisphere and the cone or pyramid is a short square neck for the latter, upon each of the four sides of which a pair of eyes is graved which typify omniscience. The hemisphere is called the garbha; the neck, gala; and the cone or pyramid, chúrámani. The Nepaulese are sufficiently familiar with Chaityas in the sense of tomb temples, or mausolea, or covers of relics (Dagopa): but all their *principal* edifices of this nature are dedicated to the self-existent, first, supreme Buddha, and to his five celestial æons. Chaityas are frequently combined with small hollow temples, of which they form the superstructure: besides which many sacred edifices of Hindoo form are used by the Buddhists for enshrining their mortal Buddhas, as well as any of the numberless gods and goddesses of their ample Pantheon. The followers of Buddha are divided into regular and secular—a division exactly equivalent to the Grihastha Asrama and Vairágí or Sannyásí Asrama of the Hindoos—but *not* equivalent to Laics and Clerics. The regulars are all monastic, as solitaries or as cœnobites, living in deserts or in monasteries (Viháras). Their collective name is Bandya (person entitled to reverence); and they are divided into four orders, called Bhikshu or 'mendicants,' Srávaka or readers, Chailaka or 'the scantly robed,' and Arhata or Arhanta or 'Adepts.' They are all monks, and constitute the *congregation* of the *faithful*, or only *real* Buddhists; the seculars having always been regarded as little better than heretics, until political ambition began to qualify the high-toned enthusiasm of the primitive saints; and until very many having come in who could not all live in idleness, these were allowed to follow the various business of the world, their instruction being provided for by the monks, some of whom thus became invested with a partially clerical character which they exercised under the names of Achárya and Vajra Achárya or 'teacher and powerful teacher.' The monasteries or conventual dwellings of the regular Buddhists are called Vihára in Sanskrit, Bahi and Báhál in Newari. They are usually large open quadrangles of a regular form, but sometimes irregular, and built round a Chaitya, or a Kútágár temple, (the latter sacred to Mánushi, the former to Dhyáni Buddhas). Every great church was formerly conventual, and the four orders had each their separate Viháras, of which there are still fifteen in the city of Patan alone, though the Nepaulese have long since abandoned the monastic institutes of their creed, and hence these monasteries are now secularized, but still exclusively appropriated to the Bandya or tonsured Buddhists. The head of a Vihára is called Náyaka, but his power appears to have been much more limited than that of the Abbots and Priors of European monachism, and since this decay of the monastic institutes in Nepaul it has become at all events strikingly so. Still, however, it is the Náyakas alone who confer the rank and character of Bandya, and every Bandya is ostensibly attached to some convent or other, even though he do not dwell in any, as many now do not. Any person may become a Bandya by submitting to tonsure and taking the usual vows of celibacy, poverty, and humility, and all these monks are alike distinguished by a peculiar dress and equipment, which as well as the ceremony of induction will be found described in the sequel.

The following list of Buddhas completes all I have at present to offer on the subject. Two lists were prepared for me some time ago by an old Bauddha of Nepaul, with whom I have long cultivated an acquaintance; but they were then laid aside for future examination and explanation when opportunity should serve. I have accordingly had them compared, under my own eyes, with the scriptures whence they were extracted, and the comparison has suggested the following brief elucidatory remarks.

In the first place, the lesser list has proved to be superfluous, all its names being contained in the larger one. In the next place, the whole number of Buddhas in the greater catalogue has been found to amount to 131, and not to 145, as stated elsewhere; the same name being repeated, in some instances, two and three times, by reason of this catalogue consisting of literal extracts from several independent works. And I have thought it better to leave it in *statu quo*, than to omit sundry names of one series because they occur in another. Such omission might have interfered with some established contiguity of time, place, or circumstances, in regard to the Buddhas, with which we are not acquainted; and with respect to the repetitions, they may be seen in the list, at a glance, by the references attached to them. There is one deviation from the catalogues as found in the works whence they are drawn, and it is this. After the names of the six great Mánushi Buddhas (No. 50 to 56) the name of Sákya Sinha, the seventh and last, is given in my list, though not found at that place in the Lalita Vistara: possibly because Sákya had not, when that work was compiled, attained Nirvána and become a Tathágata in the proper sense. His name, though occurring before, is, notwithstanding, reinserted in my catalogue in that place, in order to make up the complement of the now famous 'Sapta Buddha Mánushi,' or seven mortal Buddhas. Before each distinct series of names, the work from which it is derived, is uniformly noted.

In the works cited, many more names, besides those given in the catalogue, are to be found, and from the whole of the books which have been procured and transmitted to Calcutta, hundreds of new names might be drawn.

In the Samádhi Rája,* Sarvártha Siddha (Sákya before he became a Buddha,) is asked by Maitreya and Vajra Páni, how he acquired Samádhi Jnána. In reply, he begins by naming 120 Tathágatas, who instructed him therein in his former births; and at the conclusion of his enumeration of Buddhas, Sarvártha Siddha observes, 'he has given so many names *exempli gratiâ*, but that his instructors were really no less in number than eighty *crores!*

There is a verse in the Aparimitá Dháraní (to be found in many other, and higher authorities) purporting that "the Buddhas who have been, are, and will be, are more numerous than the grains of sand on the banks of the Ganges." Some of these Buddhas sprang, divinely not generatively, from other Buddhas; some from Akása, and some from the Lotos. These are evident nonentities, in regard to chronology and history. Yet it is often most difficult to distinguish them from their more substantial compeers, the origin of the latter having been frequently traced up to heaven by the vanity of superstition, while its grovelling genius no less frequently drew down the lineage of the former to earth. Again, among the Buddhas confessedly of mortal mould, there are three wide degrees, that of the Pratyéka Buddha, that of the Srávaka Buddha, and that of the Máháyánika Buddha. But the two former are regarded, even by their worshippers, as little more than mere men of superior sanctity; and as infinitely inferior to the Máhá-yánika Buddhas, such as Sákya and his six great predecessors. We have, however, multitudes even of this highest degree; and, besides, the title belongs not only to the supreme Mánushi Tathágatas, but also to all the Dhyánís indiscriminately. Upon the whole, then, it seems peculiarly desirable, in the present state of our information, to keep a steady eye upon the authoritative assertion of the old scriptures, that Sákya is the *seventh*, and *last* of the Buddhas. Why seven have been selected for such especial honour it seems impossible to explain on historical grounds. Four of them belong to the present cycle of ages thence called the golden *æra* or Bhadra Kalpa: the three first to the precedent Kalpa. A Kalpa is an indefinite period, and I think it may be safely asserted that all of the so-called mortal Buddhas save the last are mythological shadows. At all events it has frequently occurred to me to doubt the historical existence of Sákya's six predecessors; for I have not failed to remark that while the Buddhist writings make ample mention of Sákya's births (505), sayings, and doings, and while they ascribe to him the effectual authorship of *all* the scriptural authorities of the sect, these writings are nearly silent with respect to the origin and actions of the six Buddhas who went before him; nor are any doctrines or dogmas referred to them in the authorities in question. To go farther into this matter would lead me beyond the bounds I have prescribed to myself on the present occasion. What I have said will suffice to shew why the catalogue of Buddhas has been so long withheld, and perhaps would justify the withholding of it still. In the forthcoming scriptures the form perpetually occurs 'so said Sákya,' and this is the reason why the works are ascribed to him, though they took their written shape from his favourite disciples Kásyapa, Ananda, and Upáli.

* I have this list before me extracted from the Samádhi Rája; but I do not think it worth while to add it to the lists already given.

LIST OF TATHAGATAS, COMPILED FROM THE LALITA VISTARA, KRIYA SANGRAHA AND RAKSHA BHAGAVATI.

LALITA VISTARA, 1ST SECTION.

1 Padmottara.
2 Dharmaketu.
3 Dípankara.
4 Gunaketu.
5 Mahákara.
6 Rishideva.
7 Srítejas.
8 Satyaketu.
9 Vajrasanhata.
10 Sarvábhibhú.
11 Hemavarna.
12 Atyuchchagámí.
13 Pravaraságara.
14 Pushpaketu.
15 Vararúpa.
16 Sulochana.
17 Rishigupta.
18 Jinavaktra.
19 Unnata.
20 Pushpita.
21 Urnatejas.
22 Pushkala.
23 Surasmí.
24 Mangala.
25 Sudarsana.
26 Mahásinhatejas.
27 Sthitabuddhidatta.
28 Vasantagandhi.
29 Satyadharmavipulakírttí.
30 Tishya.
31 Pushya.
32 Lokasundara.
33 Vistírnabheda.
34 Ratnakírttí.
35 Ugratejas.
36 Brahmatejas.
37 Sughosha.
38 Supushpa.
39 Sumanojnaghosha.
40 Sucheshtárúpa.
41 Prahasitanetra.
42 Gunarásí.
43 Meghaswara.
44 Sundaravarna.
45 Ayustejas.
46 Salilagajagámí.
47 Lokábhiláshita.
48 Jitasatru.
49 Sampújita.
50 Vipasyí.*
51 Sikhí.*
52 Viswabhú.*
53 Krakutsanda.*
54 Kanakamuni.*
55 Kásyapa.*
56 Sákyamuni.*

LALITA VISTARA, 13TH SECTION.

57— 1 Amoghadarsí.
58— 2 Vairochana.
59— 3 Dundubhíswara.
60— 4 Dharmeswara.
61— 5 Samantadarsí.
62— 6 Mahárchiskandhí.
63— 7 Dharmadhwaja.
64— 8 Jnánaketu.
65— 9 Ratnasikhí.
66—10 Padmayoni.
67—11 Sarvábhibhú. (See No. 10.)
68—12 Ságara.
69—13 Padmagarbha.
70—14 Sálendraraja.
71—15 Pushpita. (See No. 20.)
72—16 Yasodatta.
73—17 Jnánameru.
74—18 Satyadarsí.

* The seven famous mortal Buddhas.

75—19 Nágadatta.
76—20 Atyuchchagámí. (See No. 12)
77—21 Mahávyúha.
78—22 Rasmíráj.
79—23 Sákyamuni. (See No. 56.)
80—24 Indraketu.
81—25 Súryánana.
82—26 Sumati.
83—27 Nágábhibhú.
84—28 Bhaishajyaráj.
85—29 Sinhaketu.
86—30 Gunágradhárí.
87—31 Kásyapa. (See No. 55.)
88—32 Archihketu.
39—33 Akshobhyaráj.
90—34 Tagarasikha.
91—35 Sarvagandhi.
92—36 Mahápradípa.
93—37 Padmottara (See No. 1.)
94—38 Dharmaketu. (See No. 2.)

LALITA VISTARA, 20TH SECTION.

95— 1 Vimalaprabhása.
96— 2 Ratnárchih.
97— 3 Pushpávalivanarájikusumitábhijna.
98— 4 Chandrasúryajihmíkaraprabha.
99— 5 Gunarjaprabhása.
100— 6 Ratnayashti.
101— 7 Meghakútábhigarjitaswara.
102— 8 Ratnachhatrábhyudgatávabhása.
103— 9 Samantadarsí.
104—10 Ganendra.

KRIYA SANGRAHA.

105— 1 Vairochana.*† (See No. 58.)
106— 2 Mahoshnísha.
107— 3 Sitátapatroshnísha.
108— 4 Tejorási.
109— 5 Vijayoshnísha.
110— 6 Vikiranoshnísha.
111— 7 Udgatoshnísha.
112— 8 Mahodgatoshnísha.
113— 9 Vijayoshnísha. (See No. 163.)
114—10 Akshobhya. (See No. 85.)
115—11 Vajrasatwa.†
116—12 Vajraraja.
117—13 Vajrarága.
118—14 Vajrasádhu.
119—15 Ratnasambhava.
120—16 Vajraratna.
121—17 Vajrasúrya.
122—18 Vajraketu.
123—19 Vajrahása.
124—20 Amitábha.†
125—21 Vajradharma.
126—22 Vajratíkshna.
127—23 Vajraketu.
128—24 Vajrabhásha.
129–25 Amoghasiddha.†
103—26 Vajrakarma.
131—27 Vajraraksha.
132—28 Vajrayaksha.
133—29 Vajrasandhi.

* This name, although a repetition, is numbered; because the personage here indicated by the name *Vairochana*, is really *Vairochana Avatára, Manjusrí*. The six celestial *Buddhas of Nepaul* will be recognised in this list; but commenting were endless. The six are those marked thus †, Vairochana being assumed to be V. proper, and not Manjusrí.

RAKSHA BHAGAVATI.

134—1 Ratnákara.
135—2 Asokasrí.
136—3 Ratnárchih. (See No. 90.)
137—4 Jayendra.
138—5 Padmottarasrí. (See No. 1.)
139— 6 Súryamandalaprabhásottama.
140— 7 Ekachhatra.
141— 8 Samádhihastyuttarasrí.
142— 9 Padmasrí.
143—10 Nandasrí.

II. Sketch of Buddhism.

From Bauddha writings of Nepaul.

Soon after my arrival in Nepaul (1821), I began to devise means of procuring some accurate information relative to Buddhism: for, though the regular investigation of such a subject was foreign to my pursuits, my respect for science in general led me cheerfully to avail myself of the opportunity afforded, by my residence in a Bauddha country, for collecting and transmitting to Calcutta the materials for such investigation. There were, however, serious obstacles in my way, arising out of the jealousy of the people in regard to any profanation of their sacred things by an European, and yet more, resulting from the Chinese notions of policy adopted by this Government. I nevertheless persevered; and time, patience, and dexterous applications to the superior intelligence of the chief minister, at length rewarded my toils.

My first object was to ascertain the existence or otherwise of Bauddha scriptures in Nepaul; and to this end I privately instituted inquiries in various directions, in the course of which the reputation for knowledge of an old Bauddha residing in the city of Pátan, drew one of my people to his abode. This old man assured me that Nepaul contained many large works relating to Buddhism; and of some of these he gave me a list. When we became better acquainted, he volunteered to procure me copies of them. His list gradually enlarged as his confidence increased; and at length, chiefly through his kindness, and his influence with his brethren in the *Bauddha* faith, I was enabled to procure and transmit to Calcutta a large collection of important *Bauddha* scriptures.*

Meanwhile, as the *Pátna Bauddha* seemed very intelligent, and my curiosity was excited, I proposed to him (about 1823) a set of questions, which I desired he would answer from his books. He did so; and these questions and answers form the text of this paper. Having in his answers quoted sundry *slokas* in proof of his statements; and many of the scriptures whence these were taken being now in my possession, I was tempted to try the truth of his quotations. Of that, my research gave me in general satisfactory proof. But the possession of the books led to questions respecting their relative age and authority; and, tried by this test, the *Bauddha's* quotations were not always so satisfactory. Thus one step

* Nearly all were eventually procured, chiefly, and in the first place solely, for Calcutta. They were deposited first with the Librarian of the College of Fort William, then with the Asiatic Society, but were for years utterly neglected, and still are so I fancy; so also the copies sent to London and Oxford. Those sent to France met with a far different reception; see Burnouf.

led to another, until I conceived the idea of drawing up, with the aid of my old friend and his books, a sketch of the terminology and general disposition of the external parts of Buddhism, in the belief that such a sketch, though but imperfectly executed, would be of some assistance to such of my countrymen as, with the books only before them, might be disposed to enter into a full and accurate investigation of this almost unknown subject.

When, however, I conceived that design, I little suspected where it would lead me; I began ere long to feel my want of languages, and (to confess the truth) of patience, and almost looked back with a sigh to the tolerably full and tolerably accurate account of Buddhism which I had obtained so long ago, and with little comparative labour, from my old friend's answers to my queries. I also saw certain notices of Buddhism coming from time to time before the world, ushered by the talents and industry of Klaproth and Rémusat; and, so far as I had opportunity to learn what these notices contained, it seemed that the answers to my questions furnished much ampler and more accurate views of the subject than these distinguished men could extract from their limited sources of information.

I add here a very considerable list of the *Bauddha* scriptures in general, extracted for me from those still existing in Nepaul, without further observation on it than that its accuracy may be relied on, and that its contents are so far from being local to Nepaul, that the largest portion of the books neither are, nor ever were procurable in this valley.

The *Bauddhas* were used, in old time, to insert at the end of any particular work, lists of the names of many of their sacred writings; and to this usage of theirs am I indebted for the large catalogue which I have obtained.

LIST OF SANSKRIT BAUDDHA WORKS.

1. PURANAS OR EXOTERIC WORKS.

1 Satasáhasrika Prajná Páramitá.
2 Pancha Vinsati Sáhasrika Prajná Páramitá.
3 Ashtádasa Sáhasrika Prajná Páramitá.
4 Ashta Sáhasrika Prajná Páramitá.
5 Sapta Sati Prajná Páramitá.
6 Prajná Páramitá Vyákhyá.
7 Ganda Vyúha.* .... Bhadráchárí.
8 Dasa Bhúmeswara.
9 Samádhi Rája.†
10 Lankávatára.
11 Saddharma Pundaríka. .... Bhadráchárí.
12 Lalita Vistara.
13 Tathágata Guhyaka, or Guhya Samádhi (Tantra).
14 Suvarna Prabhása.

* Ascribed to Arya Sanga, and teaches the Yogácharya branch of the Maháyána.

† This book and the Buddhávatamsaka and the Ratnakúta are works ascribed to Nágárjuna, a transcendentalist after whom the western barrier mountain of the Valley of Nepaul is named.

15 Mahávastuavadána.	Samajátaka. Kinnaríjátaka
	Dípankaravastu. Birkúsávadána.
16 Divyávadána.	Sárdúlakarnávadána.
17 Satakávadána.	Opakhádhávadána.
	Barikávadána.
	Ráshtra Pálávadána.
18 Bhadrakalpávadána.	Birkúsávadána.
	Kinnaríjátaka.
19 Asokávadána.	Bodhi Charyávatára.
	Sapta Kumárikávadána.
	Durgati Parishodhana.
	Ahorátri vrata.
	Kártika Máhátmya.
	Chaitya Pungava.
20 Vichitra Karnikávadána.	
21 Dwávinsatyavadána.	
22 Ratnamálávadána, or Ratnávadána.	Suchandrávadána.
23 Avadána Kalpalatá.	
24 Sugatávadána.	
25 Dharma Kosha.	
26 Dharma Sangraha.	
27 Vinaya Sútra.‡	
28 Maháyána Sútra.	
29 Maháyána Sútrálankára.	
30 Gosringa Vyákhyána.	
31 Sáláchakrávadána.	
32 Játakávadána.	
33 Játaka Málá.	Viswántarajátaka.
34 Mahá Játaka Málá.	
35 Swayambhú Purána Kalpa.	
36 Swayambhú Purána Mahatá.	
37 Swayambhú Purána Madhyama.	
38 Swayambhú Purána.	Manichúrávadána.
39 Karanda Vyúha.	
40 Gunakaranda Vyúha.	
41 Sukhávatí Vyúha.	
42 Karuná Pundaríka.	
43 Lalita Vistara, or Tathágata Janmávadána.	
44 Laukika Lankávatára.	
45 Chaitya Máhátmya.	
46 Kalpadrumávadána	Kavikumárávadána.

‡ Only trace of Vinaya *eo nomine*, though this be one grand division of the books of the Ceylonese and Tibetans. But Burnouf I think observes that the Vinaya class of books in those places is represented by the Avadána, its equivalent in Nepaul.

Uposhadhávadána.

47 Dharma Kosha Vyákhyá.

48 Avadána Sárasammuchaya. .... Sumagadhávadána.
Sahakopadesávadána.
Kapisávadána.
Kathinávadána.
Pindápátrávadána.

49 Vratávadána Málá. .... Nandimukha.
Sughoshávadána.
Dhímatyavadána.
Sringabherí, &c.

50 Anumána khanda.
51 Adikarma pradípa.
52 Sádhana yuga Tippaní.
53 Manju Srí Párájika.*
54 Vajra Satwa Párájika.
55 Lokeswara Párájika.
56 Chhando Mrityulatá.
57 Suvarnavarnávadána.
58 Tárá Satanáma.
59 Buddha Síkshá Samuchchaya.
60 Pancha Rakshá.
61 Buddhokta Sansáramaya.
62 Laksha Chaitya Vratánusansá.
63 Pratimoksha Sútra.
64 Vajra Súchi.
65 Buddha Charita Kávya.
66 Gautama Kávya.
67 Punya Pratisáha Kávya.
68 Lokeswara Sataka Kávya.
69 Sragdhará Kávya.
70 Vidagdhamukhamandana Kávya.

2. TANTRAS OR ESOTERIC WORKS.†

71 Pramodya Maháyuga Tantra.
72 Paramártha Sevá Tantra.
73 Pindí Krama Tantra.
74 Sampútodbhava Tantra.
75 Hevajra Tantra.
76 Buddha Kapála Tantra.
77 Samvara Tantra, or Samvarodya.
108 Vajravíra Tantra.
109 Vajra Satwa Tantra.
110 Maríchi Tantra.
111 Tárá Tantra.
112 Vajradhátu Tantra.
113 Vimalaprabhá Tantra.
114 Manikarniká Tantra.

* Nos. 53, 54, and 55 are Vinaya as to matter. Gogerly says 52 related to the law for expulsion from the congregation.

† See Asiatic Researches, vol. v., p. 62 and note.

78 Váráhí Tantra, or Váráhí Kalpa.
79 Yogámbara Tantra.
80 Dákiní Jála Tantra.
81 Sukla Yamári Tantra.
82 Krishna Yamári Tantra.
83 Píta Yamári Tantra.
84 Rakta Yamári Tantra.
85 Syáma Yamári Tantra.
86 Kriyá Sangraha Tantra.
87 Kriyá Kanda Tantra.
88 Kriyá Ságara Tantra.
89 Kriyá Kalpa Druma Tantra.
90 Kriyárnava Tantra.
91 Abhidánottara Tantra.
92 Kriyá Samuchchaya Tantra.
93 Sádhana Málá Tantra.
94 Sádhana Samuchchaya Tantra.
95 Sádhana Sangraha Tantra.
96 Sádhana Ratna Tantra.
97 Sádhana Paríkshá Tantra.
98 Sádhana Kalpalatá Tantra.
99 Tatwa Jnána Siddhi Tantra.
100 Jnána Siddhi Tantra.
101 Guhya Siddhi Tantra.
102 Udyána Tantra.
103 Nágárjuna Tantra.
104 Yogapítha Tantra.
105 Píthávatára Tantra.
106 Kálavíra Tantra, or Chanda Roshana.
107 Mahá Kála Tantra.

115 Trilokyavijaya Tantra.
116 Sampúta Tantra.
117 Marma Kálika Tantra.
118 Kuru Kula Tantra.
119 Bhúta Dámara.
120 Kála Chakra Tantra.
121 Yoginí Tantra.
122 Yoginí Sanchára Tantra.
123 Yoginí Jála Tantra.
124 Yogámbarapítha Tantra.
125 Uddámara Tantra.
126 Vasundhará Sádhana Tantra.
127 Nairátma Tantra.
128 Dákárnava Tantra.
129 Kriyá Sára Tantra.
130 Yamántaka Tantra.
131 Manju Srí Kalpa Tantra.
132 Tantra Samuchchaya Tantra.
133 Kriyá Vatansa Tantra.
134 Tantra Sloka Sangraha.
135 Hayagríva Tantra.
136 Sankírna Tantra.
137 Námasangíti Vyákhyá Tantra.
138 Amrita Karnika náma Sangíti Tíká.
139 Gúdhotpáda náma Sangíti Tíká.
140 Máyá jála Tantra.
141 Jnánodaya Tantra.
142 Vasanta Tilaka Tantra.
143 Nispanna Yogámbara Tantra.
144 Dháraní Sangraha. { Pancha Buddha Dháraní — Pratyangira Dháraní, Saptavara Dháraní, with hundreds more, the work being a collection of them all. }

N. B.—Names on the *right* are portions of the work written opposite them on the left; priorly they had been treated as separate works.

The whole of the above are classed under the two important heads of Exoteric and Esoteric, the subdivisions not being noted. This list has been corrected since the paper to which it was originally attached was written.

In a clever paper in the first and second numbers of the Calcutta Quarterly Oriental Magazine, (Review of the Bombay Literary Transactions), it is said that one of the distinctions between Jainism and Buddhism is, that the *Jaina* statues are all naked, and the *Bauddha* statues all clothed. The pictures were sent to prove that this notion was false. The *Bauddha* images are called *Digam-*

bara,* a name heretofore fancied to be peculiar to Jainism; this is another error, and were this the place for dissertation, I could bring forward many other presumptions in favour of the notion that the *Jainas* are sectarian *Bauddhas*,† who dissented from their *Bauddha* brethren merely in carrying to a gross excess, and in promulgating publicly, certain dangerous dogmas, which the more prudent Buddhists chose to keep veiled from all but the initiated. The Nepaul Buddhists are very jealous of any intrusion into their esoteric dogmas and symbols; so much so, that though I have been for seven years enquiring after these things, my old *Vajra Achárya* friend only recently gave me a peep at the esoteric dogmas; and my *Chitrakára*, (*Bauddha* though he be,) has only within these last twelve months brought me some esoteric pictures: nor probably should I have got at these secret things at all, if I had not been able to examine the *Bauddha* books, in some small degree, myself; and if a *Bhótiya* had not put into my hands a picture containing one of these naked saints. With these decisive means of questioning in my power, I at last got my *Bauddha* assistants to draw up the veil of the sanctuary, to bring me copies of the naked saints, and to tell me a little of the naked doctrines.

Every part of each image is significant; the differences between the five are marked, first, by the different position of the hands (which is called the *mudrá*); secondly, by the variety of the supporters, called *váhanas*; thirdly, by the variety of the cognizances or *chínas* placed between the supporters; and fourthly (where painting and colours are used), by difference of colour. *Vairochana's* appropriate colour is white; *Akshobhya's*, blue; *Ratna-Sambhava's*, yellow, or golden; *Amitábha's*, red; and *Amogha-Siddha's*, green.‡

There are a few matters connected with the following sketch of Buddhism which it may be advisable to state here; and in the first rank stands the authority upon which I have assigned the meaning of intellectual essence to the word *Buddha*, and that of material essence to the word *Dharma*. The *Bauddhas* define the words thus: '*Bodhanátmaka iti Buddha; Dháran-átmaka iti Dharma.*' About the former of these definitions there can be no difficulty; there may concerning the latter. To the word *Dhárana*, or holding, containing, sustaining (from the root *dhri*), I have assigned a material sense; first, because it is opposed to *bodhana;* secondly, because the goddess Dharma, the *právrittika* personification of this principle, is often styled, in the most authentic books, *Prákriteswari* 'the material goddess,' or 'goddess of matter;' and thirdly, because this goddess is, (under the names Dharma, Prajna, Arya Tara, etc.) in very many passages of old *Bauddha* works, described as the material cause of all things; conformably, indeed, with that bias towards materialism, which our heretofore scanty knowledge of Buddhism has led us to assign to the *Saugata* faith.

* See J.R.A.S. ii. 1, 140. † See Digambar and Yogámbar.

‡ For the positions of these Buddhas in Chaitya temples see further on; Akshobhya is enshrined on the east side, Ratna Sambhava on the south, Amitábha on the west, and Amogha Siddha on the north. Vairochana is seldom found, but if he be, his station is immediately to the right of Akshobhya. Amogha Siddha has always a canopy of snakes. For Nágapúja in Nepaul see further on.

Sangha, the third member of the Triad, belongs not to the exalted state of *nirvritti*, in which no sect of *Bauddhas* admits more than two principles of all things, or mind and matter, *Buddha* and *Dharma*. *Sangha* is defined *Samudáyi átmaka iti Sangha*, 'the multitudinous essence;' because multitude is held to be as strong a characteristic of *pravritti*, or 'the palpable world,' as unity is of the world of *nirvritti*, or 'abstraction.'

In note 31, I have distinctly rejected the fifth order of *Bandyas*,* or *Vajra Achâryas*, in opposition to my old *Bauddha* friend's statement in the text of the Sketch. There can be no doubt that my friend is mistaken: for in many high authorities, the four original and true orders of *Bandyas* are called by the collective name of the *Chatur Varna*, and are therein described without mention of the *Vajra Achâryas*. It may serve to explain my friend's statement to tell you that he is himself a *Vajra Achârya;* and that as the genuine monachism of Buddhism has long since passed away in Nepaul, sundry local books have been composed here by *Vajra Achâryas*, in which they have made their own modern order coequal with the four ancient orders; and my old friend would hold these modern Nepaul books sufficient warrant for the rank ascribed to *his own class*. I have lately spoken to him on this subject, and he has confessed that there is no *old* authority for his fifth order of *Bandyas*. In my note I have endeavoured carefully to separate Buddhism *as it is* (in Nepaul) and Buddhism *as it ought to be*, *quoad* this point of classification. If you look into Kirkpatrick's and Buchanan's works on Nepaul, you will see how they have been puzzled with the difference of things as they are from what they ought to be, in those casual and erroneous hints which they have afforded on the subject of Buddhism.

In note 15, I have stated that the *Kármikas* and *Yátnikas* entertained tolerably just views on the grand subject of free-will and necessity; and I believe I am therein essentially correct: for how otherwise are we to understand their confession of faith, 'the actions of a man's prior births are his destiny?' Exclude the metempsychosis, which is the vehicle of the sense of this passage, and we have our old adage, 'Conduct is fate:' a law of freedom surely.

Still, were I cross-examined, I might be forced to confess, that the ideas which the *Kármikas* and *Yátnikas* entertain of free-will, seem to resemble rather the qualifications of our Collins and Edwards, than the full and absolute freedom of Clarke and the best European philosophers.

The *Kármikas* and *Yâtnikas* seem to have been impressed with the *fact* of man's free-will, but to have been perplexed in reconciling such a notion with the general spirit and tendency of the old *Swábhávika* philosophy. But in the result, the *Kármikas* and *Yátnikas* seem to have adhered to free-will, though perhaps in the qualified sense above mentioned.

Question I.

How and when was the world created?

* Bandya is the original and correct form of the Chinese Bonze and Mongolian Bandida, as Arhata or Arhanta is of the Indo-Chinese Rahatun.

ANSWER.

According to the *Sambhu Purána*, in the beginning all was void *(súnya)*. The first light that was manifest was the word *Aum;* and from this *Aum* the alphabet was produced—called *Mahá Varna*, the letters of which are the seeds of the universe. (See note 1.) In the *Guna Káranda Vyúha* it is written, when nothing else was, SAMBHU was; that is the self-existent *(Swayambhú);* and as he was before all, he is also called A'DI-BUDDHA. He wished from one to become many, which desire is denominated *Prajna*. BUDDHA and PRAJNA united became PRAJNA UPAYA, as SIVA SAKTI, or BRAHMA MAYA. (See note 2.) In the instant of conceiving this desire, five forms or beings were produced, called the five BUDDHAS (see note 3), whose names are as follows: VAIROCHANA, AKSHOBHYA, RATNA-SAMBHAVA, AMITABHA, AMOGHA-SIDDHA. Each of these BUDDHAS, again, produced from himself, by means of *Dhyána*, another being called his *Bodhi-Satwa*, or son. VAIROCHANA produced SAMANTA-BHADRA; AKSHOBHYA, VAJRA-PANI; RATNA-SAMBHAVA, RATNA-PANI; AMITABHA, PADMA-PANI; and AMOGHA-SIDDHA, VISWA-PANI.

Of these five *Bodhi-Satwas*, four are engrossed with the worship of Sambhu *(Swayambhú)*, and nothing more is known of them than their names; the fifth, Padma-Páni, was engaged by Sambhu's command, in creation (see note 4): and having by the efficacy of Sambhu's *Dhyána*, assumed the virtues of the three *Gunas*, he created Brahmá, Vishnu, and Mahesa, and delegated to them respectively, creation, preservation, and destruction. Accordingly, by Padma-Páni's commands, Brahmá set about creating all things; and the *Chatur-yoni* (or oviparous, viviparous, etc.,*) came into existence by Brahmá. The creation of Brahmá, Vishnu, and Mahesa by Padma-Páni, is confirmed by the *sloka* (see note 5), the meaning of which is, Kamáli (Padma-Páni,) produced Brahmá for creating, Vishnu for preserving, and Mahesa for destroying. And the creation of Brahmá is six-sorted, *viz.*, *Déva*, *Daitya*, *Mánusha*, etc.; and, for the *Dévas*, Brahmá made heaven; and for the *Daityas*, *Pátála;* and the four remaining kinds he placed between these two regions and upon the earth.

With respect to the mansions *(Bhuvanas)* of the universe, it is related, that the highest is called *Agnishtha Bhuvaná;* and this is the abode of Adi-Buddha. And below it, according to some accounts, there are ten; and according to others, thirteen *Bhuvánas* (see note 6); named, *Pramóditá*, *Vimalá*, *Prabhákarí*, *Archishmatí*, *Sudúrjayá*, *Abhimukhi*, *Dúrangamá*, *Achalá*, *Sádhúmati*, *Dharma-mégha* (x), *Samanta-prabhá*, *Nirúpamá*, *Jnyánavatí* (xiii).† These thirteen *Bhuvanas* are the work of Adi-Buddha: they are the *Bódhi-Satwa Bhuvanas;* and whoever is a faithful follower of Buddha will be translated to one of these mansions after death.

* By et cætera *always* understand *more Bráhmanorum.*

† Aknishtha or Agnishtha is not named in the Dasa Bhuvana, and neither therein nor here is any mention made of the abodes of the five Dhyani Buddhas; and not Achala but Samanta Bhadra is the tenth Bhuvana. Nirupama, Achala, and Jnyanavati are the three extra Bhuvanas.

Below the thirteen *Bódhi-Satwa Bhuvanas* are eighteen *Bhuvanas*, called collectively *Rúpyavachara*. These are subject to Brahmá, and are named individually: Brahma-káyiká, Brahma-púróhitá, Brahma-prashádyá, Mahá Brahmaná, Paritábhá, Apramánábhá, Abháswará, Parita-subhá, Subhákishná, Anabhraká, Púnya-prasavá, Vrihat-phúlá, Arangi-satwá, Avríhá, Apáyá, Sudrishá, Sudarsaná, and Sumúkhá. Pious worshippers of Brahmá shall go to one of these eighteen *Bhuvanas* after death.

And below the eighteen mansions of Brahmá, are six others subject to Vishnu, called collectively *Kámávachará*, and separately as follows: *Chatúr-Mahá-rája-Káyiká, Trayastrinsá, Túshita, Yamá, Nirmánavatí, Paranirmitá-Vasavartí*. And whosoever worships Vishnu with pure heart shall go to one of these.

And below the six *Bhuvanas* of Vishnu are the three *Bhuvanas* of Mahadeva, called generally *Arúpyavachará*, and particularly as follows: *Abhógá-Nitya-yatnópagá, Vijnyá-yatnópagá, Akinchanya-yatnópagá*, and these are the heavens designed for pious *Siva-Márgís*. Below the mansions enumerated, are Indra Bhuvana, Yama Bhuvana, Sûrya Bhuvana, and Chandra Bhuvana; together with the mansions of the fixed stars, of the planets, and various others which occupy the places down to the *Agni Bhuvana*, also called *Agni-kunda*. And below *Agni-kunda* is *Vayu-kunda;* and below *Vayu-kunda* is *Prǐthví*, or the earth; and on the earth are seven *Dwipas, Jambu Dwipa*, etc.; and seven *Ságaras* or seas, and eight *Parvatas* or mountains (see note 7), *Suméru Parvata*, etc. And below *Prǐthví* is *Jala-kunda*, or the world of waters; and the earth is on the waters as a boat. And below the *Jála-kunda* are seven *Pátálas*, as *Dharaní*, etc.: six of them are the abodes of the *Daityas;* and the seventh is *Naraka*, consisting of eight separate abodes: and these eight compose the hell of sinners; and from the eighteen *Bhuvanas* of Brahmá down to the eight chambers of *Naraka*, all is the work of Manjusrí. Manjusrí is by the Bauddhas esteemed the great architect, who constructs the mansions of the world by Adi-Buddha's command, as Padma-Páni, by his command, creates all animate things.

Thus Manjusrí (see note 8) is the Visva-karmá of the *Bauddhas;* and is also the author of the sixty-four *Vidyás*.

Question II.

What was the origin of mankind?

Answer.

It is written in the narrative portion of our *Tantras*, that originally the earth was uninhabited. In those times the inhabitants of *Abháswará Bhuvana* (which is one of the *Bhuvanas* of Brahmá) used frequently to visit the earth, and thence speedily to return to *Abháswará*. It happened at length, that, when a few of these beings, who, though half males and half females, had never yet, from the purity of their minds, conceived the sexual desire, or even noticed their distinction of sex, came, as usual, to the earth, Adi-Buddha suddenly created in them so violent a longing to eat, that they ate some of the earth, which had the taste of almonds, and by eating it they lost their power of flying back to their *Bhuvana*, and so they remained on the earth. They were now constrained to eat the fruits of the earth for sustenance; and from eating these fruits they conceived the sexual desire, and began to associate together: and from that time, and in that manner, the origin of mankind commenced from the union of the sexes. (See note 9.)*

When the beings above-mentioned came last from *Abháswará*, Mahá Samvata was their leader, and he was the first king of the whole earth.

In another *Tantra* it is written that Adi-Buddha is the immediate creator of all things in heaven and earth.

With respect to time, we conceive the *Satya-yuga* to be the beginning of time, and the *Kali-yuga* the end of it: and the duration of the four *yugas*, the particulars of which are found in the Brahmanical scriptures, have no place in our's in which it is merely written that there *are* four *yugas;* and that in the first, men lived 80,000 years; in the second, 10,000; in the third 1,000: and the fourth is divided into four periods; in the first of which, men will live 100 years; in the second, fifty years; in the third, twenty-five years; and in the fourth, when the close of the *Kali-yuga* is approaching, seven years only; and their stature will be only the height of the thumb; and then all things will be destroyed, and Adi-Buddha alone remain: and this period of four *yugas* is a *Pralaya*. Adi-Buddha will then again create the four *yugas*, and all things else to live in their duration, which when completed, all things will be again destroyed, and thus there will be seventy-one *pralayas*, or completions of the four *yugas*, when *Mahá Pralaya* will arrive. How many revolutions of the four *yugas* (*i.e.* how many *pralayas*) have now passed, and how many remain to revolve, is nowhere written.

QUESTION III.

What is matter, and what spirit?

ANSWER.

Body (see note 10), which is called *Sárira* and *Deha*, was produced from the five elements; and soul, which is called *prána* and *jíva*, is a particle of the essence of Adi-Buddha. Body, as created out of the elements, perisheth: soul, as a particle of the divine spirit, perisheth not; body is subject to changes—to be fat and lean, etc.; soul is unchangeable. Body is different in all animals; soul is alike in all, whether in man or any other creature. But men have, besides *prána*, the faculty of speech, which other animals have not; according to the *sloka*, of which the meaning is this: "*Deha* is derived from the five *Bhútas*, and *Jíva* from the *Angas* of *Swayambhú*." (See note 11.)

QUESTION IV.

Is matter an independent existence, or derived from God?

ANSWER.

Body, according to some, depends upon the inhaling and exhaling of the *Prána-Váyu;* and this inhalation and exhalation of the breath is by virtue of the soul *(prána)*, which virtue, according to some, is derived from God, and according to

* See Turnour's and Csoma de Körös versions of this legend, in the *Journal of the Asiatic Society of Bengal.*

others (see note 12), is inherent in itself: there is much diversity of opinion on this subject. Some of the *Buddha-márgís* contend that *deha* (the body) is *Swábhávaka; i. e.*, from the copulation of males and females, new bodies proceed; and they ask who makes the eyes, the flesh, the limbs, etc. of the fœtus in the mother's womb? *Swábháva!* And the thorns of the desert, who points them? *Swábháva!* And the timidity of the deer kind, and the fury of the ravenous beasts, whence are they? from *Swábháva!*

And this is a specimen of their reasoning and proofs, according to a *sloka* of the Buddha-Charita-Kávya. (See note 13.) Some again say, that *deha* and *sansára* are Aiswarika (see note 14), *i. e.*, produced by Iswara, or Adi-Buddha, according to another *sloka*.

Some again call the world and the human body *Kármika, i. e.*, that *Karma* is the cause of this existence of *deha* and *sansára;* and they liken the first *deha* to a field *(kshetra)*, and works, to a seed. And they relate, that the first body which man received was created solely by Adi-Buddha; and at that time works affected it not: but when man put off his first body, the next body which he received was subject to *Karma*, or the works of the *first* body (see note 15); and so was the next, and all future ones, until he attained to *Mukti* and *Moksha;* and therefore they say, that whoever would be free from transmigration must pay his devotions to Buddha, and consecrate all his worldly goods to Buddha, nor ever after suffer such things to excite his desires. And, in the Buddha-Charita-Kávya it is written, that with respect to these points, Sákya expressed the following opinion: "Some persons say that *Sansára* is *Swábhávaka*, some that it is *Kármika*, and some that it is *Aiswarika* and *Atmaka;* for myself, I can tell you nothing of these matters. Do you address your meditation to Buddha; and when you have attained *Bodhijnána*, you will know the truth yourselves."

Question V.

What are the attributes of God?

Answer.

His distinctive attributes are many; one of which is, that he is *Panchajnánátmaka* (see note 16), or, in his essence are five sorts of *jnána*, possessed by him alone, and which are as follows: first, *Suvisuddha-Dharma-Dhátuja;* second, *Adarsanaja;* third, *Pratyavekshanaja;* fourth, *Sámtaja;* fifth, *Anushthánaja.* The first created beings, Vairochana, etc., were in number five, owing to these five *jnánas;* and in each of these five Buddhas is one of the *jnánas.* Another of Adi-Buddha's attributes is the faculty of individualizing, and multiplying himself, and again individualizing himself at pleasure: another is, possessing the qualities of passion and clemency.

Question VI.

Is the pleasure of God derived from action or repose?

Answer.

There are two modes of considering this subject: first, according to *nirvritti;* and secondly, according to *pravritti.*

Nirvritti (see note 17) is this: to know the world to be a mere semblance, unreal, and an illusion; and to know God to be one: and *Pravritti* is the opposite of this sublime science, and is the practice and notions of ordinary men. Therefore, according to *nirvritti*, Adi-Buddha is the author and creator of all things, without whom nothing can be done; whose care sustains the world and its inhabitants; and the moment he averts his face from them they became annihilated, and nothing remains but himself. But some persons, who profess *nirvritti*, contend that the world with all it containeth is distinct from Adi-Buddha: yet the wise know this to be an error. (See note 18.)

Adi-Buddha, though he comprehends all living things, is yet one. He is the soul, and they are but the limbs and outward members, of this monad Such is *nirvritti*, which, being deeply studied, is found to be unity; but *pravritti*, which is multiplicity, may be distinguished in all things. And in this latter view of *pravritti*, Adi-Buddha may be considered a king, who gives orders; and the five Buddhas, and other divinities of heaven, his ministers, who execute his orders; and we, poor mortals, his subjects, servants, and slaves. In this way the business of the world is distributed among the deities, each having his proper functions; and Adi-Buddha has no concern with it. Thus the five Buddhas give *mukti* (see note 19) and *moksha* to good men: Brahmá by the orders of Padma-Páni, performs the part of creator; Vishnu, by the same orders, cherishes all beings; and Mahá Deva, by the same orders, destroys; Yama takes cognizance of sins, and punishes sinners; Indra and Varuna give rain; and the sun and moon fructify the earth with their rays; and so of the rest.

Question VII.

Who is Buddha? Is he God, or the creator, or a prophet or saint; born of heaven, or of a woman?

Answer.

Buddha means, in Sanskrit, 'the wise;' also, 'that which is known by wisdom;' and it is one of the names which we give to God, whom we also call Adi-Buddha, because he was before all, and is not created, but is the creator: and the *Pancha Dhyáni Buddhas* were created by him, and are in the heavens. Sákya, and the rest of the seven human Buddhas are earth-born or human. These latter, by the worship of Buddha, arrived at the highest eminence, and attained *Nirvána Pada* (*i. e.* were absorbed into Adi-Buddha). (See note 20.) We therefore call them all Buddhas.

Question VIII.

What is the reason for Buddha being represented with curled locks?

Answer.

Adi-Buddha was never seen. He is merely light. (See note 21.) But in the pictures of Vairochana, and the other Buddhas, we have the curled hair; and since in limbs and organs we discriminate thirty-two points of beauty *(lakshanas)*, such as expansion of forehead, blackness of the eyes, roundness of the head, elevation of the nose, archedness of the eyebrows; so also the having curled locks is one of the points of beauty, and there is no other reason for Buddha's being represented with curled locks. (See note 22.)

QUESTION IX.

What are the names of the *great* Buddha? Does the *Newárí* language admit the word Buddha, or any substitute for it? and what is the *Bhotiya* name for Buddha?

ANSWER.

The names of Adi-Buddha are innumerable: Sarvajna, Sugata, Buddha, Dharma-Rája, Tathágata, Bhagaván, Sámanta-Bhadra, Márájita, Lokajita, Jina, Anádinidhana, Adi-Buddha, Nirandhaka, Jnánaikachakshu, Amala, Jnána-Múrti, Vacheswara, Mahá-Vádi, Vádirata, Vádipungava, Vádisinha, and Parajáta. Vairochana, and the other five Buddhas, have also many names. Some of Vairochana's are as follows: Mahá-Dípti, Jnána-Jyotish, Jagat-Pravritti, Mahátejas, &c.; and so of the other four. Padma-Páni also has many names, as Padma-Páni, Kamalí, Padma-Hasta, Padma-Kara, Kamala-Hasta, Kamalákara, Kamala-Páni, Aryávalokiteswara, Aryávalokeswara, Avalokiteswara, and Loka-Nátha* (See note 23.) Many of the above names are intercommunicable between the several persons to whom they are here appropriated. Buddha is a Sanskrit word, not *Newárí*: the *Bhotiya* names I do not know; but I have heard they call Sákya Sinha, Sungi Thuba: *Sungi* meaning the deity, and *Thúba*† his *Alaya* or *Vihára.*

QUESTION X.

In the opinion of the *Banras*, did God ever make a descent on earth? if so, how often; and what is the Sanskrit and *Newárí* name of each *Avatára?*

ANSWER.

According to the scriptures of the *Buddhamárgis*, neither Adi-Buddha nor any of the *Pancha Dhyáni Buddhas* (see note 24), ever made a descent; that is to say, they were never conceived in mortal womb; nor had they father or mother: but certain persons of mortal mould have by degrees attained to such excellence of nature and such *Bodhijnána*, as to have been gifted with divine wisdom, and to have taught the *Bodhi-charya* and *Buddhamárga*: and these were seven, named Vipasyí, Sikhí, Viswabhú, Krakutchanda, Kanaka muni, Kásyapa, Sákya Sinha.

In the *Satya-yuga* were three: Vipasyí, who wasborn in *Vindumatí Nagara*, in the house of Vindumán Rája; Sikhí, in *Urna Desa*; and Viswabhú, in *Anupamá Desa*, in the house of a *Kshatriya*: in the *Tretáyuga*, two persons became Buddhas; one Krakutchanda, in *Kshemavatí Nagara*, in the house of a Brahman; the other Kanaka Muni, in *Subhavatí Nagara*, in the house of a Brahman: and in the *Dwápara-yuga*, one person named Kásyapa, in *Váránasí Nagara*, in the house

* We do not find Matsyendra among these synonymes though he be now usually identified with Padma Páni. For Avalokiteswara see Fahian, p.p. 115-117.

† Sanskriticè Sthúpa, a tomb, temple. But Csoma de Körös gives Sangè Thubba as his name only.

[The name is Sangs-rGyas Thub-pa, from Sang-jäy T'ub-pa, and means: 'the Holy One, the Conqueror.' J.S.]

of a Brahman: and in the *Kali-yuga*,* Sákya, then called Sarvártha Siddha (see note 25), in the house of Suddhodana Rájá, a *Sákyavansí*, in the city of Kapilavastu, which is near Gangáságara,† became Buddhas. Besides these seven, there are many illustrious persons; but none equal to these. The particular history of these seven, and of other Buddhas, is written in the Lalita Vistara. (See note 25.)

Question XI.

How many *Avatáras* of Buddhas have there been, according to the Lamas?

Answer.

They agree with us in the worship of the seven Buddhas, the difference in our notions being extremely small; but the Lamas go further than this and contend that they themselves are *Avatáras*. I have heard from my father, that, in his time, there were five Lamas esteemed divine: the names of three of them I have forgotten, but the remaining two are called Shamurpa and Karmapa.

Question XII.

Do the Lamas worship the *Avatáras* recognized by the *Newárs?*

Answer.

The Lamas are orthodox *Buddhamárgís*, and even carry their orthodoxy to a greater extent than we do. Insomuch, that it is said, that Sankara Achárya,‡ *Siva-Márgí*, having destroyed the worship of Buddha and the scriptures containing its doctrine in Hindustan, came to Nepaul, where also he effected much mischief; and then proceeded to Bhot. There he had a conference with the grand Lama. The Lama, who never bathes, and after natural evacuations does not use topical ablution, disgusted him to that degree, that he commenced reviling the Lama. The Lama replied, "I keep my inside pure, although my outside be impure; while you carefully purify yourself without, but are filthy within:" and at the same time he drew out his whole entrails, and shewed them to Sankara; and then replaced them again. He then demanded an answer of Sankara. Sankara, by virtue of his *yoga*, ascended into the heavens; the Lama perceiving the shadow of Sankara's body on the ground, fixed a knife in the place of the shadow; Sankara directly fell upon the knife, which pierced his throat and killed him instantly. Such is the legend or tale that prevails, and thus we account for the fact that the *Buddhamárgí* practice of Bhot is purer, and its scriptures more numerous, than ours.

Question XIII.

What is the name of your sacred writings,§ and who is their author?

* This allotment into four *yugas* is apochryphal. The three first Buddhas belong to the penultimate Kalpa, and the four last to the present, or Bhadra Kalpa.

† Near or in Oude, or Rohilkhand, according to other works. Kapila was on the Bhágírathí, near Kailás, say the Tibetan authorities.

‡ He flourished in the ninth century, or about 1,000 years back. This we learn from the Brahmans, and the date is important as it agrees with the era of that persecution which led the Southerners to seek protection in Nepaul and Tibet.

§ See pp. 36-39 for a corrected list of the Sanskrit literature of Buddhism.

ANSWER.

We have nine *Puránas*, called "the nine *Dharmas.*" (See note 26.) A Purána is a narrative or historical work, containing a description of the rites and ceremonies of Buddhism, and the lives of our chief Tathágatas. The first Dharma is called Prajná Páramitá, and contains 8,000 slokas. This is a *Nyáya Sástra*, or work of a philosophic character, capable of being understood only by men of science; the second is named Ganda Vyúha,* of 12,000 slokas, which contains the history of Sudhana Kumára, who made sixty-four persons his *gurus*, from whom he acquired *Bodhijnána;* the third, is the Samádhi Rája, of 3,000 slokas, in which the nature and value of *japa* and *tapas* are explained; the fourth is the Lankávatára, of 3,000 slokas, in which is written how Rávana, lord of Lanká, having gone to Malayagiri mountain, and there heard the history of the Buddha, from Sákya Sinha, obtained *Bodhijnána.* The fifth, which is called Tathágata Guhya, is not to be found in Nepaul;** the sixth, is the Saddharma Pundaríka which contains an account of the method of building a *chaitya* or *Buddha-mandala*, and the mode and fruits of worshipping it. (Chaitya is the exclusive name of a temple dedicated to Adi-Buddha or to the Pancha Dhyáni Buddha; and whatever temple is erected to Sákya, or other Mánushi Buddhas, is called *Kútágár*): || the seventh, is the Lalita Vistara, of 7,000 slokas, which contains the history of the several incarnations of Sákya Sinha Bhagaván, and an account of his perfections in virtue and knowledge, with some notices of other Buddhas. The eighth, is the Suvarna Prabhá, containing, in 1,500 slokas, an account of Saraswatí, Lakshmí and Prithiví; how they lauded Sákya Sinha Bhagaván; and how he, in return, gave each of them what she desired. The ninth, is the Dasa Bhúmeswara, of 2,000 slokas, containing an account of the ten *Bhuvanas* of Buddha. All these *Puránas* we received from Sákya Sinha, and esteem them our primitive scriptures because before the time of Sákya our religion was not reduced to writing, but retained in memory; the disadvantages of which latter method being evident to Sákya, he secured our institutes by writing them. Besides these Puránas, we received *Tantras* and *Dháranís* from Sákya Sinha. Tantra is the name of those books in which *Mantras* and *Yantras* are written, explanatory of both of which we have very many works. Three of them are famous: first, Máyá Jála, of 16,000 slokas; second, Kála Chakra, of 6,000; third, Sámbhú Udaya, of 1,000. The Dháranís were extracted from the Tantras, and are similar in nature to the Guhya, or mysterious rites, of the Siva-Márgís. A Dháraní is never less than eight slokas or more than 500; in the beginning and middle of which are written the "Víja Mantra," and at the end, the "Phúl Stotra," or the Máhátmya, *i.e.*, what desire may be accomplished or what business achieved by the perusal of that Dháraní; such, for example, as obtaining children—advantage over an enemy—rain—or merely the approbation of Buddha. There are probably a thousand Dháranís.

* See note at page 137.

** This is a very holy Tantra. It was kept from me long, but at last I got it.

|| Kútágár is the name of the class of temples inferior to Chaityas, as now employed in Nepaul. Besides the Chaityas, the Nepaulese have temples, dedicated equally to the *Dii minores* of the *Bauddhas*, and to many of the (adopted) deities of the Brahmans.

Question XIV.

What is the cause of good and evil?

Answer.

When Padma-Páni, having become *Tri-guna-Atmaka,* that is, having assumed the form of Satyaguna, Rajo-guna, and Tamo-guna, created Brahmá, Vishnu, and Mahesa; then from Saṭya-guna, arose spontaneously (Swábhávaka), *punya* or virtue, and from Tamo-guna, *pápa* or evil, and from Rajo-guna, the mean of the two, which is neither all good nor all evil: for these three *gunas* are of such a quality that good acts, mixed acts, and bad acts, necessarily flow from them. Each of these *karmas* or classes of actions is divided into ten species, so that *pápa* is of ten kinds, first (see note 27) murder; second, robbery; third, adultery, which are called *káyaka* or bodily, *i. e.,* derived from *Káya;* fourth, lying; fifth, secret slander; sixth, reviling; seventh, reporting such words between two persons as excite them to quarrels; and these four pápas are called váchaka, *i. e.,* derived from speech; eighth, coveting another's goods; ninth, malice; and tenth, disbelief of the scriptures and immorality; and these three are called mánasa, *i. e.,* derived from *mānas* 'the mind.' The ten actions opposite to these are good actions; and the ten actions, composed, half and half, of these two sorts, are mixed actions.

Question XV.

What is the motive of your good acts—the love of God—the fear of God—or the desiring of prospering in the world?

Answer.

The primary motive for doing well, and worshipping Buddha, according to the scriptures, is the hope of obtaining *Mukti* and *Moksha,* becoming Nirvána, and being freed from transmigrations: these exalted blessings cannot be had without the love of God; therefore they, who make themselves accepted of God, are the true saints, and are rarely found; and between them and Buddha there is no difference, because they will eventually become Buddhas, and will obtain Nirvána Pada, *i. e., mukti* (absorption,) and their *jyotish* (flame, essence), will be absorbed into the *jyotish* of Buddha; and to this degree Sákya and the others of the "Sapta-Buddha" (see note 28) have arrived, and we call them Buddhas, because, whoever has reached this state is, in our creed, a Buddha. Those persons who do good from the fear of hell, and avoid evil from the desire of prospering in the world, are likewise rarely found, and their degree is much above that of the class of sinners. Their sufferings in Naraka will be therefore lessened; but they will be constrained to suffer several transmigrations, and endure pain and pleasure in this world, till they obtain Mukti and Moksha.

Question XVI.

Will you answer, in the world to come, to Adi-Buddha for your acts in this world, or to whom will you answer? and what rewards for good, and pains for evil, will you reap in the next world?

Answer.

How can the wicked arrive at Buddha? (see note 29.) Their wicked deeds will hurry them away to Naraka; and the good-will, by virtue of their good acts, be transported to the Bhuvanas of Buddha, and will not be there interrogated at all; and those who have sometimes done good and sometimes evil, are destined to a series of births and deaths on earth, and the account of their actions is kept by Yama Rája.

Question XVII.

Do you believe in the metempsychosis?

Answer.

Yes. For it is written in the Játaka Málá, and also in the Lalita Vistara, that Sákya, after having transmigrated through 501 bodies, obtained Nirvána Pada or Mukti in the last body: but so long as we cannot acquire Mukti, so long we must pass through births and deaths on earth. Some acquire Moksha after the first birth, some after the seventy-seventh, and some after innumerable births. It is no where written that Moksha is to be obtained after a prescribed number of births; but every man must atone for the sins of each birth by a proportionate number of future births; and when the sins of the body are entirely purified and absolved, he will obtain absorption into Adi-Buddha.

Question XVIII.

What and from whence are the Newárs, from Hindust'han or Bhot? (see note 30,) and what is the word Newár, the name of a country or a people?

Answer.

The natives of the valley of Nepaul are Newárs. In Sanskrit the country is called Naipála,* and the inhabitants Naipálí; and the words Newár and Newárí are vulgarisms arising from the mutation of P to V, and L to R. Thus too the word Bandya, the name of the Buddhamárgí sect (because its followers make bandana, *i. e.*, salutation and reverence to the proficients in Bodhijnána), is metamorphosed by ignorance into Bánra, a word which has no meaning.

Question XIX.

Do the Newárs follow the doctrine of caste or not?

Answer.

As inhabitants of one country they are one—but in regard to caste, they are diverse.

Question XX.

How many castes are there amongst the Bánras?

Answer.

Bánra, according to the true reading, is Bandya, as explained above. According to our Puránas, whoever has adopted the tenets of Buddha, and has cut off the lock from the crown of his head, of whatever tribe or nation he be, becomes thereby a Bandya (see note 31). The Bhotiyas, for example, are Bandyas because they follow the tenets of Buddha, and have no lock on their heads. The Bandyas are divided into two classes; those who follow the Váhya-charya, and those who

* From Né, 'the sender to Paradise,' who is Swayambhú Adi-Buddha, and pāla, 'cherished.' The Brahmans derive the word Nepaul from Né or Neyum, the proper name of a Patriarch or Muni.

adopt the Abhyantara-charya—words equivalent to the Grihastha ásrama and Vairágí ásrama of the Bráhmanas. The first class is denominated Bhikshu; the second, Vajra Achárya.* The Bhikshu cannot marry; but the Vajra Achárya is a family man. The latter is sometimes called, in the vernacular tongue of the Newárs, Gúbhál, which is not a Sanskrit word. Besides this distinction into monastic and secular orders, the Bandyas are again divided, according to the scriptures, into five classes: first, Arhat; second, Bhikshu; third, Srávaka; fourth, Chailaka; fifth, Vajra Achárya. The Arhat is he who is perfect himself, and can give perfection to others; who eats what is offered to him, but never asks for anything. The Bhikshu, is he who assumes a staff and beggar's dish (bhikshari and pinda pátra), sustains himself by alms, and devotes his attention solely to the contemplation (dhyána) of Adi-Buddha, without ever intermeddling with worldly affairs. The Srávaka is he who devotes himself to hearing the Bauddha scriptures read or reading them to others; these are his sole occupations, and he is sustained by the small presents of his audiences. The Chailaka is he who contents himself with such a portion of clothes (chilaka) as barely suffices to cover his nakedness, rejecting everything more as superfluous. The Bhikshu and the Chailaka very nearly resemble each other, and both (and the Arhat also) are bound to practice celibacy. The Vajra Achárya is he who has a wife and children, and devotes himself to the active ministry of Buddhism. Such is the account of the five classes found in the scriptures; but there are no traces of them in Nepaul.† No ōne follows the rules of that class to which he nominally belongs. Among the Bhotiyas there are many Bhikshus, who never marry; and the Bhotiya Lamas are properly Arhats. But all the Nepaulese Buddhamárgís are married men, who pursue the business of the world, and seldom think of the injunctions of their religion. The Tantras and Dháranís, which ought to be read for their own salvation, they read only for the increase of their stipend and from a greedy desire of money. This division into five classes is according to the scriptures; but there is a popular division according to Vihárs, and these Vihárs being very numerous, the separate congregations of the Bandyas, have been thus greatly multiplied.‡ In Pátan alone there are fifteen Vihárs. A temple to Adi-Buddha, or to the five Dhyáni-Buddhas, called a Chaitya, is utterly distinct from the Vihár, and of the form of a heap of rice or Dhanyarasya-ákar. But the temples of Sákya and the other of the "Sapta Buddha Mánushi," as well as those of other chief saints and leaders of Buddhism are called Vihárs. The names of the fifteen Vihárs of Pátan are as follows: Tankal-Vihár, Tú-Vihár, Hak-Vihár, Bhû-

* See further on.

† In Nepaul at present the Bandyas are divided popularly into Vajra Achárya, Sákya Vansí, Bhikshu or Bikhu, and Chiva-bare. The last derive their name from living in a Vihár which has a Chaitya, *vulgo* Chíva, in its midst. Others say that Chíva or Chivakabare is a corruption of Chailaka Bandya Potius, Bandyas wearing the Chívara, a part of the monastic dress, a sense which would make the term signify Bandyas adhering to their vows.

‡ Some years ago there were 5,000 Bandyas in the Valley of Nepaul out of a population of some 250,000.

Vihár, Haran-Varna-Mahá-Vihár,† Rudra-Varna-Mahá-Vihár,‡ Bhikshu-Vihár, Sákya-Vihár, Guhya-Vihár, Shí-Vihár, Dhom-Vihár, UnVịhár, etc. (see note 32). In short, if any Bandya die, and his son erect a temple in his name, such structure may be called such an one's (after his name) Vihár. With this distinction, however, that a temple to an eminent saint is denominated Mahá Vihár—one to an ordinary mortal, simply Vihár.§

NOTES.

(1) Here a Sloka of the Sámbhú Purána is quoted in the original paper; and it was my first intention to have repeated it on the margin of the translation; but, upon reflection, I believe it will be better to observe, that the Sámbhú Purána is a work peculiar to Nepaul. Many other Bauddha scriptures, however, which are not local, and are of high authority, symbolize the forming and changing powers of nature by the letters of the alphabet; and ascribe the pre-eminence among these letters to A, U, and M—making the mystic syllable *óm*, which is not less reverenced by Bauddhas than by Bráhmanas. A, the Bauddhas say, is the Víja Mantra of the person Buddha; U, the Víja Mantra of the person Dharma; and M, that of the person Sangha—and these three persons form the Buddhist Triad.

The Bauddhas, however, differ in their mode of classing the three persons. According to the Aiswarikas, the male, Buddha, the symbol of generative power, is the first member; the female, Dharmá, the type of productive power, is the second; and Sangha, their son, is the third, and represents *actual* creative power, or an *active* creator and ruler, deriving his origin from the union of the essences of Buddha and Dharmá. Sangha, according to all the schools, though a member, is an inferior member, of the triad.||

(2) Another sloka is here quoted; but it will not justify the language of the text, in which there is some confusion of the opposite doctrines of the Aiswarikas and Swábhávikas. In the triad of the latter, the female, Dharmá (also called Prajná), the type of productive power, is the first member; Upáya, or Buddha, the symbol of generative power, the second; and Sangha the third; their son as before, and the active author of creation; or rather the type of that spontaneous creation, which results necessarily from the union of the two principles of nature before mentioned.

Buddha and Prajná united become Upáya Prajná; or *vice versa*, according to the school, and *never* as in the text. (For some further remarks upon these chief objects of Bauddha worship, see Notes 12 and 29.)

I take this early opportunity to remark that candid criticism will compare, and not contrast, the statements made in Notes 10, 12, 17, 20, and 29, especially with reference to the Swábhávika doctrine. (See Note 16.)

† *Vulgo* Kon. ‡ *Vulgo* Uku. Throughout classical and vulgar names are mixed.

§ *Bahi* and *Bāhā* or *Bahāl* are the vulgar names for great and common Vihárs, or Vihárs with a Chaitya, and those with a Kutágár only, erected in the midst of them. Temples to Manushi Buddhas and other Deities are called Kutágár commonly, though Kutágár temples sometimes enshrine Dhyáni Buddhas. A Vihár may be built round either.

|| See Wilson's Essays and Lectures, ii. 23 ff.

(3) The deduction of the five Dhyáni Buddhas, and the five Dhyáni Bodhisattwas, from Adi-Buddha, according to the Aiswarika Bauddhas, will be stated farther on. It is a celestial or divine creation, and is here improperly mixed with the generative creations, theistic and atheistic, of various doctors.

(4) See Note 23.

(5) The sloka quoted is from the Púja Kánda, which is a mere manual of worship, of recent origin, and probably local to Nepaul. It professes, however, to be a faithful compilation from the Guna-Káranda Vyúha, and Káranda Vyúha. The latter of these is a work of respectable authority, and contains the following partial justification of the language of the Púja Kánda. (Sákya, speaking to his disciple Sarvanivarana Vishkambhí, says,) "In the very distant times of Vipasyí Buddha I was born as the son of Suganda Mukha, a merchant: in that birth I heard from Vipasyí the following account of the qualities of Aryávalokiteswara (Padma Páni). The sun proceeded from one of his eyes: and from the other, the moon; from his forehead Mahádeva; from between his shoulders, Brahmá; from his chest, Vishnu; from his teeth, Sarasvatí; from his mouth, Váyu; from his feet, Prithví; from his navel, Varuna." So many deities issued from Aryávalokiteswara's body. This passage is expanded in the Guna-Káranda Vyúha, wherein it is added, that when Aryávalokiteswara had created Brahmá, Vishnu, and Mahesa, they stood before him, and he said to the first, "be thou the lord of Satyaguna and create:" and to the second, "be thou the lord of Rajoguna and preserve;" and to the third, "be thou the lord of Tamoguna and destroy." The Guna-Káranda Vyúha, is however a mere amplification of the Káranda Vyúha, and of much less authority. In a passage of the Sáraka Dhára—which is not one of the sacred writings of Nepaul, but a work of high authority, written by Sarvajna Mitrapáda, a Bauddha ascetic of Cashmeer—the Hindu deities are made to issue from the body of the supreme Prajná just as, according to the Káranda Vyúha, they proceed from that of Padma Páni.

(6) The authority for these ten mansions is the Dasa Bhúmeswara, one of the nine great works spoken of in the answer to the thirteenth question; and which treats professedly of the subject. The thirteen mansions are, however, mentioned in sundry works of high authority; and the thirteen grades of the superior part of the Chaitya (or proper Bauddha temple) are typical of the thirteen celestial mansions alluded to in the text. The most essential part of the Chaitya is the solid hemisphere; but the vast majority of Chaityas in Nepaul have the hemisphere surmounted by a pyramid or cone, called Chúdámani, and invariably divided into thirteen grades.

(7) All this, as well as what follows, is a mere transcript from the Brahmanical writings. There is, nevertheless, authority for it in the Bauddha scriptures. The Bauddhas seem to have adopted without hesitation the cosmography and chronology of the Brahmans, and also a large part of their pantheon. They freely confess to have done so at this day. The favourite Brahmanical deities accepted by the Buddhists are, of males: Mahá Kála, Indra, Ganesa, Hanumán,

and the triad. Of females: Lakshmí and Sarasvatí. The Hindu triad are considered by the Buddhists as the mere servants of the Buddhas and Bodhisattwas, and only entitled to such reverence as may seem fit to be paid to faithful servants of so high masters. Of the origin of these deities, according to the Bauddha books, I have already given one account, and referred to another. The notions of the three gunas and of the creation, etc., by the Brahmanic triad as the delegates of the Bodhisattwas, I look upon as modern inventions. According to genuine Buddhism, the Bodhisattwas are, each in his turn, the active agents of the creation and government of the world.

(8) An important historical person, and the apparent introducer of Buddhism into Nepaul. (See note 30).

(9) This is a most curious legend. I have not yet seen the Tantra whence it professes to be extracted, and suspect that the legend was stolen from our Bible, by some inhabitant of Nepaul, who had gathered a confused idea of the Mosaic history of the origin and fall of mankind from the Jesuit missionaries, formerly resident in this valley; or perhaps the legend in question was derived from some of those various corrupt versions of the biblical story which have been current among the Jews and Moslems of Asia for many centuries.

(10) This limited reply is the fault of my friend and not of his books. Matter is called Prakriti by the Buddhists, as well as by the Brahmans.* The Swábhávika school of Bauddha philosophy (apparently the oldest school) seems to have considered matter as the sole entity, to have ascribed to it all the attributes of deity, and to have assigned to it two modalities; one termed *nirvritti*, and the other *pravritti*. (See note 12.) To speak more precisely, the above is rather the doctrine of the Prájnika Swábhávikas than of the simple Swábhávikas: for the former unitize the active and intelligent powers of nature, the latter do not unitize them; and prefer to all other symbols of those dispersed powers of nature the letters of the alphabet generally, and without much regard to the pre-eminence of A, U, and M. Indeed, it is probable that the mystic syllable AUM is altogether a comparatively recent importation into Buddhism. The Lotos is a very favourite type of creative power with all the Bauddhas; and accordingly representations of it occur in a thousand places, and in as many forms, in the Bauddha sculptures and architecture.

(11) The sloka quoted is from a modern little manual of Pújá. I have not seen any adequate original authority; but the Aiswarika Buddhists, who maintain an eternal, infinite, intellectual Adi-Buddha, in all probability made the human soul an emanation from him; and considered Moksha a remanation to him.

(12) The Swábhávikas, the name assumed by one of the four schools of Bauddha philosophy, and apparently the oldest, are divided into two sects; one called Swábhávikas simply, the other Prájnika Swábhávikas. The former maintain that an eternal revolution of entity and non-entity is the system of nature, or of matter,

* Dharmma, or that which sustains, is the Bauddha equivalent for the Brahmanical Mátra, or that which measures all qualities in space, the English 'matter.'

which alone exists. The Prájnikas deify matter as the sole substance, and give it two modes, the abstract and concrete: in the former, they unitize the active and intelligent powers held to be inherent in matter, and make this unit deity. Such is the abstract or proper mode, which is unity, immutability, rest, bliss. The second is the contingent or concrete mode, or that of actual, visible, nature. To this mode belong action, multiplicity, change, pain. It begins by the energies of matter passing from their proper and eternal state of rest into their contingent and transitory state of action; and ends when those energies resume their proper modality. The proper mode is called *nirvritti;* the contingent mode *pravritti.* The powers of matter cannot be described in their proper state of abstraction and unity. In the latter state, all the order and beauty of nature are images of their quality; they are also symbolized by the *Yoni,* and personified as a female divinity called Adi-Prajná and Adi-Dharmá. Man's *summum bonum* is to pass from the transmigrations incident to the state of *pravritti* into the eternal rest or bliss of *nirvritti.* The triadic doctrine of all the schools is referable solely to *pravritti.* In the state of *nirvritti,* with some of the Aiswarikas, Buddha represents intellectual essence and the then sole entity; with others of the Aiswarikas, Dharma, or material essence exists *biunely* with Buddha in *nirvritti,* the two being in that state *one.* With the Prájnikas, Prajná, in the state of *nirvritti,* is the *summum et solum numen, Diva Natura*—the sum of all the intellectual and physical forces of matter, considered as the sole entity, and held to exist in the state of *nirvritti* abstracted from palpable material substance, eternally, unchangeably, and essentially one. When this essential principle of matter passes into the state of *pravritti,* Buddha, the type of active power, first proceeds from it and then associates with it, and from that association results the actual visible world. The principle is feigned to be a female, first the mother, and then the wife, of the male, Buddha. [For a glimpse at the esoteric sense of these ænigmas, see note 29.]

[13] The work cited is of secondary authority; but the mode of reasoning exhibited in the text is to be found in all Bauddha works which treat of the Swábhávika doctrine.

[14] This is the name of the Theistic school of the Bauddha philosophers. The Sámbhú Purána and Guna-Káránda Vyúha contain the least obscure enunciation of Theism—and these books belong to Nepaul. Other Bauddha scriptures, however, which are not local, contain abundant expressions capable of a Theistic interpretation. Even those Bauddha philosophers who have insisted that matter is the sole entity, have ever magnified the wisdom and power of nature: and doing so, they have reduced the difference of theism and atheism almost to a nominal one: so, at least, they frequently affirm.

The great defect of all the schools is the want of Providence and of dominion in their *causa causarum,* though the comparatively recent Kármikas and Yátnikas appear to have attempted to remedy this defect. [See the following note.]

[15] Of two of the four schools of Bauddha philosophy, namely, the Swábhávika and Aiswarika, I have already said a few words: the two remaining schools are denominated the Kármika and Yátnika—from the words Karma, meaning

moral action; and Yatna, signifying intellectual force, skilful effort. The proper topics of these two schools seem to me to be confined to the phenomena of human nature—its free-will, its sense of right and wrong, and its mental power. To the wisdom of Swabháva, or Prajná, or Adi-Buddha, the Bauddhas, both Swábhávikas and Aiswarikas, had assigned that eternal necessary connexion of virtue and felicity in which they alike believed. It remained for the Kármikas and Yátnikas to discuss how each individual free-willed man might most surely hope to realize that connexion in regard to himself; whether by the just conduct of his understanding, or by the proper cultivation of his moral sense? And the Yátnikas seem to have decided in favour of the former mode; the Kármikas, in favour of the latter. Having settled these points, it was easy for the Yátnikas and Kármikas to exalt their systems by linking them to the throne of the *causa causarum*—to which they would be the more readily impelled, in order to remove from their faith the obloquy so justly attaching to the ancient Prájnika, and even to the Aiswarika school, because of the want of Providence and of Dominion in their first cause. That the Kármikas and Yátnikas originally limited themselves to the phenomena of human nature, I think probable, from the circumstance that, out of some forty slokas which I have had collected to illustrate the doctrines of these schools, scarcely one goes beyond the point of whether man's felicity is secured by virtue or by intellect? And that when these schools go further (as I have the evidence of *two* quotations from their books that they *sometimes* do), the trespassing on ground foreign to their systems seems obvious; thus in the *Divya Avadána*, Sákya says, "from the union of Upáya and Prajná, arose manas—the lord of the senses; and from manas or 'mind' proceeded good and evil; and this union of Upáya and Prajná is then declared to be a Karma. And in the same work, in regard to the Yátnika doctrine, it is said, "Iswara (*i. e.*, Adi-Buddha) produced Yatna from Prajná, and the cause of *pravritti* and *nirvritti** is Yatna; and all the difficulties that occur in the affairs of this world or of the next are rendered easy by Yatna." Impersonality and quiescence were the objections probably made to the first cause of the Prájnikas and Aiswarikas; and it was to remove these objections that the more recent Kármikas and Yátnikas feigned conscious moral agency *(Karma)*, and conscious intellectual agency *(Yatna)* to have been with the *causa causarum* (whether material or immaterial) from the beginning. Of all the schools, the Kármikas and Yátnikas alone seem to have been duly sensible of man's free-will, and God's moral attributes. The Kármika confession of faith is, "*Púrva janma kritam karma tad daivyam iti kathyate,*" which may be very well translated by our noble adage, "conduct is fate." Such sentiments of human nature naturally inclined them to the belief of immaterial existences, and accordingly they will be found to attach themselves in theology *chiefly* to the Aiswarika school.

(16) This is the divine creation alluded to in the third note. The eternal, infinite and intellectual Adi-Buddha possesses, as proper to his own essence, five sorts

* See note 17 for the sense of these cardinal terms.

of wisdom. From these he, by five separate acts of Dhyána, created the five Dhyáni Buddhas, to whom he gave the virtue of that *jnána* whence each derived his origin. These five Dhyáni Buddhas again created, each of them, a Dhyáni Bodhisatwa by the joint efficacy of the *jnána* received from Adi-Buddha, and of an act of his own Dhyána.

The five Dhyáni Buddhas are, like Adi-Buddha, quiescent—and the active work of creation and rule is devolved on the Bodhisatwas. This creation by Dhyána is eminently characteristic of Buddhism—but *whose* Dhyána possesses creative power? that of an eternal Adi-Buddha, say the Aiswarikas of the *Sámbhú Púrana*—that of *any* Buddha, even a *Mánushi* or mortal Buddha, say the Swábhávikas. The Bauddhas have no other notion of creation (than that by Dhyána,) which is not generative.

(17) These terms are common to all the schools of Bauddha philosophy; with the Aiswarikas, *nirvritti* is the state in which mind exists independent of matter; *pravritti*, the state in which it exists while mixed with matter. With the simple Swábhávikas the former term seems to import non-entity; the latter, entity. With the Prájnika Swábhávikas, the former term signifies the state in which the active and intellectual power of matter exists abstractedly from visible nature. The *Moksha* of the first is absorption into Adi-Buddha; of the second, absorption into Súnyatá; of the third, identification with Prajná. In a word, *nirvritti* means abstraction, and *pravritti*, concretion—from *nirvána* is formed *nirvritti*, but *pravritti* has no *pravána*.

(18) If so, I am afraid few Bauddhas can be called wise. The doctrine of the text in this place is that of the Aiswarikas, *set off to the best advantage:* the doctrine incidentally objected to is that of the Swábhávikas and Prájnikas. Sir W. Jones assures us that the Hindus "consider creation (I should here prefer the word change) rather as an energy than as a work." This remark is yet more true in regard to the *old Bauddha* philosophers: and the mooted point with them is, *what* energy creates? an energy *in*trinsic in some archetypal state of matter, or *ex*trinsic? The old Bauddha philosophers seem to have insisted that there is no sufficient evidence of immaterial entity. But, what is truly remarkable, *some* of them, at least, have united with that dogma a belief in *moral and intellectual operations;* nor is there one tenet so diagnostic of Buddhism as that which insists *that man is capable of extending his moral and intellectual faculties to infinity.* True it is, as Mr. Colebrooke has remarked, that the Hindu philosophy recognizes this dogma—coldly recognizes it, and that is all: whereas, the Bauddhas have pursued it into its most extravagant consequences, and made it the corner-stone of their faith and practice. (See note 29.)

(19) I have not yet found that these Dhyáni Buddhas of the Theistic school *do* anything. They seem to be mere personifications, according to a Theistic theory, of the active and intellectual powers of nature—and hence are called Pancha Bhúta, Pancha Indriya, and Pancha Ayatana-Akára.

It may seem contrary to this notion of the quiescence of the five Dhyáni Buddhas, that, according at least to some Nepaul works, each of them has a Sakti.

Vairochana's is Vajra-Dháteswarí; Akshobhya's, Lochaná; Ratna Sambhava's, Mámukhí; Amitábha's, Pándará; Amogha Siddha's, Tárá. But I apprehend that these Buddha-Saktis are peculiar to Nepaul; and though I have found their names, I have *not* found that they *do* any thing.

There *is* indeed a secret and filthy* system of Buddhas and Buddha-Saktis, in which the ladies act a conspicuous part; and according to which, Adi-Buddha is styled Yogámbara; and Adi-Dharma, Jnáneswarí. But this system has only been recently revealed to me, and I cannot say more of it at present.

(20) According to the Aiswarikas: the Swábhávikas say, into Akása and Súnyatá; the Prájnikas, into Adi-Prajná. The Swábhávika doctrine of Súnyatá is the darkest corner of their metaphysical labyrinth. It cannot mean strictly nothingness, since there are eighteen degrees of Súnyatá, whereof the first is Akása: and Akása is so far from being deemed nothingness that it is again and again said to be the only real substance. Language sinks under the expression of the Bauddha abstractions; but by their Súnyatá I understand sometimes the *place*, and sometimes the form, in which the infinitely attenuated elements of all things exist in their state of separation from the palpable system of nature.

N. B. The images of all the seven great Mánushi Buddhas, referred to in the answer to the seventh question, are exactly similar to that of Sákya Sinha, the seventh of them. This image very nearly resembles that of Akshobhya, the second Dhyáni Buddha. The differences are found only in the supporters, and in the cognizances† *(chinas.)* When *coloured* there is a more remarkable diagnosis, Akshobhya being blue, and Sákya and the other six Mánushis, yellow.

(21) The *Sámbhú Purána* says, *manifested* in Nepaul in the form of flame *(Jyotirúpa.)* According to the same work, Adi-Dharma's (or Prajná's) manifestation in Nepaul is in the form of water *(jala surúpa)*.

(22) This is the true solution of a circumstance which has caused much idle speculation: though the notion is, no doubt, an odd one for a sect which insists on tonsure!

(23) These are Padma Páni's names in his character of active creator and governor of the *present* world. Three Dhyáni Bodhisattwas preceded him in that character, and one (the fifth) remains to follow him.

(24) I have already stated that these deities, conformably with the quiescent genius of Buddhism, *do* nothing; they are merely the medium through which creative power is communicated to the Bodhisattwas from Adi-Buddha. It is the Bodhisattwas alone who *exercise* that power, one at a time, and each in his turn. It is a ludicrous instance of Bauddha contempt for action, that some recent writers have made a fourth delegation of active power to the three gods of the Hindu Triad.

(25) Until he attained *bodhijnána;* and even then, while yet lingering in the flesh, he got the name of Sákya Sinha. This náme has caused some speculation,

* Tántrika system.

† *Mudrás*, the name of the several (all) positions of the hands: *Chinas*, that of the cognizances placed between the supporters or vahans.

on the asserted ground of its not being Indian. The Bauddha scriptures differ as to the city in which Sákya was born; but all the places named are Indian. They also say that the Sakavansa was an Indian race or family; as was the Gautamavansa, in which also Sákya was once born.

(25 *bis*) This must be received with some allowance. The *Lalita Vistara* gives ample details of Sákya's numberless births and acts, but is nearly silent as to the origin or actions of his six great predecessors: and the like is true of many other Bauddha scriptures.

(26) These works are regularly worshipped in Nepaul as the "*Nava Dharma.*" They are chiefly of a *narrative* kind. The most important work of the *speculative* kind now extant in Nepaul is the *Rakshá Bhágavatí*, consisting of no less than 125,000 slokas. This is a work of philosophy rather than of religion, and its spirit is sceptical to the very verge of pyrrhonism. The Bauddhas of Nepaul hold it in the highest esteem, and I have sent three copies of it to Calcutta. Its substance though not its form or *reduction to writing*, are attributed (as are those of all the other Bauddha scriptures) to Sákya Sinha. Whatever the Buddhas have said, *(sugatai-desita)* is an object of worship with the Bauddhas. Sákya having systematised these words of the Buddhas, and his earliest disciples having reduced to writing, the books are now worshipped under the names of *Sútra* and *Dharma.* The aggregation of nine Dharmas is for ritual purposes; but why the nine specified works have been selected to be thus peculiarly honoured I cannot say. They are probably the oldest and most authentic scriptures existing in Nepaul, though this conjecture is certainly opposed to the reverence expressed for the *Rakshá Bhágavatí*, by the Buddhists. That work, (as already stated) is of vast extent, containing no less than 125,000* slokas, divided into five equal parts or *khands*, which are known by the names of the five *Páramitás* and the five *Rakshás.*

(27) The three first sins should be rendered, all destruction of life, all taking without right, and all sexual commerce whatever. The ten are the cardinal sins of Buddhism, and will bear a very favourable comparison with the five cardinal sins of Brahmanism.

(28) The Buddhas mentioned in the Bauddha scriptures are innumerable. Many of them, however, are evident non-entities in regard to history. Even the Buddhas of mortal mould are vastly numerous, and of various degrees of power and rank. These degrees are three, entitled, *Pratyeka*, *Srávaka*, and *Mahá Yánika.* Sákya Sinha is often said to be the *seventh* and *last* Mánushi Buddha who has yet reached the supreme grade of the Mahá Yánika. In the *Lalita Vistara*, there is a formal enumeration of the perfections in knowledge and virtue requisite for attaining to each of these three grades—a monstrously impracticable and impious array of human perfectibility! The three grades are known by the collective name of "*Tri Yána.*"

(29) Genuine Buddhism never seems to contemplate any measures of acceptance

* See list of books at pp. 36-39. The *Prajná Páramitá* is found in five different degrees of development; of these the second, though distinct from, is often blended with the first.

with the deity; but, overleaping the barrier between finite and infinite mind, urges its followers to aspire by their own efforts to that divine perfectibility of which it teaches that man is capable, and by attaining which man becomes God—and thus is explained both the quiescence of the imaginary *celestial*, and the plenary omnipotence of the real Mánushi Buddhas—thus too we must account for the fact, that genuine Buddhism has no priesthood; the saint despises the priest: the saint scorns the aid of mediators, whether on earth or in heaven: "conquer (exclaims the adept or Buddha to the novice or Bodhi-Sattwa)—conquer the importunities of the body, urge your mind to the meditation of abstraction, and you shall, in time, discover the great secret *(Súnyatá)* of nature: know this, and you become, on the instant, whatever priests have feigned of Godhead—you become identified with Prajná, the sum of all the power and all the wisdom which sustain and govern the world, and which, as they are manifested *out* of matter, must belong solely *to* matter; not indeed in the gross and palpable state of *pravritti*, but in the archetypal and pure state of *nirvritti.* Put off, therefore, the vile, *právrittika* necessities of the body, and the no less vile affections of the mind *(Tapas);* urge your thoughts into pure abstraction *(Dhyána)*, and then, as assuredly you *can*, so assuredly you *shall*, attain to the wisdom of a Buddha *(Bodhijnána)*, and become associated with the eternal unity and rest of *nirvritti*." Such, I believe, is the esoteric doctrine of the Prájnikas—that of the Swábhávikas is nearly allied to it, but more timid and sceptical; they too magnify the wisdom and power of nature so abundantly diffused throughout *pravritti*, but they seem not to unitize that wisdom and power in the state of *nirvritti*, and incline to conceive of *nirvritti*, as of a state of things concerning which nothing can be predicated; but which, even though it be nothingness *(Súnyatá)*, is at least a blissful *rest* to man, otherwise doomed to an eternity of transmigrations through all forms of visible nature: and while the Swábhávikas thus underrated the *nirvritti* of the Prájnikas, it is probable that they compensated themselves by magnifying, more than the Prájnikas did, that *právrittika* omnipotence of which the wise man *(Buddha)* is capable, *even upon earth.* It has been already stated that the second person of the Prájnika Triad is denominated Buddha and Upáya; of which terms the esoteric sense is this: Every man possesses in his understanding, when properly cultivated according to the rules of Buddhism, the means or expedient *(Upáya)* of discovering the supreme wisdom of nature *(Prajná)*, and of realizing by this discovery, in his own person, a plenary omnipotence or divinity! which begins even while he yet lingers in the flesh (in *pravritti);* but which is not fully accomplished till he passes, by the body's decay, into the eternal state of *nirvritti*.

And as the wisdom of man is, in its origin, but an effluence of the Supreme wisdom (*Prajná*) of nature, so is it perfected by a refluence to its source, but without loss of individuality: whence Prajná is feigned in the exoteric system to be both the mother and the wife of all the Buddhas, "*jananí sarva Buddhánám*," and "*Jina-sundarí;*" for the efflux is typified by a birth, and the reflux by a marriage.

The Buddha is the adept in the wisdom of Buddhism (*Bodhijnána*) whose first duty, so long as he remains on earth, is to communicate his wisdom to those who are willing to receive it. These willing learners are the "Bodhisattwas," so called from their hearts being inclined to the wisdom of Buddhism, and "Sanghas," from their companionship with one-another, and with their Buddha or teacher, in the *Viháras* or cœnobitical establishments.

And such is the esoteric interpretation of the third (and inferior) member of the Prájnika Triad. The Bodhisattwa or Sangha continues to be such until he has surmounted the *very last* grade of that vast and laborious ascent by which he is instructed that he can "scale the heavens," and pluck immortal wisdom from its resplendent source: which achievement performed, he becomes a Buddha, that is, an Omniscient Being, and a *Tathágata**—a title implying the accomplishment of that gradual increase in wisdom by which man becomes immortal or ceases to be subject to transmigration. These doctrines are very obscurely indicated in the Bauddha scriptures, whose words have another, more *obvious*, and very different sense; nor, but for the ambition of the *commentators* to exhibit their learning, would it be easy to gather the esoteric sense of the words of most of the original scriptures. I never was more surprised than when my old friend recently (after a six years' acquaintance) brought to me, and explained, a valuable comment upon a passage in the *Prajná Páramitá*. Let me add in this place, that I desire all searchers after the doctrine of Bodhijnyána to look into the Bauddha scriptures, and judge for themselves; and to remember, meanwhile, that I am not a Sanskrit scholar, and am indebted for all I have gathered from the books of the Buddhists to the mediation of my old Bauddha friend, and of my *Pandit*.

(30) Their physiognomy, their language, their architecture, civil and religious, their notions in regard to women, and several less important traits in their manners and customs, seem to decide that the origin of the *greater part* of the Newárs must be assigned to the north; and in the *Sámbhú Purána*, a Bauddha teacher named Manju Ghosha, and Manju Nátha and Manjusrí, is stated to have led a colony into Nepaul from China;† to have cleared Nepaul of the waters which then covered it; to have made the country habitable; to have built a temple to Jyotí-rúp-Adi-Buddha; and established Dharmákara (whom he brought with him) as first Raja of Nepaul. But I nevertheless suppose (upon the authority of tradition) that Nepaul received *some* colonists from India; and that *some* of the earliest propagators of Buddhism in Nepaul came to the valley *direct* from India. Be that as it may, the Indian origin of Nepaulese Buddhism (whether it reached the valley direct, or *viâ* Bhot or China) seems to be unquestionable from the fact that all the great *Saugata* scriptures of Nepaul are written in the *Sanskrit language*. From the gradual decay of literature and of a knowledge of Sanskrit among the Newárs has resulted the practice, now very common, of translating *ritual* works into the vernacular tongue; and also the usage of

* *Tathá*, 'thus, absolutely, verily;' and *gata*, 'got, obtained;' the thing got being cessation from versatile existence, alias, *nirvána pada*.

† See Fahien, pp. 112-115 for Manjusrí. The place named is Pancha Sírsha Parvata, which the comment says is in China. The words are both Sanskrit.

adding to the original Sanskrit of *such* works comments in the vulgar language. The great scriptures however have never been subjected to the former process; seldom to the latter; for owing to Sanskrit having always been considered by the Buddhists of Nepaul the language of literature, they have neglected to cultivate their vernacular tongue; nor does there exist to this day a dictionary or grammar of the Newárí language.

(31) Of course therefore the Bauddhas of Nepaul have not *properly* any diversity of caste; that is, any indelible distinction of ranks derived from birth, and necessarily carried to the grave. *Genuine* Buddhism proclaims the equality of all followers of Buddha—seems to deny to them the privilege of pursuing worldly avocations, and abhors the distinction of clergy and laity. All proper Bauddhas are Bandyas; and all Bandyas are equal as brethren in the faith. They are properly all ascetics or monks—some solitary, mostly cœnobitical. Their convents are called *Viháras.* The rule of these Viháras is a rule of freedom; and the door of every Vihára is always open; both to the entrance of new comers, and to the departure of such of their old inmates as are tired of their vows.§ Each Vihára has a titular superior called Náyaka,|| whose authority over his brethren depends only on their voluntary deference to his superior learning or piety. Women are held equally worthy of admission with men, and each sex has its Viháras.

The old Bauddha scriptures enumerate four sorts of Bandyas, named: Arhan, Bhikshu, Srávaka, and Chailaka, who are correctly described in the text; and from that description it will be seen that there is no essential distinction between them, the Arhan being only segregated from the rest by his superior proficiency in Bodhijnána. Of these the proper institutes of Buddhism, there remains hardly a trace in Nepaul. The very names of the Arhan and Chailaka have passed away—the names, and the names only, of the other two exist; and out of the gradual, and now total, disuse of monastic institutes, an exclusive minister of the altar, denominated *Vajra Achárya,* has derived his name, office, and existence in Nepaul, not only without sanction from the Bauddha scriptures, but in direct opposition to their spirit and tendency.

Nepaul is still covered with Viháras; but these ample and comfortable abodes have long resounded with the hum of industry and the pleasant voices of women and children. The *superior* ministry of religion is now solely in the hands of the Bandyas, entitled, *Vajra-Achárya* in Sanskrit; *Gúbhál* in Newárí: the *inferior* ministry, such Bhikshus as still follow religion as a lucrative and learned profession, are competent to discharge. And these professions of the Vajra-Achárya, and of the Bhikshu, have become by usage hereditary, as have all other avocations and pursuits, whether civil or religious, in Nepaul. And as in the modern corrupt Buddhism of Nepaul there are exclusive ministers of religion or *priests,* so are there many Bauddhas who retain the lock on the crown of the head, and are

§ "Once a priest for ever a priest" is a maxim which Buddhism utterly eschews.

|| *Náyaka,* the superior of a convent, is *Khanpo* in Tibet, *Thero* in Ceylon *Bandya* is *Bonze* in Japan, *Bandida* in Altaia; and *Arhat* is *Rahatun* in Indo-China. I demur to the frequent use of the word priest as the equivalent of any of these terms.

not Bandyas. These improper Bauddhas are called *Udás*, *Japu*, *Kami*, etc., according to their various avocations and crafts; the Udás are traders; the Japu, agriculturists; the Kami, craftsmen. They comprise the untonsured class: they never dwell in the Viháras; look up to the Bandyas with a reverential respect derived from the misapplication of certain ancient tenets; and follow those trades and avocations which are comparatively disreputable (among which is *foreign* commerce); while the Bandyas, who have abandoned the profession of religion, practise those crafts which are most esteemed. Agriculture is equally open to both; but is, in fact, chiefly followed by the untonsured class, who have thus become, in course of time, more numerous than the Bandyas, notwithstanding the early abandonment by the Bandyas of those monastic vows which their faith enjoins, the resort of the greater part of them to the active business of the world, and their usurpation of all the liberal, and many of the mechanical, arts of their country. The Vajra-Achárya and Bhikshu are the religious guides and priests of both Bandyas and non-Bandyas.* All Bandyas, whatever be the profession or trade they hereditarily exercise, are still equal; they intermarry, and communicate in all the social offices of life—and the like is true of all of the other classes—but between the one class and the other, growing superstition has erected an insuperable barrier. To the above remarks it may be well to add, that Buddhists, of some one or other of the above denominations, comprise the vast majority of the Newar race, and that the minority, are mostly Saivas and Sáktas; but in a sense peculiar to themselves, and with which my subject does not entitle me here to meddle.

(33) The names are almost all barbarous; that is, not derived from Sanskrit, but from Newárí. I have not thought it worth while to enumerate any more of these examples. The Vihára is built round a large quadrangle, or open square, two stories high; the architecture is Chinese. Chaitya properly means a *temple of* Buddha, and Vihára, an abode of cœnobitical *followers* of Buddha.† In the open square in the midst of every Vihára, is placed a Chaitya or a Kutagar—but those words always bear the senses here attached to them; and Vihára can never be construed temple—it is a convent, or monastery, or religious house, but never *templum* DEI *vel* BUDDHÆ. At the base of the hemisphere of *every Nepaul Chaitya* are placed the images of the Dhyáni Buddhas. The Chaitya has often been *blended* with sundry structures, more or less appropriate to Buddhism.

To conclude: with respect to the notes—that portion of this sketch, which is my own—no one can be more sensible than I am that the first half contains a sad jumble of cloudy metaphysics. How far the sin of this indistinctness is mine, and how far that of my original authorities, I cannot pretend to decide; but am ready to take a large share of it to myself. In regard to *this*, the most

* Bandya has no correlative term, like Laicus of Clerus; one of many arguments in favour of the nonadmittance of that distinction by Buddhism, as elsewhere attempted to be shown: see Fahian pp. 12, 172, 175, and 289, for sundry notices of so-called *Clerus et Laicus*. Those passages seem to prove that the distinction is foreign to genuine Buddhism.

† Fergusson, tree and serpent worship, p. 79.

speculative part of Buddhism, it is sufficient happiness for me to have discovered and placed within the reach of my countrymen the *materials* for more accurate investigation, by those who have leisure, patience, and a knowledge of languages for the undertaking; and who, with competent talents, will be kind enough to afford the world the benefit of so irksome an exercise of them.

But I trust that the *latter* half of the notes, which embraces topics more practical and more within the range of the favourite pursuits of my leisure, will not be found wanting in distinctness; and I can venture confidently to warrant the *accuracy* of the information contained in it.

QUOTATIONS FROM ORIGINAL SANSKRIT AUTHORITIES.

Several distinguished orientalists having, whilst they applauded the novelty and importance of the information conveyed by my Sketch of Buddhism,‡ called upon me for proofs, I have been induced to prepare for publication the following translation of significant passages from the ancient books of the Saugatas, which still are extant in Nepaul in the original Sanskrit.

These extracts were made for me (whilst I was collecting the works* in question) some years ago by Amrita Nanda Bandya, the most learned Buddhist then, or now, living in that country; they formed the materials from which chiefly I drew my sketch; and they would have been long since communicated to the public, had the translator felt sufficiently confident of his powers, or sufficiently assured that enlightened Europeans could be brought to tolerate the '*ingens indigestaque moles*' of these 'original authorities;' which however, in the present instance, are original in a higher and better sense than those of Csoma de Körös or of Upham. Without stopping to question whether the sages who formed the Bauddha system of philosophy and religion used Sanskrit or high Prákrit, or both, or seeking to determine the consequent pretension of Upham's authorities to be considered original,† it may be safely said, that those of Csoma de Körös can support no claims of the kind.

‡ Transactions of the Royal Asiatic Society, London; —necnon, Transactions of Bengal Society, vol. xvi.

* The collection comprises, besides sixty volumes in *Sanskrit*, procured in *Nepaul*, the very names of which had previously been unknown, some 250 volumes in the language of *Tibet*, which were obtained from *Lassa* and *Digarchi*. But for the existence of the latter at Calcutta, Csoma de Körös's attainments in *Tibetan* lore had been comparatively useless. The former or Sanskrit books of *Nepaul* are the authorities relied on in this paper. One complete set has been presented to the Indian Home Government, another procured for the Asiatic Society, and most of the Sanskrit series for the Libraries of Paris and of Oxford. Since the first collection was made in *Nepaul*, very many new works in the Sanskrit language have been discovered and are yet daily under discovery. The probability now is, that the entire *Kahgyur* and *Stangyur* may be recovered in the original language. The whole series has been obtained in that of Tibet, 327 large volumes.

† Upham's authorities, however, even if allowed to be original, appear to consist entirely of childish legends. I allude to the three published volumes. The received hypothesis, *viz.*, that the philosophers of *Ayodhyá* and *Magadha*, (the acknowledged founders of *Buddhism*) postponed the use of Sanskrit to that of Prakrit, in the original exposition of their subtle system appears to me as absurd as it does probable that their successors, as *Missionaries*, resorted to Prakrit versions of the original Sanskrit authorities, in propagating the system in the remotest parts of the continent and in

The native works which the latter gentleman relies on are avowedly Tibetan translations of my Sanskrit originals, and whoever will duly reflect upon the dark and profound abstractions, and the infinitesimally-multiplied and microscopically-distinguished personifications of Buddhism, may well doubt whether the language of *Tibet* does or can adequately sustain the weight that has been laid upon it.

Sanskrit, like its cognate Greek, may be characterised as a speech "capable of giving a soul to the objects of sense, and a body to the abstractions of metaphysics." But, as the Tibetan language can have no pretensions to a like power, those who are aware that the Saugatas taxed the whole powers of the Sanskrit to embody in words their system, will cautiously reserve, I apprehend, for the Bauddha books still extant in the classical language of India, the title of original authorities. From such works, which, though now found only in Nepaul, were composed in the plains of India before the dispersion of the sect, I have drawn the accompanying extracts; and though the merits of the "doing into English" may be small indeed, they will yet, I hope, be borne up by the paramount and (as I suspect) unique authority and originality of my "original authorities," a phrase which, by the way, has been somewhat invidiously, as well as laxly, used and applied in certain quarters.

It is still, I observe, questioned amongst us, whether Brahmanism or Buddhism be the more ancient creed, as well as whether the latter be of Indian or extra Indian growth. The Buddhists themselves have no doubts upon either point. They unhesitatingly concede the palm of superior antiquity to their rivals and persecutors, the Brahmans; nor do they in any part of the world hesitate in pointing to India as the cradle of their faith.

Formerly we might be pardoned for building fine-spun theories of the exotic origin of Buddhism upon the supposed African locks of Buddha's images: but surely it is now somewhat too late,* in the face of the abundant direct evidence which we possess against the exotic theory, to go in quest of presumptions to the time-out-of-mind illiterate Scythians,† in order to give to them the glory of

Ceylon. On this ground, I presume the Prakrit works of Ceylon and Ava to be translations, not originals :—a presumption so reasonable that nothing but the production from Ceylon or Ava of original Prakrit works, comparable in importance with the Sanskrit books discovered in Nepaul, will suffice to shake it in my mind. Sir W. Jones had a copy of the *Lalita Vistara* whence he quotes a description of Dharma as Diva Natura. Sir W. Jones I believe to be the author of the assertion, that the *Buddhists* committed their system to high Prakrit or Pali : and so long at least as there were no Sanskrit works of the sect forthcoming, the presumption was not wholly unreasonable. It is, however, so now. And Sir W. Jones was not unaware that *Magadha* or *Bihar* was the original head-quarters of *Buddhism*, nor that the best Sanskrit lexicon extant was the work of a *Bauddha;* nor that the *Brahmans* themselves acknowledged the pre-eminent *literary* merits of their heterodox adversaries. But for his *Brahmanical* bias therefore, Sir William might have come at the truth, that the *Bauddha* philosophers employed the classical language.

* Recent discoveries make it more and more certain, that the cave temples of the Western Coast and its vicinity, are *exclusively Bauddha*. Every part of India is illustrated by splendid remains of Buddhism.

† The Uighurs of Bish Balig had letters derived from the Nestorian Christians. Thence Sramanism and Christian monachism may have met on the common ground of monachism. Sramanism is nothing more than Tantrika Buddhism.

originating a system built upon the most subtle philosophy, and all the copious original records of which are inshrined in Sanskrit,‡ a language which, whencesoever primevally derived, had been, when Buddhism appeared, for ages proper to the Indian continent.

The Buddhists make no serious pretensions to a very high antiquity: never hint at an extra Indian origin.

Sákya Sinha is, avowedly, a Kshatriya; and, if his six predecessors had really any historical existence, the books which affirm it, affirm too, that all the six were of Brahmanical or Kshatriya lineage.§ Saugata books treating on the subject of caste never call in question the antique fact of a fourfold division of the Hindu people, but only give a more liberal interpretation to it than the current Brahmanical one of their day.‖ The Chinese, the Mongols, the Tibetans, the Indo-Chinese, the Japanese, Ceylonese, and other Indian Islanders, all point to India as the father-land of their creed. The records of Buddhism in Nepaul and in Tibet, in both of which countries the people and their mother-tongues are of the Mongol stock, are still either Sanskrit or avowed translations from it by Indian *pandits*. Nor is there a single record or monument of this faith in existence which bears intrinsic or extrinsic evidence of an *extra* Indian origin.**

The speculations of a writer of Sir W. Jones's day (Mr. Joinville), tending to prove, argumentatively, from the characters of Buddhism and Brahmanism, the superior antiquity of the former, have been lately revived (see Asiatic Journal, No. CLX.) with applause. But besides that fine drawn presumptions are idle in the face of such a mass of direct evidence as we now possess, the reasonings of Joinville appear to me altogether based on errors of fact. Buddhism (to hazard a character in few words), is monastic asceticism in morals, philosophical scepticism in religion; and whilst ecclesiastical history all over the world affords abundant instances of such a state of things resulting from gross abuse of the reli-

‡ The difference between high Prakrit and Sanskrit could not affect this question, though it were conceded that the founders of Buddhism used only the former and not the latter—a concession however, which should not be lightly made, and to which I wholly demur. In fact, it now appears that they used both languages, but Sanskrit only in the philosophical or speculative series of their Sastras.

§ The Brahmanical or Kshatriya family from which each of these Buddhas sprung is expressly and carefully stated by the Bauddha writers, a fact which I hold to be decisive of this dispute, since if we would carry the etymon of Buddhism beyond the last of these seven Buddhas, we cannot surely think of carrying it beyond the first of them.

‖ See the Bauddha disputation on caste, Royal Asiatic Society's Transactions.

** See Crawfurd's remarks on the purely Indian character of all the great sculptural and architectural monuments of Buddhism in Java. Also Barrow's remarks to the same effect in his travels in China. The Chinese *Pu-sa* is *Visvarúpyá Prajná* or the polyform type of "Diva Natura." See Oriental Quarterly Magazine, No. xvi. pp. 218—222, for proofs of the fact that numberless Bauddha remains have been mistaken for Brahmanical by our antiquaries, and even by the natives. In the same work I have proved this in reference to Crawfurd's Archipelago, Oriental Quarterly, No. xvi. pp. 232, 235.

Yet, no sooner had I shown, from original authorities, how thoroughly *Indian* Buddhism is, than it was immediately exclaimed, "Oh! this is *Nepaulese* corruption! these are merely popular grafts from Brahmanism." The very same character belongs to the oldest monuments of Buddhism, extant in India and beyond it; and I have traced that character to the highest scriptural authorities.

gious sanction, that ample chronicle gives us no one instance of it as a primitive system of belief. Here is a legitimate inference from sound premises. But that Buddhism was, in truth, a reform or heresy, and *not* an original system, can be proved by the most abundant direct evidence both of friends and of enemies. The oldest Saugata works incessantly allude to the existing superstition as the *Máracharya* or way of the evil one,†† contradistinguishing their reformation thereof as the *Bodhicharya* or way of the wise; and the Brahmanical impugners of those works (who, upon so plain a fact, could not lie), invariably speak of Buddhism as a notorious heresy.

An inconsiderable section of the Saugatas alone, ever held the bald doctrine of mortal souls: and the Swábhávika denial of a creation of matter by the fiat of an absolutely immaterial being, springs not out of the obesity of barbarian dulness, but out of the over-refinement of philosophical ratiocination. Joinville's idea of the speculative tenets of Buddhism is utterly erroneous. Many of them are bad indeed: but they are of philosophy "all compact," profoundly and painfully subtle, sceptical too, rather than atheistically dogmatic.

At the risk of being somewhat miscellaneous in this preface, I must allude to another point. The lamented Abel Rémusat sent me, just before he died, a copy of his essay on the Saugata doctrine of the Triad; and Mr. Upham, I find, has deduced from Rémusat's interpretation of that doctrine, the inference (which he supports by reference to sundry expressions in the sacred books of Ceylon), that I am in error in denying that Buddhism, in its first, and most characteristic form, admits the distinction of *Clerus et Laicus.* It is difficult expressly to define that distinction; but it may be seen in all its breadth in Brahmanism and in Popery; whilst in Islamism, and in the most enthusiastic of the Christian sects, which sprang out of the Reformation, it is wholly lost. According to my view, Apostolic Christianity recognised it not;* the congregation of the faithful, the Church, was a society of peers, of brethren in the faith, all essentially equal, in gifts, as in place and character. On earth, there were no indispensable mediators, no exclusive professional ones; and such alone I understand to be priests.† Again, genuine monachism all over the world, I hold to be, in its own nature, essentially opposed to the distinction of clergyman and layman, though we all know that monastic institutions no sooner are rendered matters of public law and of extensive popular prevalence, than, *ex vi necessitatis,* the distinction in question is superinduced upon them, by the major part of the monks laicising, and the rest *becoming clergy.*‡ There are limits to the number of those whom the public can

†† Námuchi by name, chief of the Kakodemons.

* I would not be understood to lay stress on his opinion, which is merely adduced to illustrate my argument.

† For example, the Anglican church holds that there is no virtue in any sacerdotal function not performed by the successors of the apostles, who are the only clergy.

‡ History informs us that, soon after monachism supervened upon our holy and eminently social religion, there were in Egypt as many monks almost as peasants. Some of these monks necessarily laicised, and the rest became clergy. The community of the *Gosains* and several others, of strictly ascetical origin, now in India, exhibit the same necessary change after the sects had become numerously followed.

support in idleness; and whoso would eat the bread of the public must perform some duty to the public. Yet who can doubt that the true monk, whether cœnobite or solitary, is he who abandons the world to save his *own* soul; as the true clergyman is he who mixes with the world to save the souls *of others?*§ The latter in respect to the people or laics has a distinctive function, and, it may be, also an exclusive one: the former has no function at all. Amongst entirely monastic sects, then, the exclusive character of priest is objectless and absurd; and who that has glanced an eye over ecclesiastical history knows not that in proportion as sects are enthusiastic, they reject and hate, (though nothing tainted with monachism) the exclusive pretensions of the clergy! Whoever has been able to go along with me in the above reflections can need only to be told that primitive Buddhism was entirely monastic, and of an unboundedly enthusiastical genius,‖ to be satisfied that it did not recognise the distinction in question. But if, being suspicious of the validity of argumentative inferences, he demand of me simple facts, here they are. In the *Sata Sáhasrika, Prajná Páramitá*, or *Raksha Bhágavatí*, and also in the nine *Dharmás* (the oldest and highest written authorities), it is affirmed more or less directly, or is clearly deducible from the context, in a thousand passages (for the subject is not expressly treated), that the only true followers of Buddha are monks; the majority being cœnobites, the rest, solitaries. The fullest enumeration of these followers *(Bhikshu, Srávaka* or *Srámana,* Chailáka*, and *Arhata* or *Arhana* or *Arhanta)* proves them to have been all monks, tonsured, subject to the usual vows, (nature teaching to all mankind that wealth, women and power, are the grand tempters,) resident in monasteries (*Vihára*) or in deserts, and essentially peers, though of course acknowledging the claims of superior wisdom and piety. The true church, the congregation of the faithful, (called from this very circumstance *Sangha*,) is constantly said to consist of such only; and I am greatly mistaken indeed if the church in this sense be synonymous with the clergy;§§ or, if the primitive church of Buddha recognized an absolutely distinct body such as we (*i. e.*, Catholics, Lutherans, and Kirkmen) ordinarily mean when we speak of the latter. The first mention of an exclusive, professional, active, minister of religion, or priest, in the Bauddha books, is in those of a comparatively recent date, and not of scriptural authority. Therein the *Vajra Achárya* (for so he is called) first appears arrayed with the ordinary attributes of

§ See Guizot's *Civilization of Europe*. ii. 61-63, & i. 86.

‖ Its distinguishing doctrine is that finite mind can be enlarged to infinite; all the schools uphold this towering tenet, postponing all others to it. As for the scepticism of the Swábhávikas relative to those transcendent marvels, creation and providence, it is sufficient to prove its remoteness from "flat Atheism," simply to point to the *coexistence* of the cardinal tenet first named.

* Srámana includes the whole, and is equally ascetic; Srámaní feminine, equal to monk and nun. Sákya is often called the great Srámana.

§§ Bunsen's controversy with Gladstone, and his work on the constitution of the church (published in 1847) set this matter clearly in the light in which I viewed it; Bunsen insists on the congregational church as the only true one, says the clergy church is pregnant with priestcraft and essentially untenable, contends that the future church must be of the former kind, and adds that the reformation virtually extinguished the clergy church. So Sákya argued and instituted in opposition to the cleric exorbitances of the Brahmans.

a priest. But his character is anomalous, as is that of everything about him; and the learned Bauddhas of Nepaul at the present day universally admit the falling off from the true faith. We have in these books, *Bhikshus*, *Srávakas*, *Chailakas*, and *Sákya Vansikas*,* bound by their primitive rules for ten days (in memory of the olden time) and then released from them; tonsured, yet married; ostensibly monks, but really citizens of the world.

From any of the above the *Vajra Achárya* is drawn indiscriminately; he keeps the keys of the no longer open treasury; and he is surrounded by *untonsured* followers, who now present themselves for the first time. I pretend not to trace with historical nicety all the changes which marked the progress of Buddhism as a public institute and creed of millions up to the period of the dispersion: but I am well aware, that the primitive doctrines were not, because they could not be, *rigidly* adhered to, when what I hold to have been at first the closet speculation of some philosophers, had become the dominant creed of large kingdoms. That the latter character was, however, assumed by Buddhism in the plains of India for centuries† before the dispersion, seems certain; and, as many persons may urge that the thing in question is the dominant public institute, not the closet speculation, and that whatever discipline prevailed before the dispersion must be held for primitive and orthodox, I can only observe that the ancient books of the Saugatas, whilst they glance at such changes as I have adverted to, do so in the language of censure; and that, upon the whole, I still strongly incline to the opinion that genuine or primitive Buddhism (so I cautiously phrased it originally) rejected the distinction of *Clerus et Laicus;* that the use of the word priest by Upham, is generally inaccurate; and that the *Sangha* of the Buddhist triad ought to have been invariably rendered by Rémusat into 'congregation of the faithful' or 'church,' and never into 'clergy' or 'priesthood.' Rémusat indeed seems to consider (*Observations*, 28-29, and 32,) these phrases as synonymous; and yet the question which their discrimination involves is one which, in respect to our own religion, has been fiercely agitated for hundreds of years; and still, by the very shades of that discrimination, chiefly marks the subsisting distinction between the various Churches of Christ!

*An inscription at Karli identifies the splendid *Sálivâhana* with the head of the Saka tribe, which is that of Sákya Sinha. The Sákya-Vansikas, or people of the race of Sákya, appeared in Nepaul as refugees from Brahman bigotry, some time after Buddhism had been planted in these hills. Sákya is universally allowed to have been the son of king Sudhodana, sovereign of Magadha, or Bihar (Kosala says Wilson, who calls it a dependancy of Bihar). He is said to have been born in the "Sthána of Kapila Muni," at Gangá Ságara, according to some; in Oude, as others say. His birth place was not necessarily within his father's kingdom. He may have been born when his father was on a pilgrimage to the shrine of the saint Kapila. Sákya died, according to my authorities, in Assam, and left one son named Rahula Bhadra. (See Csoma de Körös in No. 20 of Journal of Bengal Asiatic Society for origin of Sákya-Vansika. Their primitive sect was Tatta, their next Kapila in Oude, whence they migrated into Nepaul.) The Sakas were Kshatriyas of the solar line, according to *Bauddha* authorities: nor is it any proof of the contrary that they appear not in the Brahmanical genealogies. See note in the sequel.

† Even if we begin with Asoka we can hardly assign less than six to eight centuries for Buddhist predominance, nor less than about double that duration for more or less of prevalence in the plains of India. (See note at page 76.)

Following the authority he has relied on, Mr. Upham was at liberty, therefore, to adopt a sense which would consist with *my* interpretation of phrases such as he alluded to, and which, of course, I found copiously scattered over the works I consulted.

I always rendered them advisedly into English, so as to exclude the idea of a priesthood, because I had previously satisfied myself, by separate inquiry and reflection that that cardinal tenet was repugnant to the genius of the creed, and repudiated by its primitive teachers. This important point may have been wrongly determined by me; but assuredly the determination of it upon such grounds as Mr. Upham's is perfectly futile. Such words as *Arhanta* and *Bandya*, (which, by the way, are the correct forms of the Barmese *Rahatun* and the Japanese *Bonze*,) no more necessarily mean priest, clergy, than do the Latin *fideles* and *milites* as applied to Christianity, as little can such a sense be ascribed to the word *Bhikshu*, which means 'mendicant friar;' and as for the word *Sangha*, it is indisputable that it does *not* mean *literally* priest,** and that it *does* mean *literally* 'congregation.'

If, as Rémusat and Upham* appear to insist is the case, every monastic follower of Buddha be a priest, then Bandya or Bonze† must be rendered into English by the word 'clergyman.' But there will still remain as much difference between Bandya and Sangha as, in Christian estimation, between an ordinary parson of the present day, and one of the inspired primitive professors. Of old, the spirit descended upon all alike; and Sangha was this hallowed and gifted congregation. But the glory has passed away, and the term been long sanctified and set apart. So has, in part, and for similar reasons, the word Arhata. But Bandya, as a generic title, and Bhikshu, Srávaka, and Chailaka,‡ as specific ones, are still

** Observations, p. 63.

* *Bhikshu* now appears to be the word rendered priest by us in Ceylon. But it is unquestionably mendicant, holy beggar, as *Thero* is *Náyaka* or Superior and *Upásika* Servitor, of a Convent. See Fahian, 12, 172, 234.

† The possible meaning of this word has employed in vain the sagacity of sundry critics. In its proper form of *Bandya (Vandya)*, it is pure Sanskrit, signifying a person entitled to reverence, and is derived from *Vandana*.

Equally curious and instructive is it to find in the *Sanskrit* records of *Buddhism* the solution of so many enigmas collected by travellers from all parts of Asia; *e grege*, Elphinstone's mound is a genuine Chaitya, and its proper name is *Manikálaya*, or the place of the precious relic. The mound is a tomb temple. The "*tumuli eorum Christi altaria*" of the poet, is more true of *Buddhism* than even of the most perverted model of Christianity; the *cause* being probably the same, originally, in reference to both creeds, *viz.*, persecution and martyrdom, with consequent divine honours to the sufferers. The *Bauddhas*, however, have in this matter gone a step further in the descending scale of representative adoration than the Catholics; for they worship the mere image of that structure which is devoted to the enshrining of the relics of their saints; they worship the architectural model or form of the *Chaitya*.

The *Chaitya* of Sambhuná h in *Nepaul* is affirmed to cover *Jyoti rúpya* Swayambhu, or the self-existent, in the form of flame: nor was there ever anything *exclusive* of theism in the connection of tomb and temple: for *Chaityas* were always dedicated to the *Celestial Buddhas*, not only in *Nepaul*, but in the plains of India, as the *Chaityas* of *Sanchi*, of *Gyá*, and of *Bág*, demonstrate. The *Dhyáni Buddhas* appear in the oldest monuments of the continent and islands.

‡ Buddhist monachism agrees surprisingly with Christian, whether owing to Nestorian infusion among the Uighurs or otherwise. Thus there are several orders of monks in both; in the former mendicant saints, naked or scantly clothed saints, and learned

every-day names of every-day people, priests, if it must be so, but as I conceive, ascetics or monks merely. In the thick night of ignorance and superstition which still envelopes Tibet, the people fancy they yet behold Arhatas in the persons of their divine Lámas. No such imagination however possesses the heads of the followers of Buddha in Nepaul, Ceylon, or Indo-China; though in the last mentioned country the *name* Arhata is popularly applied to the modern order of the clergy, an order growing there, as in Nepaul, (if my opinions be sound) out of that deviation from the primitive genius and type of the system which resulted necessarily from its popular diffusion as the rule of life and practice of whole nations.

In conclusion I would observe, that, in my apprehension, Rémusat's interpretation of the various senses of the Triadic doctrine is neither very complete, nor very accurate. In a religious point of view, by the first member is understood the founder of the creed, and all who, following his steps, have reached the full rank of a Maháyánika Buddha; by the second, the law or scriptures of the sect; and by the third, the congregation of the faithful, or primitive church, or body of original disciples, or any and every assemblage of true, *i. e.*, of monastical observers of the law, past or present.

In a philosophical light, the precedence of Buddha or of Dharma indicates the theistic or atheistic school. With the former, Buddha is intellectual essence,§ the efficient cause of all, and underived. Dharma is material essence,‖ the plastic cause, and underived, a co-equal biunity with Buddha; or else the plastic cause, as before, but dependent and derived from Buddha. Sangha is derived from, and compounded of Buddha and Dharma, is their collective energy in the state of action; the immediate operative cause of creation, its type or its agent.* With the latter or atheistic schools, Dharma is *Diva natura*, matter as the sole entity, invested with intrinsic activity and intelligence, the efficient and material cause of all.

Buddha is derivative from Dharma, is the active and intelligent force of nature, first put off from it and then operating upon it. Sangha is the *result* of that operation; is embryotic creation, the type and sum of all specific forms, which are spontaneously evolved from the union of Buddha with Dharma.*† The

saints like the Franciscans, Dominicans, etc., and all of both creeds are usually social, though hermits also be found.

§ *Bodhanátmaka iti Buddha*, 'the intellectual essence is *Buddha*.'

‖ *Dharanátmaka iti Dharma*, 'the holding, sustaining or containing substance is *Dharma*.' Again, *Prakriteswari iti Prajna*, 'the material goddess is *Prajna*,' one of the names of Dharma. The word *Prajna* is compounded of the intensive prefix *pra*, and *jnana* wisdom, or *jna*, to know. It imports the supreme wisdom of nature. *Dharma* is the universal substratum, is that which supports all form and quality in space. The *Bauddha Dharma* is the exact equivalent of the *Brahmanical Matra*. *Matra* is that which measures space; *Dharma* that which supports form and quality in space; both are very just and philosophical ideas relative to what we call matter and substance. The *substans* or supporter of all phænomena, whatever its nature, is *Dharma*.

* *Samudayâtmika iti Sangha*, 'the multitudinous essence is *Sangha*:' multitude is the diagnosis of the versatile universe, as unity is of that of abstraction.

*† *Prajnaopayatmakam Jagatah*, from *Prajna* and *Upaya*, the world. *Upaya* is the energy of *Prajna*.

above are the principal distinc tions; others there are which I cannot venture here to dwell on.

With regard to Rémusat's remark, "*on voit que les trois noms sont placés sur le même niveau, comme les trois représentations des mêmes êtres dans les planches de M. Hodgson avec cette différence que sur celles-ci, Sangha est à droite, et Dharma à gauche,*" I may just add, that the placing of *Sangha* to the right is a merely ritual technicality, conformable to the *pújá* of the *Dakshinácháras,** and that all the philosophers and religionists are agreed in postponing *Sangha* to *Dharma.*

I possess very many drawings exhibiting the arrangement mentioned by Rémusat; but all subservient to mere ritual purposes and consequently worthy of no serious attention. The *Matantara,* or variorum text of the *Pújáris* of the present day, displays an infinite variety of formula,† illustrated by corresponding sculptural and pictorial devices, embodied in those works, and transferred from them to the walls and interior of temples existing all over the valley of Nepaul.

THE SWABHAVIKA‡ DOCTRINE.

1. All things are governed or perfected by Swabháva;‡ I too am governed by Swabháva. *(Ashta Sáhasrika.)*

2. It is proper for the worshipper at the time of worship to reflect thus: I am *Nirlipta,*§ and the object of my worship is Nirlipta; I am that God *(Iswara)* to whom I address myself. Thus meditating, the worshipper should make *pújá* to all the celestials: for example, to Vajra Satwa Buddha, let him pay his adorations, first, by recollecting that all things with their *Víja Mantras* come from Swabháva in this order:—from the *víja* ‖ of the letter Y, air; from that of the letter R, fire; from that of the letter V, or B, water; and from that of the letter L, earth; and from that of the letter S, Mount Sumeru. On the summit of Sumeru is a lotus of precious stones, and above the lotus, a moon crescent, upon which sits, supremely exalted, Vajra Satwa. And as all (other) things proceed from Swabháva, so also does Vajra Satwa, thence called the self-existent.** *(Pújá Kánda.)*

3. All things and beings (in the versatile universe) which are alike perishable, false as a dream, treacherous as a mirage, proceed, according to some, from Swabháva (nature), and according to others, from God (Iswara); and hence it is said, that Swabháva and Iswara are essentially one, differing only in name.*‡ *(Ashta Sáhasrika.)*

* The theistic sects so call themselves, styling their opposites, the *Swabhavikas* and *Prajnikas, Vamacharas.* The *Pauranikas,* too, often designate the *Tantrikas* by the latter name, which is equivalent to left-handed.

† See the classified enumeration of the principal objects of *Bauddha* worship appended to this paper.

‡ *Swa,* own, and *bhava,* nature. Idiosyncrasis.

§ Intact and intangible, independent.

‖ Root, radix, seed.

** This may teach us caution in the interpretation of terms. I understand the dogma to announce, that infinite intelligence is as much a part of the system of nature as finite. The mystic allusion to the alphabet imports nothing more than its being the indispensable instrument and means of knowledge or wisdom, which the *Buddhists* believe man has the capacity of perfecting up to the standard of infinity.

*‡ See note on No. 3, on the *Yatnika* system.

4. At the general dissolution of all things, the four elements shall be absorbed in *Súnyákára-Akása* (sheer space) in this order:—earth in water, water in fire, fire in air, and air in Akása, and Akása in Súnyatá, and Súnyatá in Tathatá,* and Tathatá in Buddha, (which is Mahá Súnyatá†) and Buddha in Bhávana, and Bhávana in Swabháva. And when existence is again evolved, each shall in the inverse order, progress from the other. From that Swabháva, which communicates its property of infinity to Akása, proceeded into being, in Akása, the letter A, and the rest of the letters; and from the letters Adi-Buddha‡ and the other Buddhas; and from the Buddhas the Bodhi-Satwas, and from them the five elements, with their Víja Mantras.§ Such is the Swábhávika Sansára; which Sansára (universe) constantly revolves between Pravritti and Nirvritti, like a potter's wheel. *(Divya Avadána).*

5. Mahá Súnyatá is, according to some Swabháva, and according to others, Iswara it is like the ethereal expanse, and self-sustained. In that Mahá-Súnyatá, the letter A, with the Víja Mantra of Upáya,|| and the chief of all the Víja Mantras of the letters, became manifest. *(Rakshá Bhagavatí.)***

6. Some say creation is from God: if so, what is the use of Yatna or of Karma? *† That which made all things, will preserve and destroy them; that which governs Nirvritti governs Pravritti also. *(Buddha Charitra Kávya.)*

7. The Sandal tree freely communicates its fragrance to him who tears off its bark. Who is not delighted with its odour? It is from Swabháva. *(Kalpalatá.)*

8. The elephant's cub, if he find not leafless and thorny creepers in the green wood, becomes thin. The crow avoids the ripe mango.*‡ The cause is still Swabháva. *(Kalpalatá.)*

9. Who sharpened the thorn? Who gave their varied forms, colours, and habits to the deer kind, and to the birds? Swabháva! It is not according to the will *(ichchhá)* of any; and if there be no desire or intention, there can be no intender or designer.†* *(Buddha Charitra.)*

* *Tathata*, says the comment, is *Satya Jnyana;* and *Bhavana* is *Bhava* or *Satta, i. e.*, sheer entity.

† See note on quotation 1 of section on *Adi-Buddha*.

‡ Here again I might repeat the caution and remark at quotation 2. I have elsewhere observed that *Swabhavika* texts, differently interpreted, form the basis of the *Aiswarika* doctrine, as well as that the *Buddhas* of the *Swabhavikas*, who derive their capacity of identifying themselves with the first cause from nature, *which is that cause*, are as largely gifted as the *Buddhas* of the *Aiswarikas*, deriving the same capacity from *Adi-Buddha*, who *is that cause*. See remarks on Rémusat in the Journal of the Bengal Asiatic Society, Nos. 32, 33, and 34.

§ A. Cunningham has found this literal symbolic representation of the elements, and also that of the triad at Bhilsa. See his *Bhilsa Topes*, p. 355 f.

|| *Upáya*, the expedient, the energy of nature in a state of activity. See the note on No. 6 of the section *Adi-Sangha*.

** The *Raksha Bhágavati* is the same work as the *Prajná Paramita*.

*† See the note on quotation 9 of this head. *Yatna* and *Karma* may here be rendered by intellect and morality.

*‡ These are assumed facts in Natural History; but not correct.

†* Here is plainly announced that denial of self-consciousness or personality in the *causa causarum* which constitutes the great defect of the *Swabhavika* philosophy: and if this denial amount to atheism, the *Swabhavikas* are, for the most part, atheists:

10. The conch, which is worthy of all praise, bright as the moon, rated first among excellent things, and which is benevolent to all sentient beings, though it be itself insensate, yields its melodious music, purely by reason of Swabháva. *(Kalpalatá.)*

11. That hands and feet, and belly and back, and head, in fine, organs of whatever kind, are found in the womb, the wise have attributed to Swabháva; and the union of the soul or life *(Atmá)* with body, is also Swabháva. *(Buddha Charitra Kávya.)*

12. From Swabháva (nature) all things proceeded; by Swabháva all things are preserved. All their differences of structure and of habits are from Swabháva: and from Swabháva comes their destruction. All things are regulated *(suddha)* by Swabháva. Swabháva is known as the Supreme. *(Pújá Kánda*, from the *Rakshá Bhagavatí*, where the substance is found in sundry passages.)

13. *Akása* is Swábhávika, because it is established, governed perfected *(suddha)*, by its own force or nature. All things are absorbed in it: it is uncreated or eternal; it is revealed by its own force; it is the essence *(Atma**)* of creation, preservation, and destruction; it is the essence of the five elements; it is infinite; it is intellectual essence (*Bodhanátmika*). The five colours are proper to it; and the five Buddhas; and the letters. It is Súnyatá; self-supported; omnipresent: to its essence belong both Pravritti and Nirvritti. This Akása, which is omnipresent, and essentially intellectual,* because infinite things are absorbed into it, is declared to be infinite. From the infinite nature of this Akása were produced all moving things, each in its own time, in due procession from another, and with its proper difference of form and habits. From the secret† nature of Akása pro-

their denial also of a moral ruler of the universe being a necessary sequel to it. Excepting, however, a small and mean sect of them, they all affirm eternal necessary, entity; nor do any of them reject the soul's existence beyond the grave, or the doctrine of atonement. Still Newton's is, upon the whole, the right judgment, '*Deus sine providentia et dominio nihil est nisi fatum et natura.*' The *Swabhavika* attempts to deify nature are but a sad confusion of cause and effect. But, in a serious religious point of view, I fail to perceive any superiority possessed by the immaterial pantheism of the Brahmanists over the material pantheism of the *Buddhists.* Metempsychosis and absorption are common to both. Both admit eternal necessary, entity or a substans for phænomena; both admit intellect; both deny two classes of phænomena as well as two substantes for them; both affirm the homogeneousness and unreality of all phænomena, and lastly, both leave the personality and active dominion of the *causa causarum* in obscurity.

** One comment on the comment says, *Atma* here means *sthan* or *alaya, i. e.*, the *ubi* of creation, etc.

* *Akása* is here understood as synonymous with *Súnyatá*, that is, as the elemental state of all things, the universal *ubi* and *modus* of primal entity, in a state of abstraction from all specific forms: and it is worthy of note, that amidst these primal principles, intelligence has admission. It is therefore affirmed to be a necessary ens, or eternal portion of the system of nature, though separated from self-consciousness or personality. In the same manner, *Prajná*, the sum of all things, *Diva natura*, is declared to be eternal, and essentially intelligent, though a material principle.

† Secret nature of *Akasa*, that is, *Akasa* or *Ether* has no sensible cognizable properties such as belong to the ordinary elements. The gradual evolution of all things in Pravritti and their revolution into Nirvritti being perpetual, seem to prove that the Buddhist Súnyatá is not nothingness, but rather the utterly inscrutable character of the ultimate *semina rerum.*

ceeded likewise, together with the *Vija Mantra* of each one, air with its own mobility; and from air, fire with its own heat; and from fire, water with its intrinsical coldness; and from water, earth with its own proper solidity or heaviness; and from earth, Mount Sumeru with its own substance of gold, or with its own sustaining power (*Dhátwátmika);* and from Sumeru, all the various kinds of trees and vegetables; and from them, all the variety colours, shapes, flavours, and fragrances, in leaves, flowers, and fruits. Each derived its essential property (as of fire to burn) from itself; and the order of its procession into existence from the one precedent, by virtue of Swabháva,‡ operating in time. The several manners of going peculiar to the six classes of animate beings (four-legged, two-legged, etc.) and their several modes of birth, (oviparous, etc.§) all proceeded from Swabháva. From the Swabháva of each mansion or habitat (*Bhuvana*) resulted the differences existing between the several abodes of all the six orders of animate beings. The existence of the fœtus in the womb proceeds from the Swabháva of the union of male and female; and its gradual growth and assumption of flesh, bones, skin, and organs, is caused by the joint energy of the Swabháva of the fœtus, and that of time, or the Swabháva of the fœtus, operating in time. The procession of all things from birth, through gradual increase, to maturity, and thence, through gradual decay, to death, results spontaneously from the nature of each being; as do the differences appropriated to the faculties of the senses and of the mind, and to those external things and internal, which are perceived by them. Speech and sustenance from dressed food in mankind, and the want of speech and the eating of grass in quadrupeds, together with the birth of birds from eggs, of insects from sweat, and of the Gods (*Devatás*) without parentage of any sort: all these marvels proceed from Swabháva. (Comment on the *Pújá Kánda*, quotation 12.)

THE AISWARIKA* SYSTEM.

1. The self-existent God is the sum of perfections, infinite, external, without members or passions; one with all things (in Pravritti), and separate from all things (in Nirvritti), infiniformed and formless, the essence of Pravritti and of Nirvritti†. (*Swayambhú Purána.*)

‡ By virtue of Swabháva and of time says another comment; thus time stands out like space, as a something superior to all phænomena, and both are quasi deified by Buddhists and by Brahmanists.

§ By etcætera, understand *always* more Brahmanorum. That *Buddhism* forms an integral part of the Indian philosophy is sufficiently proved by the multitude of terms and classifications common to it, and to Brahmanism. The theogony and cosmogony of the latter are expressly those of the former, with sundry additions only, which serve to prove the posteriority of date, and schismatical secession, of the *Buddhists.* M. Cousin, in his course of philosophy, notices the absence of a sceptical school amongst the Indian philosophers. *Buddhism,* when fully explained, will supply the desideratum; and I would here notice the precipitation with which we are now constantly drawing general conclusions relative to the scope of Indian speculation, from a knowledge of the Brahmanical writings only—writings equalled or surpassed in nnmber and value by those of the *Buddhists, Jains,* and other dissenters from the existing orthodox system of *Vyasa* and *Sankara Acharya.*

* From Iswara, 'God.'

† *Pravritti,* the versatile universe; *Nirvritti,* its opposite, this world and the next. *Pravritti* is compounded of *Pra,* an intensitive, and *vritti,* action, occupation, from the root *va,* to blow as the wind; *Nirvritti,* of *Nir,* a privative, and *vritti,* as before.

2. He whose image is Súnyatá, who is like a cypher|| or point, infinite, unsustained (in Nirvritti), and sustained (in Pravritti), whose essence is Nirvritti, of whom all things are forms (in Pravritti), and who is yet formless (in Nirvritti), who is the Iswara, the first intellectual essence, the Adi-Buddha, was revealed by his own will. This self existent is he whom all know as the only true Being; and, though the state of Nirvritti be his proper and enduring state, yet, for the sake of Pravritti, (creation), having become Pancha-jnánátmika, he produced the five Buddhas thus:—from Suvisuddhadharma-dhátuja-jnána, Vairochana, the supremely wise, from whom proceed the element of *Akása*, the organ of sight, and colours; and from Adarsana-jnána, Akshobhya, from whom proceed the element of air, the organ of hearing, and all sounds; and from Pratyavekshana-jnána, Ratna Sambhava, from whom proceed the element of fire, the organ of smell, and all odours; and from Sánta-jnána, Amitábha, from whom proceed the element of water, the organ of taste, and all savours; and from Krityanushtha-jnána, Amogha Siddha, from whom proceed the element of earth, the organ of touch, and all the sensible properties of outward things dependent thereon. All these five Buddhas are Pravritti-karmánas, or the authors of creation. They possess the five *jnánas*, the five colours, the five *mudrás*, and the five vehicles.* The five elements, five organs of sense, and five respective objects† of sense, are forms of them.‡ And these five Buddhas each produced a Bodhi-Satwa, (for the detail, see Asiatic Society's Transactions, vol. xvi.) The five Bodhi-Satwas are Srishti-karmánas, or the immediate agents of creation; and each, in his turn, having become Sarvaguna, (invested with all qualities, or invested with the three *gunas*,) produced all things by his fiat. (Comment on quot. 1.)

3. All things existent (in the versatile universe) proceed from some cause (*hetu*): that cause is the Tathágata§ (Adi-Buddha); and that which is the cause of

|| This is the symbol of the Triad and of the Saktis.

* See Appendix A.

† If Manas, as the sum of the faculties of sense, be excluded, we may render the passage as here; else we must say elements, organs, and objects.

‡ The five *Dhyáni Buddhas* are said to be *Pancha Bhúta*, *Pancha Indriya*, and *Pancha Ayatana âkâra*. Hence my conjecture that they are mere personifications, according to a theistic theory, of the phænomena of the sensible world. The sixth *Dhyáni Buddha* is, in like manner, the icon and source of the sixth sense, and its object, or *Manas* and *Dharma*, *i. e.*, the percipient principle, soul of the senses, or internal sense, and moral phænomena. *Manas* is the *Bhûta*, *Dhârana* the *Indriya*, and *Dharma* the *Ayatana*, or mind, mental apprehension and the appropriate objects of such apprehension, or all things. Mind is the seat of consciousness and perception; whatever its essence, and is the effective cause of all sensation and perception.

§ This important word is compounded of *Tathá*, thus, and *gata*, gone or got, and is explained in three ways. First, thus got or obtained, *viz.*, the rank of a *Tathágata*, obtained by observance of the rules prescribed for the acquisition of perfect wisdom, of which acquisition, total cessation of births is the efficient consequence. Second, thus gone, *viz.*, the *mundane* existence of the *Tathágata*, gone so *as never to return*, mortal births having been closed, and *Nirvritti* obtained, by perfection of knowledge. Third, gone in the same manner as it or they (birth or births) came; the sceptical and necessitarian conclusion of those who held that both metempsychosis and absorption are beyond our intellect (as objects of knowledge), and independent of our efforts (as objects of desire and aversion—as contingencies to which we are liable); and that *that* which causes births, causes likewise *(proprio vigore)* the ultimate cessation of them.

(versatile) existence is the cause of the cessation or extinction of all (such) existence: so said Sákya Sinha. (*Bhadra Kalpávadána.*)*

4. Body is compounded of the five elements: soul, which animates it, is an emanation from the self-existent. (*Swayambhú-Purána.*)

5. Those who have suffered many torments in this life, and have been burned in hell, shall, if they piously serve the *Tri Ratna* (or *Triad*), escape from the evils of both. (*Avadána Kalpalatá.*)

6. Subandhu (a Rája of Benares) was childless. He devoted himself to the worship of Iswara (Adi-Buddha;) and by the grace of Iswara a sugar-cane was produced from his semen, from which a son was born to him. The race remains to this day, and is called Ikshava Aku. (*Avadána Kalpalatá.*)

7. When all was void, perfect void, [Súnya, Mahá Súnya] the triliteral syllable *Aum* became manifest, the first created, the ineffably splendid, surrounded by all the radical letters (*Vija Akshara,*) as by a necklace. In that *Aum*, he who is present in all things, formless and passionless, and who possesses the *Tri Ratna*, was produced by his own will. To him I make adoration. (*Swayambhú-Purána.*)

THE KARMIKA† SYSTEM.

1. From the union of Upáya and Prajná,‡ arose Manas, the lord of the senses, and from Manas proceeded the ten virtues and the ten vices; so said Sákya Sinha. [*Divya Avadána*].

The epithet *Tathágata,* therefore, can only be applied to *Adi-Buddha,* the self-existent, who is never incarnated, in a figurative, or at least a restricted, sense;—cessation of human births being the essence of what it implies. I have seen the question and answer, 'what is the *Tathágata?* It does not come again,' proposed and solved by the *Rakshá Bhágavati,* in the very spirit and almost in the words of the *Vedas.* One of a thousand proofs that have occurred to me how thoroughly Indian *Buddhism* is. *Tathágata,* 'thus gone, or gone as he came,' as applied to Adi-Buddha, alludes to his voluntary secession from the versatile world into that of abstraction, of which no mortal can predicate more than that his departure and his advent are *alike* simple results of his volition. Some authors substitute this interpretation, exclusively applicable to *Adi-Buddha,* for the third sceptical and general interpretation above given. The synonym *Sugata,* or 'well gone, (or well got, that is, happily got so as never to be lost—or virtually got, that is, by rigid observance of the laws or rules prescribed,) for ever quit of versatile existence,' yet further illustrates the ordinary meaning of the word *Tathágata,* as well as the ultimate scope and genius of the *Buddhist* religion, of which the end is, freedom from metempsychosis; and the means, perfect and absolute enlightenment of the understanding, and consequent discovery of the grand secret of nature. What that grand secret, that ultimate truth, that single reality, is, whether all is God, or God is all, seems to be the sole *propositum* of the oriental philosophic religionists, who have all alike sought to discover it by taking the high *priori* road. That God is all, appears to be the prevalent and dogmatic determination of the Brahmanists; that all is God, the preferential but sceptical solution of the *Buddhists;* and, in a large view, I believe it would be difficult to indicate any further essential difference between their theoretic systems, both, as I conceive, the unquestionable growth of the Indian soil, and both founded upon transcendental speculations, conducted in the very same style and manner. See Guizot's *Civilization,* ii. 386. India long long preceded Europe in the paths of transcendental philosophy.

* Since ascertained that this passage was misquoted for me, and that it is in fact equivalent to the Sarnath inscription, which should be rendered thus, "Of all things cause-produced the causes hath the Tathágata explained. The great Sramana hath likewise explained the causes of the extinction of all things." For these causes of existence and non-existence see the next section.

† From *Karma,* morality, the moral law of the universe.

‡ See the note on quotation 6 of the section *Adi Sangha.* Also the note on quotation 1 of the *Yátnika* system.

2. The being of all things is derived from belief, reliance, [*pratyaya,*] in this order: from false knowledge, delusive impression; from delusive impression, general notions; from them, particulars; from them, the six seats [or outward objects] of the senses; from them, contact; from it, definite sensation and perception; from it, thirst or desire; from it, embryotic [physical] existence; from it, birth or actual physical existence; from it, all the distinctions of genus and species among animate things; from them decay and death, after the manner and period peculiar to each. Such is the procession of all things into existence from Avidyá, or delusion: and in the inverse order to that of their procession, they retrograde into non-existence. And the egress and regress are both Karmas,* wherefore this system is called Kármika. (Sákya to his disciples in the *Rakshá Bhágavatí.)*

3. The existence of the versatile world is derived sheerly from fancy or imagination, or belief in its reality; and this false notion is the first Karma of Manas, or first act of the sentient principle, as yet unindividualized (?) and unembodied. This belief of the unembodied sentient principle in the reality of a mirage is attended with a longing after it, and a conviction of its worth and reality; which longing is called *Sanskára* and constitutes the second† Karma of Manas. When Sanskára becomes excessive, incipient individual consciousness arises [third Karma]: thence proceeds an organised and definite, but archetypal body, the seat of that consciousness, [fourth Karma]: from the last results the existence of [the six sensible and cognizable properties of] natural‡ objects, moral and physical, [fifth

* The *Dasa Karma* are, 1 *Sanskára,* 2 *Vijnána,* 3 *Námarúpa,* 4 *Shadáyatana,* 5 *Vedaná,* 6 *Trishná,* 7 *Upádaná,* 8 *Bhava,* 9 *Játi,* 10 *Jarámarana.*

† The first, not second; ten in all.

‡ So I render, after much inquiry, the *Shadáyatana,* or six seats of the senses external and internal; and which are in detail as follows: *Rúpa, Sabda, Ganda, Rasa, Sparsa, Dharma.* There is an obvious difficulty as to *Sparsa,* and some also as to *Dharma.* The whole category of the *Ayatanas* expresses *outward* things: and after much investigation, I gather, that under *Rúpa* is comprised not only colour, but form too, so far as its discrimination (or, in *Kármika* terms, its existence) depends on sight; and that *all* other *unspecified* properties of body are referred to *Sparsa,* which therefore includes not only temperature, roughness, and smoothness, and hardness, and its opposite, but also gravity, and even extended figure, though not extension in the abstract.

Here we have not merely the secondary or sensible properties of matter, but also the primary ones; and, as the *existence* of the *Ayatanas* or outward objects perceived, is said to be derived from the *Indriyas,* (or from *Manas,* which is their collective energy,) in other words, to be derived from the sheer exercise of the percipient powers the Kármika system amounts to idealism. Nor is there any difficulty thence arising in reference to the *Kármika* doctrine, which clearly affirms that theory by its derivation of all things from *Pratyaya* (belief), or from *Avidyá* (ignorance). But the *Indriyas* and *Ayatanas,* with their *necessary connexion,* (and, possibly, also, the making *Avidyá* the source of all things,) belong likewise to one section at least of the *Swábhávika* school; and, in regard to it, it will require a nice hand to exhibit this Berkleyan notion existing co-ordinately with the leading tenet of the *Swábhávikas.* In the way of explanation I may observe, first, that the denial of material entity involved in the *Indriya* and *Ayatana* theory (as in that of *Avidyá*) respects solely the versatile world of *Pravritti,* or of specific *forms merely,* and does not touch the *Nirvrittika* state of formative power and of primal substance, to which latter, in that condition, the qualities of gravity, and even of extended figure, in any sense cognizable by human faculties, are denied, at the same time, that the real and even eternal existence of a substance, in that state, is affirmed.

Second, though *Dharma,* the sixth *Ayatana,* be rendered by virtue, the appropriated

Karma.] When the archetypally embodied sentient principle comes to exercise itself on these properties of things, then definite perception or knowledge is produced, as that this is white, the other, black; this is right, the other wrong, [sixth Karma.] Thence arises desire or worldly affection in the archetypal body, [seventh Karma,] which leads to corporeal conception, [eighth,] and that to physical birth, [ninth.] From birth result the varieties of genus and species distinguishing animated nature, tenth Karma,] and thence come decay and death in the time and manner peculiar to each, [eleventh and final Karma]. Such is the evolution of all things in Pravritti; opposed to which is Nirvritti; and the recurrence of Nirvritti is the sheer consequence of the abandonment of all absurd ideas respecting the reality and stability of Pravritti, or, which is the same thing, the abandonment of Avidyá; for, when Avidyá is relinguished or overcome, Sanskára and all the rest of the Karmas or acts of the sentient principle, vanish with it; and also, of course, all mundane things and existences, which are thence only derived. Now, therefore, we see that Pravritti or the versatile world is the consequence of affection for a shadow, in the belief that it is a substance; and Nirvritti is the consequence of an abandonment of all such affection and belief. And *Prävritti* and *Nirvritti*, which divide the universe, are Karmas; wherefore the system is called Kármika. [Comment on quotation 2.]

4. Since the world is produced by the Karma of Manas, or sheer act of the percipient principle, it is therefore called Kármika. The manner of procession of all things into existence is thus: from the union of Upáya* and of Prajná, Manas proceeded; and from Manas, Avidyá; and from Avidyá, Sanskára; and from Sanskára, Vijnána; and from Vijnána, Námarúpa; and from Námarúpa, the Shad Ayatana;† and from them, Vedaná; and from it, Trishná; and from it, Upádána;

object of the internal sense, it must be remembered, that most of the *Swábhávikas*, whilst they deny a moral ruler of the universe, affirm the existence of morality as a part of the system of nature. Others again (the minority) of the Swábhávikas reject the sixth *Indriya*, and sixth *Ayatana*, and, with them, the sixth *Dhyáni Buddha*, or *Vajra Satwa*, who, by the way, is the *Magnus Apollo* of the *Tántrikas*, a sect the mystic and obscene character of whose ritual is redeemed by its unusually explicit enunciation and acknowledgment of a "God above all."

The published explanations of the procession of all things from *Avidyá* appear to me irreconcilably to conflict with the ideal basis of the theory.

* See Fahian, 159 and 291. See also Gogerly, p. 15, his enumeration is precisely ours, though his explanation differs, and is I think unintelligible, as is also Colebrooke's. See Ceylon Journal, No. 1.

† That is; colour, odour, savour, sound, the properties dependent on touch, (which are hardness, and its opposite, temperature, roughness and smoothness, and also, I believe gravity and extended figure,) and lastly, right and wrong. They are called the seats of the six senses, the five ordinary, and one internal. In this quotation I have purposely retained the original terms. Their import may be gathered from the immediately preceding quotations and note, which the curious may compare with Mr. Colebrooke's explication. See his paper on the *Bauddha* philosophy, *apud* Trans. Roy. As. Society, quarto vol. The following are the details of the three catagories, *viz*:—

Bhútas.	*Indriyas.*	*Ayatanas.*
Earth	Skin	Tangible properties.
Water	Palate	Savours.
Fire	Nose	Odours.
Air	Ear	Sounds.
Akása	Eye	Colours, forms.

and from it, Bhava; and from it Játi; and from it, Jarámarana. And from Játi-rúpya Manas, [*i. e.*, the sentient principle in organized animate beings] emanated the ten virtues and ten vices. And as men's words and deeds partake of the character of the one or the other, is their lot disposed; felicity being inseparably bound to virtue, and misery to vice, by the very nature of Karma.

Such is the procession of all things into existence from Manas through Avidyá: and when Avidyá ceases, all the rest cease with it. Now, since Avidyá is a false knowledge, and is also the medium of all mundane existence, when it ceases, the world vanishes; and Manas, relieved from its illusion, is absorbed into Upáya Prajná.† Pravritti is the state of things under the influence of Avidyá; and the cessation of Avidyá is Nirvritti; Pravritti and Nirvritti are both Karmas. [Another comment on Quot. 2.]

5. The actions of a man's former births constitute his destiny.‡ [*Punya Paroda.*]

6. He who has received from nature such wisdom as to read his own heart, and those of all others, even he cannot erase the characters which *Vidhátri*§ has written on his forehead. [*Avadána Kalpalatá.*]

7. As the faithful servant walks behind his master when he walks, and stands behind him when he stands, so every animate being is bound in the chains of Karma. (*Avadána Kalpalatá.*)

8. Karma accompanies everyone, everywhere, every instant, through the forest, and across the ocean, and over the highest mountains, into the heaven of *Indra*, and into *Pátála* (hell); and no power can stay it. (*Avadána Kalpalatá.*)

9. Kanála, son of king Asoka, because in one birth he plucked out the golden eyes from a *Chaitya*,* had his own eyes plucked out in the next; and because he in that birth bestowed a pair of golden eyes on a *Chaitya*, received himself in the succeeding birth eyes of unequalled splendour. (*Avadána Kalpalatá.*)

10. Sákya Sinha's son, named Rahula Bhadra, remained six years in the womb of his mother Yasodhará. The pain and anxiety of mother and son were caused by the Karmas of their former births. (*Avadána Kalpalatá.*)

11. Although I had acquired (Sákya speaks of himself) a perfect body, still, even in this body, defect again appeared; because I had yet to expiate a small residue of the sins of former births. (*Lalita Vistara.*)

Bhútas.	*Indriyas.*	*Ayatanas.*
Manas	Perception or conscious sensation.	The sum of all phænomena which are homogeneous and result from Manas,

and include thought, considered as one of the phænomena of *Diva Natura*, or thought, that is, human perception regarded as the sole measure of all things, the sole reality.

† The *Vámácháras* say, into *Prajná Upáya*: see note on quotation 6 of the section *Adi Sangha*.

‡ *Daivya*, identified with *Adi Buddha* by the theistic, and with Fate, by the atheistic doctors. The precise equivalent of the maxim itself is our 'conduct is fate.'

§ *Brahma*, but here understood to be *Karma*.

* *Chaitya* is the name of the tomb temples or relic-consecrated churches of the *Buddhists*. The essential part of the structure is the basal hemisphere: above this a square neck or *Gala* always supports the acutely conical or pyramidal superstructure; and on all four sides of that neck two eyes are placed, which are typical of omniscience. Wherever the hemisphere is found, it is indisputable evidence of *Buddhism*, *e. g.*, 'the topes' of *Manikyála* and of *Peshâwar*. In niches at the base of the hemisphere are

THE YATNIKA* SYSTEM.

1. Iswara (Adi-Buddha) produced Yatna from Prajná;§ and the cause of Pravritti and Nirvritti is Yatna; and all the difficulties that occur in the affairs of this world and the next are vanquished by Yatna (or conscious intellectual effort.) (*Divya Avadána.*)

2. That above mentioned Iswara, by means of Yatna, produced the five Jnánas, whence sprang the five Buddhas. The five Buddhas, in like manner, *i. e.*, by means of Yatna, produced the five Bodhisatwas; and they again, by the same means, created the greater Devatás from their bodies, and the lesser ones from the hairs of the bodies. In like manner, Brahma created the three *Lokas*† and all moving and motionless things. Among mortals, all difficulties are overcome by Yatna; for example, those of the sea by ships, those of illness by medicine, those of travelling by equipages—and want of paper, by prepared skin and bark of trees. And as all our worldly obstacles are removed by Yatna, so the wisdom which wins Nirvritti for us is the result of Yatna; because by it alone are charity and the rest of the virtues acquired. Since therefore all the goods of this world and of the next depend upon Yatna, Sákya Sinha wandered from region to region to teach mankind that cardinal truth. (Comment on quotation 1.)

3. That Adi-Buddha, whom the Swábhávikas call Swabháva, and the Aiswarikas, Iswara,§ produced a Bodhisatwa, who, having migrated through the three worlds, and through all six forms of animate existence, and experienced the goods and evils of every state of being, appeared, at last, as Sákya Sinha, to teach mankind the real sources of happiness and misery, and the doctrines of the four schools of philosophy;‖ and then, by means of Yatna, having obtained Bodhijnána, and having fulfilled all the Páramitás (transcendental virtues,) he at length became Nirvána. (*Divya Avadána.*)

4. Sákya Sinha, having emanated from that self-existent, which, according to some, is Swabháva, and, according to others, is Iswara, was produced for the purpose of preserving all creatures. He first adopted the Pravritti Márga (secular

frequently enshrined four of the five *Dhyáni Buddhas*, one opposite to each cardinal point. *Akshobhya* occupies the eastern niche; *Ratnasambhava*, the southern; *Amitábha* the western; and *Amoghasiddha*, the northern. *Vairochana*, the first *Dhyáni Buddha* is supposed to occupy the centre, invisibly. Sometimes, however, he appears visibly, being placed at the right-hand of *Akshobhya*.

* From *Yatna*, 'intellect, intellectual force and resource.'

† The celestial, terrene, and infernal divisions of the versatile universe.

§ This, as I conceive, is an attempt to remedy that cardinal defect of the older *Swábhávika* school, *viz.*, the denial of personality, and conscious power and wisdom in the first cause. To the same effect is the *Kármika* assertion, that *Manas* proceeded from the union of *Upáya* and *Prajná*. *Karma* I understand to mean conscious moral effort, and *Yatna*, conscious intellectual effort. Their admission in respect to human nature implies its *free will*, as their assignation to the divine nature implies its *personality*.

§ Passages of this entirely pyrrhonic tenure incessantly recur in the oldest and highest authorities of the *Buddhists*; hence the assertion of the preface that Sugatism is rather sceptical than atheistically dogmatic.

‖ Expressly called by my Bauddha pandit the *Swábhávika, Aiswarika, Yátnika*, and *Kármika* systems; and the terms well denote the things meant to be designated: see note at p. 23.

character,) and in several births exercised Yatna and Karma, reaping the fruits of his actions in all the three worlds. He then exercised Yatna and Karma in the Nirvritti Márga (ascetical or monastic character) essaying a release from this mortal coil, fulfilling the ten virtues from the Satwa to the Dwápara Yuga, till at last, in the Kali Yuga, having completely freed himself from sublunary cares, having become a Bhikshuka,** and gone to Buddha Gayá, he rejected and reviled the Bráhmanical penance, did all sorts of true penance for six years under the tree of knowledge on the banks of the Niranjana river; conquered the Namuchimára,*|| obtained Bodhijnána, became the most perfect of the Buddhas, seated himself among the Bodhisatwas, (Ananda Bhikshu and the rest,) granted wisdom to the simple, fulfilled the desires of millions of people, and gave *Moksha** to them and to himself. *(Lalita Vistara.)*

5. A hare fell in with a tiger: by means of Yatna the hare threw the tiger into a well. Hence it appears that Yatna prevails over physical force, knowledge, and the Mantras. *(Bhadra Kalpávadána.)* 6. Nara Sinha, Rája of Benares, was a monster of cruelty. Satta Swáma Rája, by means of Yatna, compelled him to deliver up 100 Rájkumárs, whom Nara Sinha had destined for a sacrifice to the gods. *(Bhadra Kalpávadána.)*

7. Sudhana Kumára found a beautiful daughter of a horse-faced Rája named Druma. By means of Yatna he carried her off, and kept her; and was immortalized for the exploit. *(Swayambhú Purána.)*

ADI-BUDDHA.†

1. Know that when, in the beginning, all was perfect void (Mahá súnyatá,‡) and the five elements were not, then Adi-Buddha, the stainless, was revealed in the form of flame or light.

2. He in whom are the three *gunas*, who is the Mahá Múrti and the Visvarúpa (form of all things,) became manifest: he is the self-existent great Buddha, the Adinátha, the Maheswara.

** Mendicant: one of the four regular orders of the *Bauddhas*. See the preface.

*|| A *Daítya* of *Kánchanapura*, personification of the principle of evil. *Bodhijnána* is the wisdom of *Buddhism*. Ananda was one of the first and ablest of Sákya's disciples. The first *code* of *Buddhism* is attributed to him in conjunction with Kásyapa and Upáli. He succeeded the former as heresiarch.

* Emancipation, absorption. † *Adi* 'first,' *Buddha* 'wise.'

‡ The doctrine of *Súnyatá* is the darkest corner of the metaphysical labyrinth. Eighteen kinds of *Súnyatá* are enumerated in the *Rakshá Bhágavati*. I understand it to mean generally space, which some of *our* philosophers have held to be a *plenum*, others a *vacuum*. In the transcendental sense of the *Buddhists*, it signifies not merely the universal *ubi*, but also the *modus existendi* of all things in the state of quiescence and abstraction from phænomenal being. The *Buddhists* have eternised matter or nature in *that* state. The energy of nature ever *is*, but is not ever *exerted;* and when not *exerted*, it is considered to be void of all those qualities which necessarily imply perishableness, and, which is the same thing, of all those qualities which are cognisable or distinguishable, and hence the energy in that state is typed by sheer space. Most of the *Buddhists* deem (upon different grounds) all phænomena to be as purely illusory as do the Vedantists. The phænomena of the latter are sheer energies of God; those of the former are sheer energies of Nature, deified and substituted for God. See note on quot. 6 of this section *Adi Sangha*. The *Aiswarikas* put their *Adi Buddha* in place of the nature of the older *Swábhávikas*. See Journal of As. Soc. No. 33, Art. 1.

3. He is the cause of all existences in the three worlds; the cause of their well-being also. From his profound meditation (Dhyána,) the universe was produced by him.

4. He is the self-existent, the Iswara, the sum of perfections, the infinite, void of members or passions: all things are types of him, and yet he has no type: he is the form of all things, and yet formless.

5. He is without parts, shapeless, self-sustained, void of pain and care, eternal and not eternal;* him I salute. *(Káranda Vyúha.)*

6. Adi-Buddha is without beginning. He is perfect, pure within, the essence of the wisdom of thatness, or absolute truth. He knows all the past. His words are ever the same.

7. He is without second. He is omnipresent. He is the Nairátmya lion to the Kutírthya deer.† *(Náma sangíti.)*

8. I make salutation to Adi-Buddha, who is one and sole in the universe; who gives every one Bodhi-jnána; whose name is Upáya; who became manifest in the greatest Súnyatá, as the letter A. Who is the Tathágata; who is known only to those who have attained the wisdom of absolute truth. *(Náma sangíti.)*

9. As in the mirror we mortals see our forms reflected, so Adi-Buddha is known (in Pravritti) by the thirty-two lakshanas and eighty anuvinjanas. *(Náma sangíti.)*

10. As the rainbow, by means of its five colours, forewarns mortals of the coming weather, so does Adi-Buddha admonish the world of its good and evil actions by means of his five essential colours.§ *(Náma sangíti.)*

11. Adi-Buddha delights in making happy every sentient being; he tenderly loves those who serve him. His majesty fills all with reverence and awe. He is the assuager of pain and grief. *(Náma sangíti.)*

12. He is the possessor of the ten virtues; the giver of the ten virtues; the lord of ten heavens; lord of the Universe; present in the ten heavens. *(Náma sangíti.)*

13. By reason of the ten jnánas, his soul is enlightened. He too is the enlightener of the ten jnánas. He has ten forms and ten significations, and ten strengths, and ten vasitás. He is omnipresent, the chief of the Munis. *(Náma sangíti.)*

* One in *Nirvritti;* the other in *Pravritti;* and so of all the preceding contrasted epithets. *Nirvritti* is quiescence and abstraction: *Pravritti,* action and concretion. All the schools admit these two modes, and thus solve the difficulty of different properties existing in cause and in effects.

† Comment says, that *Nairátmya* is '*Sarva Dharmánám nirábhása lakshanam,*' that is, all things are unreal; and that *Tirtha* means *Moksha,* and *Kutirtha,* any perversion of the doctrine of *Moksha,* as to say it consists in absorption into *Brahma*: and it explains the whole thus, 'He thunders in the ears of all those who misinterpret *Moksha,* there is no true *Moksha* but *Súnyatá.*' Another comment gives the sense thus, dividing the sentence into two parts, 'there is no *átmá* (life or soul) without him: he alarms the wicked as the lion the deer.' The first commentator is a *Swábhávika;* the second, an *Aiswarika* one.

§ White, blue, yellow, red, and green, assigned to the five *Dhyáni Buddhas.* For a detail of the *lakshanas, anuvinjanas, balas, vasitás,* etc., of the neighbouring quotations, see Appendix A.

14. He has five bodies, and five jnánas, and five sights; is the múkat of the five Buddhas, without partner. (*Náma sangíti.*)

15. He is the creator of all the Buddhas: the chief of the Bodhisatwas are cherished by him. He is the creator of Prajná, and of the world; himself unmade. *Aliter*, he made the world by the assistance of Prajná; himself unmade. He is the author of virtue, the destroyer of all things.* (*Náma sangíti.*)

16. He is the essence of all essences. He is the Vajra-átmá (eternal being). He is the instantly produced lord of the universe; the creator of Akása. He assumes the form of fire, by reason of the Prajnarúpi-jnána, to consume the straw of ignorance. (*Náma sangíti.*)

ADI PRAJNA,† OR DHARMA.

1. I salute that Prajná Páramitá, who by reason of her omniscience causes the tranquillity-seeking Srávakas‡ to obtain absorption; who, by her knowledge of all the ways of action, causes each to go in the path suited to his genius; of whom wise men have said, that the external and internal diversities belonging to all animate nature, are produced by her; who is the mother of Buddha (Buddha Mátra), of that Buddha to whose service all the Srávakas and Bodhisatwas dedicate themselves. (*Panchavinsati Sáhasrika.*)

2. First air, then fire, then water, then earth,§ and in the centre of earth, Sumeru, the sides of which are the residence of the thirty-three millions of gods (Devatás,) and above these, upon a Lotos of precious stones, sustaining the mansion of the moon (or a moon-crescent), sits Prajná Páramitá, in the Lallita-ásan manner; ‖ Prajná, the mother of all the gods (Prasú-bhagavatáng,) and without beginning or end, (anádyanta.) (*Bhadra Kalpávadána.*)

3. I make salutation to the Prajná Deví, who is the Prajná Páramitá, the Prajnárúpya, the Nirrúpya, and the universal mother. (*Pújá kánda.*)

4. Thou Prajná, art like Akása, intact and intangible; thou art above all human wants; thou art established by thy own power. He who devoutly serves thee serves the Tathágata also. (*Ashta Sáhasrika.*)

5. Thou mighty object of my worship! thou Prajná, art the sum of all good qualities; and Buddha is the Guru of the world. The wise make no distinction between thee and Buddha. (*Ashta Sáhasrika.*)

* The comment on this passage is very full, and very curious, in as much as it reduces many of these supreme deities to mere *parts of speech*. Here is the summing up of the comment: 'He (*Adi-Buddha*) is the instructor of the *Buddhas* and of the *Bodhisatwas*. He is known by the knowledge of spiritual wisdom. He is the creator and destroyer of all things, the fountain of virtue.' Spiritual wisdom is stated to consist of *Síla, Samádhi, Prajná, Vimukti*, and *Jnána*.

† *Adi* 'first,' *Prajná* 'supreme wisdom, nature:' see p. 12.

‡ Name of one of the ascetical orders of *Buddhists*.

§ In this enumeration of material elements, *Akása* is omitted: but it is mentioned, and most emphatically, in quotation 4, as in the fifty other places quoted. In like manner, the five elements are frequently mentioned without allusion to the sixth, which however occurs in fit places. Omission of this sort is no denial.

‖ *i. e.*, one leg tucked under the seat: the other, advanced and resting on the bow of the moon-crescent.

6. O thou who art merciful to thy worshippers, the benevolent, knowing thee to be the source of Bauddha excellence, attain perfect happiness by the worship of thee! *(Ashta Sáhasrika.)*

7. Those Buddhas who are merciful, and the Gurus of the world, all such Buddhas are thy children. Thou art all good, and the universal mother (Sakalajagat Pitá Mahí.) *(Ashta Sáhasrika.)*

8. Every Buddha assembling his disciples instructs them how from unity thou becomest multiformed and many named. *(Ashta Sáhasrika.)*

9. Thou comest not from any place, thou goest not to any place. Do the wise nowhere find thee?* *(Ashta Sáhasrika.)*

10. The Buddhas, Pratyeka Buddhas, and Srávakas,† have all devoutly served thee. By thee alone is absorption obtained. These are truths revealed in all Sástras. *(Ashta Sáhasrika.)*

11. What tongue can utter thy praises, thou of whose being (or manifestation) there is no cause but thy own will. No Purána hath revealed any attribute by which thou mayest certainly be known. *(Ashta Sáhasrika.)*

12. When all was Súnyatá, Prajná Deví was revealed out of Akása with the letter U; Prajná, the mother of all the Buddhas and Bodhisatwas, in whose heart Dharma ever resides; Prajná, who is without the world and the world's wisdom, full of the wisdom of absolute truth; the giver and the *ikon* of that wisdom; the ever living (Sanátani); the inscrutable; the mother of Buddha.‡ (*Pujá kánda.*)

13. O Prajná Deví! thou art the mother (Jananí) of all the Buddhas, the grandmother of the Bodhisatwas, and great grandmother of all (other) creatures! thou art the goddess (Isání.) (*Pujá kánda.*)

14. Thou, Srí Bhágavatí Deví Prajná, art the sum of all the sciences, the mother of all the Buddhas, the enlightener of Bodhijnána, the light of the universe! (*Gunakáranda Vyúha.*)

15. The humbler of the pride of Namuchimára,§ and of all proud ones; the giver of the quality of Satya; the possessor of all the sciences; the Lakshmí; the protector of all mortals; such is the Dharma Ratna. (*Gunakáranda Vyúha.*)

16. All that the Buddhas have said, as contained in the Maháyána Sútra and the rest of the Sútras, is also Dharma Ratna.* (*Gunakáranda Vyúha.*)

* The force of the question is this, the wise certainly find thee.

† The *Buddhas* are of three grades: the highest is *Maháyána*, the medial, *Pratyeka*, and the lowest, *Srávaka*. These three grades are called collectively the *Triyána*, or 'three chariots,' bearing their possessors to transcendental glory. The *Triyána* are otherwise explained as three paths leading to different degrees of beatitude suited to the different capacities of those who propose to follow them. The *Maháyána* is the great or popular, or the great or most excellent.

‡ *Sugataja*, which the *Vámácháras* render, 'of whom *Buddha* was born;' the *Daksihnácháras*, 'born of *Buddha*, or goer to *Buddha*,' as wife to husband.

§ Bauddha personification of the principle of evil.

* Hence the scriptures are worshipped as forms of *Adi Dharma*. *Sútra* means

17. Because Buddha sits on thy brow, the splendour thence derived to thy form illuminates all the ethereal expanse, and sheds over the three worlds the light of a million of suns; the four Devatás, Brahma, Vishnu, Mahesa, and Indra, are oppressed beneath thy foot, which is advanced in the Alir-Asana. O Arya Tárá! he who shall meditate on thee in this form shall be relieved from all future births (*Saraká Dhárá.*†)

18. Thy manifestation, say some of the wise, is thus; from the roots of the hairs of thy body sprang Akása, heaven, earth, and *hades*, together with their inhabitants, the greater Devatás, the lesser, the Daityas, the Siddhas, Gandharbas, and Nágas. So too (from thy hairs,) wonderful to tell! were produced the various mansions of the Buddhas, together with the thousands of Buddhas who occupy them.‡ From thy own being were formed all moving and motionless things without exception. (*Saraká Dhárá.*)

19. Salutation to Prajná Deví, from whom, in the form of desire, the production of the world was excellently obtained,§ who is beautiful as the full moon, the mother of Adi Buddha, (Jinendra Mátra,) and wife of (the other) Buddha, who is imperishable as adamant. (*Sádhana Málá.*)

20. That Yoni, from which the world was made manifest, is the Trikonákára Yantra.‖ In the midst of the Yantrá or *trikon* (triangle) is a *bindu* (point,cypher): from that *bindu*, Adi Prajná revealed herself by her own will. From one side of the triangle Adi Prajná produced Buddha, and from another side, Dharma, and from the third side, Sangha. That Adi Prajná is the mother of that Buddha who issued from the first side; and the Dharma, who issued from the second side, is the wife of the Buddha of the first side, and the mother of the other Buddhas. (Comment on quotation 19.)

21. Salutation to Prajná Páramitá, the infinite, who, when all was void, was revealed by her own will, out of the letter U. Prajná, the Sakti of Upáya, the sustainer of all things, (Dhármikí) the mother of the world, (Jagan-mátá;) the Dhyánarúpyá, the mother of the Buddhas. The modesty of women is a form

literally thread of (discourse,) aphorism. *Sákya*, like other Indian sages, taught orally, and it is doubtful if he himself reduced his doctrines to a written code, though the great scriptures of the sect are now generally attributed to him, though in fact reduced to writing and systematized by his disciples Kásyapa, Ananda, and Upáli. Sútra is now the title of the *books* of highest authority among the *Bauddhas*.

† Composed by *Sarvajna Mitrapáda* of *Kashmir*, and in very high esteem, though not of scriptural authority.

‡ These thousands of *Buddhas* of mortal mould are somewhat opposed to the so-called simplicity of *Buddhism!!* whatever were the primitive doctrines of *Sákya* it is certain that the system attributed to him, and now found in the written authorities of the sect, is the very antipodes of simplicity.

§ *Dharmodaya-saugata Kámarupiní*, variously rendered, 'well got from the rise of virtue,' 'well got from the rise or origin of the world;' also as in text, *Dharmodaya*, the source of all things, signifies likewise the *Yoni*, of which the type is a triangle. See 20. The triangle is a familiar symbol in temples of the *Buddha Saktis*, and of the *Triad*. The point in the midst represents either *Adi-Buddha* or *Adi Prajná*, according to the theistic or atheistic tendency of his opinions who uses it. Our commentator is of the *Vámáchára* or atheistic school, and such also is his text. (See Ravenshaw in the J.R.A.S. on the *Khat Kon Yantra.*)

‖ See J.R.A.S. xiii. 1, 79, and 171.

of her, and the prosperity of all earthly things. She is the wisdom of mortals, and the ease, and the joy, and the emancipation, and the knowledge. Prajná is present everywhere. (*Sádhana Málá.*)

ADI SANGHA.*

1. That Amitábha, by virtue of his Sánta-jnána, created the Bodhi-satwa named Padma-páni, and committed to his hands the lotos.† (*Gunakáranda Vyúha.*)

2. From between his (Padma-páni's) shoulders sprang Brahmá; from his forehead, Mahá Deva; from his two eyes, the sun and moon; from his mouth, the air; from his teeth, Saraswatí; from his belly, Varuna; from his knees, Lakshmí; from his feet, the earth; from his navel, water; from the roots of his hair, the Indras and other Devatás. (*Gunakáranda Vyúha.*)

3. For the sake of obtaining Nirvritti, I devote myself to the feet of Sangha who, having assumed the three Gunas, created the three worlds. (*Pújá kánda.*)

4. He (Padma-páni) is the possessor of Satya Dharma, the Bodhi-satwa, the lord of the world, the Mahá-satwa, the master of all the Dharmas. (*Gunakáranda Vyúhá.*)

5. The lord of all worlds, (Sarvalokádhipa,) the Sri-mán, the Dharma Rája, the Lokeswara, sprang from Adi-Buddha** (Jinátmaja.) Such is he whom men know for the Sangha Ratna. (*Gunakáranda Vyúha.*)

6. From the union of the essences of Upáya and of Prajná*a proceeded the world which is Sangha.

---o---

P. S. With regard to the consistency or otherwise of the view of the subject taken in the sketch of Buddhism, with the general tenor of the foregone quota-

* *Adi* 'first,' *Sangha* 'congress, union.'

† Type of creative power. *Amitábha* is the fourth *Dhyáni* or celestial *Buddha*: *Padma-páni* is his *Æon* and executive minister. *Padma-páni* is the *præsens Divus* and creator of the *existing* system of worlds. Hence his identification with the third member of the *Triad.* He is figured as a graceful youth, erect, and bearing in either hand a *lotos* and a jewel. The last circumstance explains the meaning of the celebrated *Shadakshari Mantra,* or six-lettered invocation of him, *viz., Om! Mani padme hom!* of which so many corrupt versions and more corrupt interpretations have appeared from Chinese, Tibetan, Japanese, Mongolian, and other sources. The *mantra* in question is one of three, addressed to the several members of the *Triad.* 1. *Om sarva vidye hom.* 2. *Om Prajnáye hom.* 3. *Om mani-padme hom.* 1. The mystic triform Deity is in the all-wise (Buddha). 2. The mystic triform Deity is in Prajná (Dharma). 3. The mystic triform Deity is in him of the jewel and lotos (Sangha). But the *præsens Divus,* whether he be Augustus or *Padma-páni,* is everything with the many. Hence the notoriety of *this mantra,* whilst the others are hardly ever heard of, and have thus remained unknown to our travellers.

** From *Amitábha Buddha* immediately: mediately from *Adi-Buddha.*

*a Such is the *Aiswarika* reading. The *Prâjnikas* read 'from the union of *Prajná* and *Upáya.*'

With the former, *Upáya* is *Adi-Buddha,* the efficient and plastic cause, or only the former; and *Prajná* is *Adi Dharma,* plastic cause, a biunity with *Buddha,* or only a

tions, I would observe, that the ideal theory involved in the Prájnika-Swábhávika, and in the Kármika doctrines, was omitted by me in the sketch, from some then remaining hesitation as to its real drift, as well as its connexion with those schools, *and no other*. Upon this *exclusive* connexion I have still some doubt. For the rest, I retain unchanged the opinions expressed in the sketch, that the Kármika and Yátnika schools are more recent than the others—that they owe their origin to attempts to qualify the extravagant quietism of the primitive Swábhávikas, and even of the Aiswarikas — and that their contradistinguishing mark is the preference given by them respectively to mortals, or to intellect, with a view to final beatitude. The assertion of the Ashtasahasrika, that Swabháva or nature absolutely disposes of us, not less than the assertion of other works, that an immaterial abstraction *so* disposes of us, very logically leads the author of the Buddha Charitra to deny the use of virtue or intellect. To oppose these ancient notions was, I conceive, the especial object of those who, by laying due stress on Karma and Yatna, gave rise to the Kármika and Yátnika schools. But that these latter entertained such just and adequate notions of God's providence, or man's free will, as we are now familiar with, it is not necessary to suppose, and is altogether improbable. None such they *could* entertain if, as I believe, they adopted the more general principles of their predecessors. The ideal theory or denial of the reality of the versatile world, has, in some of its numerous phrases, a philosophical foundation; but its prevalence and popularity among the Buddhists are ascribable principally to that enthusiastic contempt of action for which these quietists are so remarkable. Their passionate love of abstractions is another prop of this theory.

product. With the latter, *Upáya* is the energy of *Prajná*, the universal material cause. The *original* aphorism, as I believe, is, '*Prajnopáyatmakam jagatah*,' which I thus translate; 'From the universal material principle, in a state of activity, proceeded the world.' This original Sútra has, however, undergone two transformations to suit it to the respective doctrines of the *Triadic Aiswarikas* and of the *Kârmikas*. The version of the former is, *Upáyaprajnátmakam sangha*, that of the latter is, *Upáyaprajnátmakam manas*. Of both, the *Upáya* is identical with *Adi-Buddha*, and the *Prajná*, with *Adi Dharma*. But the result—the unsophisticated *jagat* of the *Prâjnikas*, became *Adi Sangha*, *a creator*, with the *Aiswarikas*; and *Manas*, the sentient principle in man, the first production, and *producer of all other things*, with the *Kārmikas*. *Avidyá*, or the condition of mundane things and existences, is an illusion, alike with the *Prājnikas* and with the *Kārmikas*. But, whilst the former consider *Avidyá* the universal affection of the *material* and *immediate cause* of all things whatever; the latter regard *Avidyá* as an affection of *manas* merely, which they hold to be an *immaterial* principle and the *mediate* cause of all things else, *Adi-Buddha* being their final cause. The phænomena of both are homogeneous and unreal: but the *Prājnikas* derive them, directly, from a material source—the *Kàrmikas*, indirectly, from an immaterial fount. Our sober European thoughts and languages can scarcely cope with such extravagancies as these: but it would seem we must call the one doctrine material, the other, immaterial, idealism.

The phænomena of the *Pràjnikas* are sheer energies of matter: those of the *Kàrmikas*, are sheer (human) perceptions. The notions of the former rest on general grounds—those of the latter, on particular ones, or (as it has been phrased) upon the putting the world into a man's self: the Greek "*panton metron anthropos.*"

APPENDIX A.

DETAIL OF THE PRINCIPAL ATTRIBUTES OF ADI-BUDDHA AND OF THE EIGHTEEN SUNYATA.

THE THIRTY-TWO LAKSHANA.*

1. Chakránkitapánipádatalatá.
2. Supratishthitapánipádatalatá.
3. Jálábuddhavajrángulípánipádatalatá.
4. Mridutarunahastapádatalatá.
5. Saptochhandatá.
6. Dírghángulitá.
7. Ayatapárshnitá.
8. Rijugátratá.
9. Utsangapádatá.
10. Urdhángaromatá.
11. Aineyajunghatá.
12. Paturubáhutá.
13. Koshagatavastiguhyatá.
14. Suvarnavarnatá.
15. Suklachhavitá.
16. Pradakshinávartaikaromatá.
17. Urnálankritamukhatá.
18. Sinhapúrvárdhakáyatá.
19. Susambhritaskandhatá.
20. Chittántarangatá.
21. Rasaraságratá.
22. Nyagrodhaparimandalatá.
23. Ushníshasiraskatá.
24. Prabhútajihwatá.
25. Prastambaratá.
26. Sinhahanutá.
27. Suklahanutá.
28. Samadantatá.
29. Hansavikrántagamitá.
30. Aviraladantatá.
31. Samachatwárinsaddantatá.
32. Abhinílanetratá.

THE EIGHTY VYANJANA.

1. Atámranakhatá.
2. Snigdhanakhatá.
3. Tunganakhatá.
4. Chitrángulitá.
5. Anupúrvángulitá.
6. Gúdhasiratá.
7. Nirgranthisiratá.
8. Gúdhagulphatá.
9. Avishamapádatá.
10. Sinhavikrántagámitá.
11. Nágavikrántagámitá.
12. Hansavikrántagámitá.
13. Vrishabhavikrántagámitá.
14. Pradakshinagámitá.
15. Chárugámitá.
16. Avakragámitá.
17. Vrittagátratá.
18. Mrishtagátratá.
41. Suchisamudácháratá.
42. Vyapagatatilakálagátratá.
43. Gandhasadrisasukumárapánitá.
44. Snigdhapánilekhitá.
45. Gambhírapánilekhitá.
46. Ayatapánilekhitá.
47. Nátyáyatavachanatá.
48. Bimbapratibimbosthatá.
49. Mridujihwatá.
50. Tanujihwatá.
51. Meghagarjitaghoshatá.
52. Raktajihwatá.
53. Madhurachárumanjuswaratá.
54. Vrittadanshtratá.
55. Tíkshnadanshtratá.
56. Sukladanshtratá.
57. Samadanshtratá.
58. Anupúrvadanshtratá.

* Rémusat in his *Mélanges* applies all these to Sákya.

19. Anupúrvagátratá.
20. Suchigátratá.
21. Mridugátratá.
22. Visuddhagátratá.
23. Paripúrnavyanjanatá.
24. Prithuchárumandalagátratá.
25. Samakramatá.
26. Visuddhanetratá.
27. Sukumáragátratá.
28. Adínagátratá.
29. Utsáhagátratá.
30. Gambhírakukshitá.
31. Prasannagátratá.
32. Suvibhaktángapratyangatá.
33. Vitimirasuddhálokatá.
34. Vitungakukshitá.
35. Mrishtakukshitá.
36. Abhayakukshitá.
37. Akshobhakukshitá.
38. Gambhíranábhitá.
39. Pradakshinávartanábhitá.
40. Samantaprásádikatá.
59. Tunganásikatá.
60. Suchinásikatá.
61. Visálanetratá.
62. Chittrapakshmatá.
63. Sitásitakamaladalanetratá.
64. Ayatakrikatá.
65. Suklabhrúkatá.
66. Susnigdhabhrúkatá.
67. Pínáyatabhujalatatá.
68. Samakarnatá.
69. Anupahatakarnendriyatá.
70. Aparisthánalalátatá.
71. Prithulalátatá.
72. Suparipúrnottamángatá.
73. Bhramarasadrisakesatá.
74. Chittrakesatá.
75. Guhyakesatá.
76. Asangunitakesatá.
77. Aparushakesatá.
78. Surabhikesatá.
79. Srívatsamuktikanandyatá.
80. Vartulachihnitapánipádatalatá.

THE FIVE VARANA.

1. Sweta. 2. Níla. 3. Píta. 4. Rakta. 5. Syáma.

THE TEN PARAMITA.*

1. Dána.
2. Síla.
3. Sánti.
4. Vírya.
5. Dhyána.
6. Prajná.
7. Upáya.
8. Bala.
9. Pranidhi.
10. Jnána.

THE TEN BHUVANA.†

1. Pramuditá.
2. Vimalá.
3. Prabhákarí.
4. Archishmatí.
5. Sudurjayá.
6. Abhimukhí.
7. Durangamá.
8. Sádhumatí.
9. Samantaprabhá.
10. Dharmameghá.

* Burnouf renders the ten: Charity, Morality, Patience, Industry, Meditation, Ingenuity, Wish or Prayer, Fortitude, Foreknowledge, Method.

† Compare pp. 42 43. We have here no heaven for Adi-Buddha, nor any for any one of the five Dhyáni Buddhas.

THE TEN JNANAS.

1. Duhkhajnána.‡
2. Samúdyajnána.‡
3. Nirodhajnána.‡
4. Márgajnána.‡
5. Dharmajnána.‡
6. Arthajnána.§
7. Samvrittijnána.§
8. Parachittajnána.§
9. Kshayajnána.§
10. Anutpádajnána.§

THE TEN AKARA.

1. Prithivyákára.*
2. Jalákára.*
3. Agnyákára.*
4. Váyvákára.*
5. Akásákára.*
6. Akásanirodhákára.†
7. Váyunirodhákára.†
8. Agninirodhákára.†
9. Jalanirodhákára.†
10. Prithivínirodhákára.†

THE TEN ARTHA.||

1. Pránártha.
2. Apánártha.
3. Samánártha
4. Udánártha.
5. Vyánártha.
6. Kúrmártha.
7. Krikárártha.
8. Nágártha.
9. Devadátártha.
10. Dhananjayártha.

THE TEN BALA.

1. Sthánásthánajnánabala.
2. Karmavipákajnánabala.
3. Nánádhátujnánabala.
4. Nánávimuktijnánabala.
5. Sadindriyaparáparajnánabala.
6. Sarvatragámipratipattijnánabala.
7. Dhyánavimokshasamádhisamápattisanklesavyavadánasthánajnánabala.
8. Púrvanivásánusmritijnánabala.
9. Chyutyutpattijnánabala.
10. Asravakshayajnánabala.

THE TEN VASITA.

1. Ayurvasita.
2. Chittavasita.
3. Parishkáravasita.
4. Dharmavasita.
5. Avadhivasita.
6. Janmavasita.
7. Adhimuktivasita.
8. Pranidhánavasita.
9. Karmavasita.
10. Jnánavasita.

THE FIVE KAYA.

1. Dharmakáya.
2. Sambhogakáya.
3. Nirmánakáya.
4. Mahásukhakáya.
5. Jnánakáya.

‡ Five in Nirvritti. § Five in Pravritti.
* Evolution of the five elements in Pravritti.
† Revolution of the five elements in Nirvritti.
|| Five in Pravritti and five in Nirvritti; and so of the Bala and Vasita.

THE FIVE CHAKSHU.

1. Mánsachakshu.
2. Dharmachakshu.
3. Prajnánachakshu.
4. Divyachakshu.
5. Buddhachakshu.

THE EIGHTEEN SUNYATA.

1. Adhyátmasúnyatá.
2. Bahirdhásúnyatá.
3. Adhyátmabahirdhásúnyatá.
4. Súnyatásúnyatá.
5. Mahásúnyatá.
6. Paramárthasúnyatá.
7. Sanskritasúnyatá.
8. Asanskritasúnyatá.
9. Atyantasúnyatá.
10. Anavarágrasúnyatá.
11. Anavakárasúnyatá.
12. Prakritisúnyatá.
13. Sarvadharmasúnyatá.
14. Salakshanasúnyatá.
15. Anupalambhasúnyatá.
16. Abhavasúnyatá.
17. Subhavasúnyatá.
18. Abhavasubhavasúnyatá.

MATANTARA TWENTY SUNYATA.

19. Lakshanasúnyatá.
20. Alakshanasúnyatá.

APPENDIX B.

CLASSIFIED ENUMERATION OF THE PRINCIPAL OBJECTS OF BAUDDHA WORSHIP.

Ekámnáya.

Upáya.

Adi-Buddha.

Mahá-Vairochana.

Ekámnáyi.

Prajná.

Prajná-páramitá.

Dwyámnáya.

1. Upáya.* 2. Prajná.*

1. Prajná.† 2. Upáya.†

Tryámnáya.

1. Dharma.‡ 2. Buddha.‡ 3. Sangha.‡

2. Sangha.§ 1. Buddha.§ 3. Dharma.§

1. Buddha.§ 2. Dharma.§ 3. Sangha.§

Pancha-Buddhámnáya.

4. Amitábha. 2. Akshobhya. 1. Vairochana. 3. Ratnasambhava. 5. Amoghasiddha.‖

Pancha-Prajnámnáyi.

4. Pándurá 2. Lochaná. 1. Vajradhátwísvarí. 3. Mámakí. 5. Tárá.

* Root of theistic doctrine. † Root of atheistic doctrine.

‡ Atheistic. § Theistic; diversely so.

‖ These five are the famous Dhyáni Buddhas. A sixth is often added, or Vajra Satwa. The series of five is the common exoteric one: the sixth seems to belong rather to the esoteric system.

Pancha-Sanghámnáya.

4. Padmapáni. 2. Vajrapáni. 1. Samantabhadra. 3. Ratnapáni. 5. Viswapáni.

Pancha-Sanghámnáyí.

4. Bhrikuti-tárá. 2. Ugratárá. 1. Sitatárá. 3. Ratnatárá. 5. Viswatárá.

Matántara-Pancha-Buddhámnáya.

1. Vairochana. 2. Akshobhya. 3. Ratnasambhava. 4. Amitábha. 5. Amoghasiddha.

Matántara-Pancha-Prajnámnáyí.

1. Vajradhátwísvarí. 2. Lochaná. 3. Mámakí. 4. Pándurá. 5. Tárá.

Matántara-Pancha-Sanghámnáya.

1. Samantabhadra. 2. Vajrapáni. 3. Ratnapáni. 4. Padmapáni. 5. Viswapáni.

Matántara-Pancha-Sanghámnáyí.

1. Sitatárá. 2. Ugratárá. 3. Ratnatárá. 4. Bhrikutitárá. 5. Visvatárá.

Matántara-Pancha-Buddhámnáya.

4. Amitábha. 2. Amoghasiddha. 1. Vairochana. 3. Ratnasambhava. 5. Akshobhya.

Matántara-Pancha-Prajnámnáyí.

4 .Tárá. 2. Mámakí. 1. Vajradhátwísvarí. 3. Pándurá. 5. Lochaná.

Shad-Amnáya-Buddháh.

2. Akshobhya. 3. Ratnasambhava. 4. Amitábha. 5. Amoghasiddha.
1. Vairochana. 6. Vajrasatwa.

Shat-Prajnámnáyí.

2. Lochaná. 3. Mámakí. 4. Pándurá. 5. Tárá.
1. Vajradhátwísvarí. 6. Vajrasatwátmiká.

Shat-Sanghámnáya.

2. Vajrapáni. 3. Ratnapáni. 4. Padmapáni. 5. Viswapáni.
1. Samantabhadra. 6. Ghantápáni.

*Mánushiya-Sapta-Buddhámnáya.**

2. Sikhí. 3. Viswabhú. 4. Kakutsanda. 5. Kanakamuni. 6. Kásyapa.
1. Vipasyí. 7. Sákyasinha.

Matántara-Mánushiya-Sapta-Buddhámnáya.

4. Kakutsanda. 2. Sikhí. 1. Vipasyí. 3. Viswabhú. 5. Kanakamuni.
6. Kásyapa. 7. Sákyasinha.

Prajná-Misrita-Dyáni-Nava-Buddhámnáya.

2. Akshobhya. 1. Vairochana-Vajradhátwísvarí. 3. Ratnasambhava.
8. Pándurá. 6. Lochaná. 4. Amitábha. 5. Amoghasiddha. 7. Mámakí. 9. Tárá.

* All the Deities named above are Dhyáni, or celestial. The following are Mánushíya Dhyáni, as specified.

Dhyáni-Nava-Buddhámnáyáh.

4. Amitábha. 2. Akshobhya. 1. Vairochana. 3. Ratnasambhava. 5. Amoghasiddha.
8. Vajradharma. 6. Vajrasatwa. 7. Vajraraja. 9. Vajrakarma.

Dhyáni-Nava-Prajnámnáyí.

4. Pándurá. 2. Lochaná. 1. Vajradhátwísvarí. 3. Mámakí. 5. Tárá.
8. Dharmavajriní. 6. Vajrasatwátmiká. 7. Ratnavajriní. 9. Karmavajriní.

Dhyáni-Nava-Sanghámnáyáh.

4. Padmapáni. 2. Vajrapáni. 1. Samantabhadra. 3. Ratnapáni. 5. Viswapáni.
8. Dharmapáni. 6. Ghantápáni. 7. Manipáni. 9. Karmapáni.

Misrita-Nava-Sanghámnáyáh.

2. Maitreya. 1. Avalokiteswara. 3. Gaganaganja.
6. Manjughosha. 4. Samantabhadra. 5. Vajrapáni. 7. Sarva-nivarana-vishkambhí.
8. Kshitigarbha. 9. Khagarbha.*

Nava-Dharmámnáyáh-Paustakáh (Buddha-Dharma-sangha-Mandale Pújanakrame etan Múlam.)

2. Gandavyúha. 1. Prajná-páramitá. 3. Dasabhúmíswara.
6. Saddharmapundaríka. 4. Samádhirája. 5. Lankávatára. 7. Tathágataguhyaka.
8. Lalita-vistara. 9. Suvarna-prabhá.

Nava-Bodhisatwa-Sanghámnáyáh.

4. Sitatárá. 2. Maitráyaní. 1. Bhrikutitárá. 3. Pushpatárá. 5. Ekajátá.
8. Dípatárá. 6. Vagíswarí. 7. Dhúpatárá. 9. Gandhatárá.

Nava-Deví-Prajnámnáyí.

2. Vajravidáriní. 1. Vasundhara. 3. Ganapati-hridaya. 8. Maríchi. 4. Ushnísha-vijaya.
5. Parnasavarí. 7. Grahamátrika. 8. Pratyangirah. 9. Dhwajágrakeyurí.

Misrita-Nava-Dharmámnáyáh.

4. Pándurá. 2. Lochaná. 1. Vajradhátwíswarí. 3. Mámakí. 5. Tárá.
8. Pratyangirah. 6. Vajrasatwátmiká. 7. Vasundhara. 9. Guhyeswarí.‡

Mánushiya-Nava-Buddhámnáyáh.

4. Sikhí. 2. Ratnagarbha. 1. Dípankara. 3. Vipasyí. 5. Viswabhú.
8. Kásyapa. 6. Kakutsanda. 7. Kanakamuni. 9. Sákyasinha.

Mánushiya-Nava-Buddhámnáyáh.

1. Dípankara. 2. Ratnagarbha.§ 3. Vipasyí. 4. Sikhí. 5. Viswabhú.
6. Kakutsanda. 7. Kanakamuni. 8. Kásyapa. 9. Sákyasinha.

* Avalokiteswara is probably identical with Matsyendra náth, the introducer of Nathism into Buddhism, but not with Padma Páni, the fourth Dhyáni Bodhisatwa, though now usually so identified. Maitreya is the Buddha next to come; Manjughosha is a historical person and the apparent introducer of Saktiism into Buddhism: 4-5 are Dhyánis, shadows like the rest.

‡ Guhyeswari is now worshipped by the orthodox as the Sakti of Pasupati Nath. But the expelled Buddhists claim the goddess as their own and affirm that there is a subterranean way from their great temple of Sambhunath to hers.

§ For Ratnagarbha see Fahian, p. 116. We have here nine mortal Buddhas

Mánushíya-Nava-Prajnámnáyí.

1. Jwalavatí. 2. Lakshanavatí. 3. Vipasyantí. 4. Sikhámáliní. 5. Viswadhará. 6. Kakudvatí. 7. Kanthanamáliní. 8. Mahídhará. 9. Yasodhará.*

Nava-Bhikshu-Sanghámnáyáh.

1. Pradípeswara. 2. Ratnarája. 3. Mahámatí. 4. Ratnadhara. 5. Akásaganja. 6. Sakalamangala. 7. Kanakarája. 8. Dharmodara. 9. Ananda.

Iti-Sri-Ekámnáyádi-Navámnáya-Devatáh Samáptáh

N. B.—The authority for these details is the Dharma Sangraha, or *catalogue raisonné* of the terminology of the Bauddha system of philosophy and religion.

———o———

EUROPEAN SPECULATIONS ON BUDDHISM.||

In the late M. Abel-Rémusat's review of my sketch of Buddhism, (Journal des Savans, Mai, 1831,) with the perusal of which I have been favoured by Mr. J. Prinsep, there occurs (p. 263) the following passage: "L'une des croyances les plus importantes, et celle sur laquelle l'essai de M. Hodgson fournit le moins de lumières, est celle des avénemens ou incarnations *(avatára)*. Le nom de *Tathágata* (avenu**) qu'on donne à Sakia n'est point expliqué dans son mémoire; et quant aux incarnations, le religieux dont les reponses ont fourni la substance de ce mémoire, ne semble pas en reconnoître d'autres que celles des sept Boudhas. Il est pourtant certain qu'on en compte une infinité d'au tres; et les lamas du Tibet se considèrent eux mêmes comme autant de divinités incarnées pour le salut des hommes."

I confess I am somewhat surprised by these observations, since whatever degree of useful information relative to Buddhism my essays in the Calcutta and London Transactions may furnish, they profess *not* to give *any*, (save *ex vi necessitatis*) concerning the 'veritable nonsense' of the system. And in what light, I pray you, is sober sense to regard "une infinité" of phantoms, challenging belief in their historical existence as the founders and propagators of a given code of laws? The *Lalita Vistara* gravely assigns 505, or according to another copy, 550, *avatáras* to Sákya *alone*. Was I seriously to incline to the task of collecting and recording all that is attributed to these palpable *nonentities?* or, was it merely desired that I should explain the rationale of the doctrine

instead of seven, which latter is the usual series, *vide* the Amarakosha. The Southerns usually cite only four. All depends on the Kalpas, each has its own Buddhas, and to the last or present Kalpa belong the four of southern notoriety.

* Yasodhará was the wife of Sákya, and Rahula their son. Rahula therefore ought to have been the ninth Sangha: but he was dull and little known whilst Ananda was most famous and succeeded Sákya as Heresiarch after Kásyapa's speedy demise.

|| Printed from the Journal of the Asiatic Society of Bengal. Nos. 32, 33, and 34, A.D. 1834.

** A radical mistake; see the sequel.

of incarnation? If the latter only be the desideratum, here is a summary recapitulation of what I thought I had already sufficiently explained.

The scale of Bauddha perfectibility has countless degrees, several of which towards the summit express attributes really divine, however short of the transcendental glory of a *tathágata* in *nirvritti.* Nevertheless, these attributes appertain to persons subject to mortal births and deaths, of which the series is as little limited as is that scale of cumulative merits to which it expressly refers. But, if the scale of increasing merits, with proportionate powers in the occupiers of each grade, have almost infinite extent, and yet mortal birth cleave to every grade but the very highest, what wonder that men-gods should be common? or, that the appearance again in the flesh, of beings, who are far more largely gifted than the greatest of the *devatás*, should be called an *avātár?* Such *avatáras*, in all their successive mortal advents till they can reach the estate of a *tathágata*, are the *arhantas*, and the *bodhisatwas*, the *pratyeka* and ths *srávaka-Buddhas.* They are gods and far more than gods; yet they were originally, and still *quoād* birth and death are, mere men. When I stated that the divine Lamas of Tibet are, in fact, *arhantas:* but that a very gross superstition had wrested the just notion of the character of the latter to its own use, I thought I had enabled every reader to form a clear idea of that marvel of human folly, the immortal mortals, or present palpable divinities of Tibet! How few and easy the steps from a theory of human perfectibility, with an apparently interminable metempsychosis, to a practical tenet such as the Tibetans hold!

But Rémusat speaks of the incarnations of the *tathágatas:* this is a mistake, and a radical one. A Tathágata may be such whilst yet lingering in the flesh of that mortal birth in which he reached this supreme grade;—and here, by the way, is *another* very obvious foundation for the Tibetan extravagance—but when once, by that body's decay, the Tathágata has passed into nirvritti, he can never be again incarnated. The only true and proper Buddha is the Mahá Yánika or Tathágata Buddha. Such are all the 'sapta Buddha;' of whom it is abundantly certain that *not one* ever was, or, by the principles of the creed, could be, incarnated. Sákya's incarnations all belong to the period preceding his becoming a Tathágata. Absolute quietism is the enduring state of a Tathágata: and, had it been otherwise, Buddhism would have been justly chargeable with a more stupendous absurdity than that from which Rémusat in vain essays to clear it. 'Plusieurs absolus—plusieurs infinis' there are; and they are bad enough, though the absolute infinity be restricted to the fruition of the subject. But the case would have been tenfold worse had activity been ascribed to these beings; for we should then have had an unlimited number of infinite ruling providences! The infinite of the Buddhists is *never* incarnated; nor the finite of the Brahmans. Avatáras are an essential and consistent part of Brahmanism—an unessential and inconsistent part of Buddhism: and there is always this material difference between the avatára of the former and of the latter, that whereas in the one it is an incarna-

* Not a syllable is told of these mortal Bodhisatwas with the exception of the last, Sákya's most famous disciple.

tion of the supreme and infinite spirit, for recognised purposes of creation or rule ; in the other, it is an incarnation of a mere human spirit—(however approximated by its own efforts to the infinite) and for what purpose it is impossible to say, *consistently with the principles of the creed.* I exclude here all considerations of the *dhyáni*, or celestial Buddhas, because Rémusat's reference is expressly to the seven *mánushi* or human ones.

The word Tathágata is reduced to its elements, and explained in three ways—1st. *thus gone*, which means gone in such a manner that he (the Tathágata) will never appear again; births having been closed by the attainment of perfection. 2nd. *thus got* or *obtained*, which is to say, (cessation of births) obtained, degree by degree, in the manner described in the Bauddha scriptures, and by observance of the precepts therein laid down, in a word by *tapas* and Dhyána, or severe ascetic purity and transcendental meditation. 3rd. *thus gone*, that is, gone as it (birth) came—the pyrrhonic interpretation of those who hold that doubt is the end, as well as beginning, of wisdom; and that *that* which causes births, causes likewise the ultimate cessation of them, whether that 'final close' be conscious immortality or virtual nothingness. Thus the epithet Tathágata, so far from meaning 'come' (avenu), and implying incarnation, as Rémusat supposed, signifies the direct contrary, or 'gone for ever,' and expressly announces the *impossibility* of incarnation; and this according to all the schools, sceptical, theistic, and atheistic.

I shall not, I suppose, be again asked for the incarnations of the Tathágatas.* Nor, I fancy, will any philosophical peruser of the above etymology of this important word have much hesitation in refusing, on this ground alone, any portion of his serious attention to the 'infinite' of Bnddhist *avatáras*, such as they really are. To my mind they belong to the very same category of mythological shadows with the infinity of *distinct* Buddhas, which latter, when I first disclosed it as a fact in relation to the belief of these sectaries, led me to warn my readers "to keep a steady eye upon the authoritative assertion of the old scriptures, that Sákya is the seventh and *last* of the Buddhas,"† though I believe that Sákya's six predecessors are *voces et prœterea nihil.*

The purpose of my two essays on Buddhism was to seize and render intelligible the *leading* and *least* absurd of the opinions and practices of these religionists, in order to facilitate to my countrymen the study of an entirely new and difficult subject in those original Sanskrit authorities[a]* which I had discovered and placed

* To the question, what is the *tathágata*, the most holy of Buddhist scriptures returneth for answer, "It does not come again, it does not come again."

† Asiatic Researches, vol. xvi. p. 445.

[a]*Nearly seventy volumes in Sanskrit, and some in the language of Tibet, were sent by me to Calcutta between the years 1824 and 1830. The former had never been before heard of, nor the latter possessed, by Europeans.

[See the notices of the contents of the Tibetan works and their Sanskrit originals by M. Csoma de Körös, and by Professor H. H. Wilson in the third volume of Gleanings, and first volume of Journal As. Soc.—Ed.]

See at pp. 137-139 of vol. i. for list of Sanskrit works. Eventually I procured from Lhasa the complete Kahgyur and Stangyur in 327 large volumes. The catalogue thereof had previously been obtained, and its general character reported on before Csoma de Körös made his appearance.

within their reach, but no living interpreters of which, I knew, were accessible to them, in Bengal or in Europe.

I had no purpose, nor have I, to meddle with the interminable sheer absurdities of the Bauddha philosophy or religion; and, had I not been called upon for *proofs* of the numerous novel statements my two essays contained, I should not probably have recurred at all to the topic. But sensible of the prevalent literary scepticism of our day and race, I have answered that call, and furnished to the Royal Asiatic Society, a copious selection from those original works which I had some years previously discovered the existence of in Nepaul. I trust that a further consideration of my two published essays, as illustrated by the new paper just mentioned, will suffice to remove from the minds of my continental readers most of those doubts of Rémusat, the solution of which does not necessarily imply conversancy on my part with *details as absurd as interminable.* I cannot, however, be answerable for the mistakes of my commentators. One signal one, on the part of the lamented author in question, I have just discussed: others of importance I have adverted to elsewhere: and I shall here confine myself to the mention of one more belonging to the review from which I have quoted. In speaking of the classification of the people, Rémusat considers the *vajra áchárya* to be laics; which is so far from being true that they and they alone constitute the clergy. The *bhikshuka* can indeed perform some of the lower offices of religion: but the *vajra áchárya* solely are competent to the discharge of the higher; and, in point of fact, are the only real clergy. That the distinction of *clerus et laicus* in this creed is altogether an anomaly, resulting from the decay of the primitive asceticism of the sect, I have endeavoured to shew elsewhere, and cannot afford room for repetition in this place.

The critics generally have been, I observe, prompt to adopt my caution relative to local superstitions, as opposed to the original creed of the Bauddhas. But they have carried their caution too far, and by so doing, have cast a shade of doubt and suspicion over things sufficiently entitled to exemption therefrom. Allow me, then, to reverse the medal, and to shew the grounds upon which a great degree of certainty and uniformity may always be presumed to exist in reference to this creed, be it professed where it may.

Buddhism arose in an age and country celebrated for literature; and the consequence was, that its doctrine and discipline were fixed by means of one of the most perfect languages in the world (Sanskrit), during, or immediately after, the age of its founder.

Nor, though furious bigots dispersed the sect, and attempted to destroy its records, did they succeed in the latter attempt. The refugees found, not only safety, but protection, and honour, in the *immediately* adjacent countries, whither they safely conveyed most of their books, and where those books still exist, either in the original Sanskrit, or in most carefully made translations from it. The *Sata Sáhasrika-Prajná-Páramitá*, and the nine Dharmas, discovered by me in Nepaul, are as indisputably original evidence of Buddhism as the Vedas and Puránas are of Brahmanism. The Káhgyur of Tibet has been *proved* to have been rendered into Tibetan from Sanskrit, with pains and fidelity: and if the numerous books of the Burmese and Ceylonese be not originals, it is certain that they were translated in the earlier ages of Buddhism, and that they were rendered into a language (high Prakrit) which, from its close affinity to that of the original books of the sect, (Sanskrit,) must have afforded the translators every facility in the prosecution of their labours.

But if the Buddhists, whether of the continent or islands of India, or of the countries beyond the former, still possess and consult the primitive scriptures of their faith, either in the original language, or in careful translations, made in the best age of their church, how can Buddhism in the several countries where it is practically used as the rule of life and of faith, fail to exhibit a common character as to essentials at least. And wherefore, I would fain know, should European scholars, from their study, incessantly prate about mere local rites and opinions, constituting the substance of whatever is told to the intelligent traveller by the present professors of this faith in diverse regions—nay, constituting the substance of whatever he can glean from their books? In regard to Nepaul, it is just as absurd to insinuate, that the Prajná Páramitá, and the nine Dharmas were composed in that country, and have exclusive reference to it, as to say that the Hebrew Old, or Greek New Testament was composed in and for Italy, France, or Spain exclusively. Nor is it much less absurd to affirm, that the Buddhism of one country is essentially unlike the Buddhism of any and every other country professing it, than it would be to allege the same of Christianity.

Questionless, in the general case, documentary is superior to verbal evidence. But the superiority is not without limit: and where, on the one hand, the books referred to by our closet students are numerous and difficult, and respect an entirely new subject, whilst, on the other hand, our personal inquirers have time and opportunity at command, and can question and cross-question intelligent witnesses, and cause reference to be made to the written authorities, the result of an appeal to the living oracles will oft times prove as valuable as that of one to the dead without any other guide.

Let the closet student, then, give reasonable faith to the traveller, even upon this subject; and, whatever may be the general intellectual inferiority of the orientals of our day, or the plastic facility of change peculiar to every form of polytheism, let him not suppose that the living followers of Buddha cannot be profitably interrogated touching the creed they live and die in; and, above all, let him not presume that a religion fixed, at its earliest period, by means of a noble written language, has no identity of character in the several countries where it is now professed, notwithstanding that that identity has been guarded, up to this day, by the possession and use of original scriptures, or of faithful translations from them, which were made in the best age of this church.

For myself, and with reference to the latter point, I can safely say that my comparisons of the *existing* Buddhism of Nepaul, with that of Tibet, the Indo-Chinese nations and Ceylon, as reported by our local enquirers, as well as with that of *ancient* India itself, as evidenced by the sculptures of Gaya,* and of the cave temples of Aurungabad, have satisfied me that this faith possesses as much identity of character in all times and places as any other we know, of equal antiquity and diffusion.†

—o—

P.S.—Whether Rémusat's *avenu* be understood loosely, as meaning 'come,' or strictly, as signifying 'come to pass,' it will be equally inadmissible as the interpretation of the word *Tathágata;* because *Tathágata* is designed expressly to announce that all reiteration and contingency whatever is barred with respect to the beings so designated. They cannot come; nor can anything come to pass affecting them.§

And if it be objected, that the mere use of the word *avenu*, in the past tense, does not necessarily imply such reiteration and conditional futurity, I answer that Rémusat clearly meant it to convey these ideas, or what was the sense of calling on me for the successive incarnations of these *avenus*? It has been suggested to me that *absolu*, used substantively, implies 'activity.' Perhaps so, in Parisian propriety of speech. But I use it merely as opposed to relative with reference to *mere* mortals; and I trust that the affirmation—there are many absolutes, many infinites, who are nevertheless inactive—may at least be distinctly understood. I have nothing to do with the reasonableness of the tenet so affirmed or stated, being only a reporter.

* See the explanation of these sculptures by a Nepaulese Buddhist in the Quarterly Oriental Magazine No. xiv. pp. 218, 222.

† As a proof of the close agreement of the Bauddha systems of different countries, we may take this opportunity of quoting a private letter from Colonel Burney, relative to the 'Burmese Philosopher Prince,' Mekkhara Men, the King of Ava's uncle.

"The prince has been reading with the greatest interest M. Csoma de Körös's different translations from the Tibet scriptures in your journal, and he is most anxious to obtain the loan of some of the many Tibetan works, which the Society is said to possess. He considers many of the Tibetan letters to be the same as the Burmese, particularly the *b*, *m*, *n*, and *y*. He is particularly anxious to know if the monastery called Zedawuna still exists in Tibet, where, according to Burmese books, Godama dwelt a long time, and with his attendant Ananda planted a bough which he had brought from the great *pipal* tree, at Buddha-Gaya. The prince is also anxious to know whether the people of Tibet wear their hair as the Burmese do? how they dress, and how their priests dress and live? The city in which the monastery of Zedawuna stood, is called in the Burmese scriptures *Thawotthi*, and the prince ingeniously fancies, that Tibet must be derived from that word. The Burmese have no *s*, and always use their soft *th*, when they meet with that letter in *Pali* or foreign words—hence probably *Thawotthi* is from some Sanskrit name Sawot. I enclose a list of countries and cities mentioned in the Burmese writings, as the scene of Godama's adventures, to which if the exact site and present designation of each can be assigned from the Sanskrit or the Tibet authorities, it will confer an important favour on Burmese *literati*." It is highly interesting to see the spirit of inquiry stirring in the high places of this hitherto benighted nation. The information desired is already furnished, and as might be expected, the Burmese names prove to be copied through the *Prakrit* or *Pali*, directly from the Sanskrit originals, in this respect differing from the Tibetan, which are *translations* of the same name.

§ *Avenu* signifies *quod evenit*, *contigit*, that which hath happened.—*(Dictionnaire de Trevoux.)* *Tathágata; tathá* thus (what really is), *gata* (known, obtained.)—(Wilson's Sans. Dict.)—*Ed.*

FURTHER REMARKS ON M. REMUSAT'S REVIEW OF BUDDHISM.†

Adverting again to Rémusat's Review in the *Journal des Savans* for May, 1831, I find myself charged with another omission more important than that of all mention of the Avatars. It is no less than the omission of all mention of any other *Buddhas* than the seven celebrated Mánushis. The passage in which this singular allegation is advanced is the following: "Les noms de ces sept personnages (the '*Sapta Buddhá*') sont connus des Chinois, et ils en indiquent une *infinité d'autres* dont le Bouddhiste Nipálien *ne parle pas*."

My Essay in the London Transactions was the complement and continuation of that in the Calcutta Researches. Rémusat was equally well acquainted with *both;* and, unless he would have had me indulge in most useless repetition, he must have felt convinced that the points enlarged on in the former essay would be treated cursorily or omitted, in the latter. Why, then, did he not refer to the Calcutta paper for what was wanting in the London one? Unless I greatly deceive myself, I was the first person who shewed clearly, and *proved* by extracts from original Sanskrit works, that Buddhism recognises "une infinité" of Buddhas,—Dhyáni and Mánushi, Pratyeka, Srávaka, and Mahá Yánika.* The sixteenth volume of the Calcutta Transactions was published in 1828. In that volume appeared my first essay, the substance of which had, however, been in the hands of the Secretary nearly three years before it was published.§ In that volume I gave an *original list of nearly* 150 *Buddhas* (p. 446, 449): I observed that the Buddhas named in the Buddhist scriptures were "as numerous as the grains of sand on the banks of the Ganges;" but that, as most of them were nonentities in regard to chronology and history, the list actually furnished would probably more than suffice to gratify rational curiosity; on which account I suppressed *another long list*, drawn from the Samádhi Rája, *which was then in my hands*, (p. 444.) By fixing attention on that cardinal dogma of Sugatism, *viz.*, that man can enlarge his faculties to infinity, I enabled every inquirer to conclude with certainty that the Buddhas had been multiplied *ad libitum.* By tracing the connexion between the Arhantas and the Bodhisatwas; between the latter again, and the Buddhas of the first, second, and third, degree of eminence and power; I pointed out the *distinct steps* by which the finite becomes confounded with the infinite,—man with Buddha; and I observed in conclusion that the epithet Tathágata, a synonym of Buddha, *expressly pourtrays this transition.* (London Transactions, vol. ii. part i.) Facts and dates are awkward opponents except to those, who, with Rémusat's compatriot, dismiss them with a 'tant pis pour les faits!' For years before I published my first essay, I had been in possession of hundreds of drawings, made from the Buddhist pictures and sculptures with which this land is saturated, and which drawings have not yet been published, owing to the delay incident to procuring authentic explanations of

† Printed from the Journal of the Bengal Asiatic Society, No. 33, A.D. 1834.

* The *triyána*, or three paths to bliss (of three different degrees) suited to the respective capacities of the several followers of this creed, want elucidation. The *Maháyána* is elsewhere spoken of as the humblest path; some call it the highest.

§ According to usage in that matter provided.

them from original sources. All the gentlemen of the residency can testify to the truth of this assertion; and can tell those who would be wiser for the knowledge, that it is often requisite to walk heedfully over the classic fields of the valley of Nepaul, lest perchance you break your shins against an image of a Buddha! These images are to be met with everywhere, and of all sizes and shapes, very many of them endowed with a multiplicity of members sufficient to satisfy the teeming fancy of any Brahman of Madhya Desa! Start not, gentle reader, for it is literally thus, and not otherwise. Buddhas with three heads instead of one —six or ten arms in place of two! The necessity of reconciling these things with the so called first principles of Buddhism,* may reasonably account for delay in the production of my pictorial stores. Meantime, I cannot but smile to find myself condoled with for my poverty when I am really, and have been for ten years, *accablé des richesses !* One interesting result only have I reached by means of these interminable trifles; and that is, strong presumptive proof that the cave temples of Western India are the work of Buddhists *solely*, and that the most apparently Brahmanical sculptures of those venerable fanes are, in fact, Buddhist. A hint to this effect I gave so long ago as 1827, in the Quarterly Oriental Magazine, (No. XVI. p. 219;) and can only afford room to remark in this place, that subsequent research had tended strongly to confirm the impressions then derived from my very learned old friend Amrita Nanda. The existence of an infinite number of Buddhas; the existence of the whole Dhyáni class of Buddhas; the personality of the Triad; its philosophical and religious meanings; the classification and nomenclature of the (ascetical or true) followers of this creed; the distinction of its various schools of philosophy; the peculiar tenets of each school, faintly but rationally indicated; the connexion of its philosophy with its religion; and, as the result of all these, the means of speaking consistently upon the general subject,† are matters for the knowledge of which, if Rémusat be not wholly indebted to me and my authorities, it is absolutely certain that I am wholly *un*indebted to him and his; for till he sent me his essay on the Triad, I had never seen one line of his, or any other continental writer's, lucubrations on Buddhism.

I have ventured to advance above that in the opinion of a learned friend, the Chinese and Mongolian works on Buddhism, from which the continental *savans* have drawn the information they possess on that topic, are not *per se* adequate to supply any very intelligible views of the general subject.

As this is an assertion which it may seem desirable to support by proof, allow me to propose the following. Rémusat observes, that a work of *the first order* gives the subjoined sketch of the Buddhist cosmogony. "Tous les êtres etant contenus dans la tres pure substance de la pensée, une idée surgit inopinement et

* See Erskine's Essays in the Bombay Transactions.

† A learned friend assures me that "a world of Chinese and Mongolian enigmas have been solved by means of your general and consistent outline of the *system*, but for which outline the said enigmas would have continued to defy all the Continental Œdipuses." (Sir G. Haughton in epis. 16 January, 1832.)

produisit la fausse lumière; Quand la fausse lumière fut née, le vide et l'obscurité s'imposèrent reciproquement des limites. Les formes qui en resultèrent étant indeterminées, il y eut agitation et mouvement. De là naquit le tourbillon de vent qui contient les mondes. L'intelligence lumineuse étoit le principe de solidité, d'ou naquit la roue d'or qui soutient et protège la terre. Le contact mutuel du vent et du metal produit le feu et la lumière, qui sont les principes des changemens et des modifications. La lumière précieuse engendre la liquidité qui bouillonne à la surface de la lumière ignée, d'ou provient le tourbillon d'eau qui embrasse les mondes de toute part."

Now I ask, is there a man living, not familiar with the subject, who can extract a particle of sense from the above passage? And are not such passages, produced in illustration of a novel theme the veriest obscurations thereof? But let us see what can be made of the enigma. This *aperçu cosmogonique* of the Langyen-king, is, in fact, a description of the procession of the five elements, one from another, and ultimately from *Prajná*, the universal material principle, very nearly akin to the *Pradhána* of the Kapila Sánkhya. This universal principle has two modes or states of being, one of which is the proper, absolute, and enduring mode; the other, the contingent, relative, and transitory. These modes are termed respectively Nirvritti and Pravritti.

The former is abstraction from all effects, or quiescence: the latter is concretion with all effects, or activity.* When the intrinsic energy of matter is exerted, effects exist; when that energy relapses into repose, they exist not. All worlds and beings composing the versatile universe are cumulative effects; and though the so-called elements composing them be evolved and revolved in a given manner, one from and to another, and though each be distinguished by a given property or properties, the distinctions, as well as the orderly evolution and revolution, are mere results of the gradually increasing and decreasing energy of nature in a state of activity.‡ *Upáya*, or 'the expedient,' is the name of this energy;—increase of it is increase of phenomenal properties;—decrease of it is decrease of phænomenal properties. All phænomena are homogeneous and alike unreal; gravity and extended figure, no less so than colour and sound. Extension in the abstract is not a phænomenon, nor does it belong *properly* to the versatile world. The productive energy begins at a minimum of intensity, and increasing to a maximum, thence decreases again to a minimum. Hence *ákása*, the first product, has but one quality or property; air, the second, has two; fire, the third, has three; water, the fourth, has four; and earth, the fifth, has five.*

* See Bailly's *History of Asia*, pp. 114, 118, 124, 187, of vol. i; also pp. 130, 187. Wondrous concord of ideas! Also Goguet, 1. 170.

‡ Causes and effects, *quoad* the versatile world, cannot be truly alleged to exist. There is merely customary conjunction, and certain limited effects of proximity in the precedent and subsequent, by virtue of the one true and universal cause, viz. *Prajná*. With the primitive Swabhávikas cause is not unitised: for the rest, their tenets are very much the same with those above explained in the text; only their conclusions incline rather to scepticism than dogmatism. It may also perhaps be doubted whether with the latter school, phænomena are unreal as well as homogeneous. In the text, I would be understood to state the tenets of the Prajnikas only.

* There is always cumulation of properties, but the number assigned to each element is variously stated.

These elements are evolved uniformly one from another in the above manner, and are revolved uniformly in the inverse order.

Súnyatá, or the total abstraction of phænomenal properties, is the result of the total suspension of nature's activity. It is the *ubi*, and the *modus*, of the universal material principle in its proper and enduring state of *nirvritti*, or of rest. It is *not* nothingness, except with the sceptical few. The opposite of *Súnyatá* is *Avidyá*, which is the mundane affection of the universal principle, or the universal principle in a state of activity, that is, of *pravritti.* Avidyá is also the result of this disposition to activity; in other words it represents phænomenal entities, or the sum of phænomena, which are regarded as wholly unreal, and hence their existence is ascribed to ignorance or Avidyá. Now, if we revert to the extract from the Lang-yen-king, and remember that la pensée,* l'intelligence lumineuse,* and la lumière precieuse,* refer alike to Prajná, the material principle of all things, (which is personified as a goddess by the religionists,) we shall find nothing left to impede a distinct notion of the author's meaning, beyond some metaphorical flourishes analogous to that variety of descriptive epithets by which he has characterised the one universal principle. *Tourbillon de vent*, and *tourbillon d'eau* are the elements of air and of water, respectively; and *le principe de solidité* is the element of earth.

"Tous les êtres étant contenus dans la pure substance de Prajná une idée surgit inopinement et produisit la fausse lumière:"—that is, the universal material principle, or goddess Prajná, whilst existing in its, or her, true and proper state of abstraction and repose, was snddenly disposed to activity, or impressed with delusive mundane affection *(Avidyá.)* "Quand la fausse lumière fut née, le vide et l'obscurité s'imposèrent réciproquement des limites." The result of this errant disposition to activity, or this mundane affection, was that the universal void was limited by the coming into being of the first element, or *ákása*, which, as the primary modification of *súnyatá* (space), has scarcely any sensible properties. Such is the meaning of the passage "les formes qui en resultèrent étant indeterminées," immediately succeeding the last quotation. Its sequel again, "il y eut agitation et mouvement," merely refers to mobility being the characteristic property of that element (air) which is about to be produced. "De la naquit le tourbillon de vent, qui contient les mondes." Thence (*i.e.*, from *ákása*) proceeded the element of the circumambient air. "L'intelligence lumineuse étoit le principe de solidité, d'ou naquit la roue d'or qui soutient et protége la terre." *Prajná* in the form of light (her *pravrittika* manifestation) was the principle of solidity, whence proceeded the wheel of gold which sustains and protects the earth. Solidity, the diagnostic quality of the element of earth, stands for that element; and the wheel of gold is mount Meru, the distinctive attribute of which is protecting and sustaining power: this passage, therefore, simply announces the evolu-

* Prajná is literally the supreme wisdom, videlicet, *of nature.* Light and flame are *types* of this universal principle, in a *state of activity.* Nothing but extreme confusion can result from translating these terms *au pied de la lettre*, and without reference to their technical signification. That alone supremely governs both the literal and metaphorical sense of words.

tion of the element of earth, with its mythological appendage, mount Meru. But, according to all the authorities within my knowledge, earth is the *last* evolved of the material elements. Nor did I ever meet with an instance, such as here occurs, of the direct intervention of the first cause *(Prajná)* in the *midst* of this evolution of the elements. "Le contact mutuel du vent et du *métal* produit le feu et la lumière, qui sont les principes des changemens." The mutual contact of the elements of air and of *earth* produce fire and light, which are the principles of change. This is intelligible, allowance being made for palpable mistakes. I understand by it, merely the evolution out of the element of air of that of fire, of which light is held to be a modification. To the igneous element is ascribed the special property of heat, which is assumed by our author as the principle of all changes and transformations. Metal for earth is an obvious misapprehension of Rémusat's. Nor less so is the false allocation of this element (earth) in the general evolution of the five, and its introduction here.

"La lumière précieuse engendre la liquidité qui bouillonne à la surface de la lumière ignée, d'on provient le tourbillon d'eau qui embrasse les mondes."

Prajná (in the form of light) produces the liquidity which boils on the surface of igneous light, whence proceeds the element of water embracing the world.

This figurative nonsense, when reduced to plain prose, merely announces the evolution of the element of water from that of fire. Our terrestrial globe rests upon the waters like a boat, according to the Buddhists; and hence the allusion (embracing the world) of the text. What is deserving of notice is the direct interference, a *second time*, (and in respect to earth, a third time,) of the *causa causans* with the procession of the elements, one from another. All my authorities are silent in regard to any such repeated and direct agency; which amounts in fact, to creation properly so called—a tenet directly opposed to the fundamental doctrine of all the Swábhávikas. Certain Buddhists hold the opinion, that all material subtances in the versatile world have no existence independent of *human* perception. But that the Chinese author quoted by Mr. Rémusat was one of these idealists, is by no means certain. His more immediate object, in the passage quoted, evidently was, to exhibit the procession of the five material elements, one from another. To that I at present confine myself, merely observing of the other notion, that what has been stated of the homogeneousness and unreality of all phænomena, is not tantamount to an admission of it. The doctrine of *Avidyá*, the mundane affection of the universal principle, is not necessarily the same with the doctrine which makes the percipient principle *in man* the *measure of all things*.* Both may seem, in effect, to converge towards what we very vaguely call idealism; but there are many separate paths of inquiry by which that conclusion may be reached.

Nepaul, *August*, 1834.

* *Manas*, the sixth element, is the percipient principle in man. The Chinese author mentions it *not*, unless the passage beginning "la même force," and immediately following that I have quoted, was designed to announce its evolution. That passage as it stands, however, does not assert more than the homogeneousness of this sixth element with the other five.

I resume my notice of Rémusat's speculations on Buddhism in the Journal des Savans.

He observes, "On ne seroit pas surpris de voir que, dans ce système, la formation* et la déstruction des mondes soient présentés comme les résultats d'une révolution perpetuelle et spontanée, sans fin et sans interruption;" and afterwards remarks, "Il y a dans le fond même des idées Bouddhiques une objection contre l'éternité du monde que les théologiens de cette religion ne semblent pas avoir prévue. Si tous les êtres rentroient dans le repos réel et definitif à l'instant que les phénomènes cesseroient et disparoitroient dans le sein de l'existence absolu, on conçoit un terme où tous les êtres seroient devenus Buddha, et où le monde auroit cessé d'exister."

This Buddha, it is said, is "l'intelligence infinie, la cause souveraine, dont la nature est un effet."

Now, if there be such a supreme immaterial cause of all things, what is the meaning of alleging that worlds and beings are *spontaneously* evolved and revolved? and, if these spontaneous operations of nature be expressly allowed to be *incessant and endless*, what becomes of the apprehension that they should ever fail or cease?

As to the real definitive repose, and the absolute existence, spoken of, they are as certainly and customarily predicated of *Diva natura* by the Swábhávikas, as of God or Adi-Buddha, by the Aiswarikas; to which two sects respectively the two opposite opinions confounded by Rémusat exclusively belong.

Again, "Tout est vide, tout est délusion, pour l'intelligence suprème (Adi-Buddha, as before defined). L'Avidyá seul donne aux choses du monde sensible une sorte de réalité passagère et purement phénomenal." Avidyá, therefore, must *according to this statement*, be entirely dependant on the volition of the one supreme immaterial cause: yet immediately after, it is observed, "on voit, à travers des brouillards d'un langage énigmatique, ressortir l'idée d'une *double cause* de tout ce qui existe, savoir l'intelligence suprème (Adi-Buddha) et l'Avidyá ou matière." But the fact is, that Avidyá is not a material or plastic cause. It is not a substance, but a mode—not a being, but an affection of a being—not a cause, but *an effect*. Avidyá, I repeat, is nothing primarily causal or substantial: it is a phænomenon, or rather the sum of phænomena; and it is "made of such stuff as dreams are." In other words, all phænomena are, according to this theory, absolutely homogeneous, and utterly unreal. The Avidyálists, therefore, are so far from belonging to that set of philosophers who have inferred two distinct substances and causes from the two distinct classes of phænomena existing in the world, that they entirely deny the justice of the premises on which that inference is rested.

Rémusat next observes, "Les effets matériels sont subordonnés aux effets psychologiques"—and in the very next page we hear that "on appelle lois les rapports qui lient les effets aux causes, tant dans l'ordre physique que dans l'ordre moral,

* The question of formation is a very different one from that of continuance. Yet Rémusat would seem to have confounded the two. See the passage beginning "Mais ce qui mérite d'être remarqué."

ou, pour parler plus exactement, dans l'ordre *unique*, qui constitue l'univers."

Now, if there be really but one class of phænomena in the world, it must be either the material, or the immaterial, class: consequently, with those who hold this doctrine, the question of the dependence or independence of mental upon physical phænomena, must, in one essential sense, be a mere *façon de parler*. And I shall venture to assert, that with most of the Buddhists—whose cardinal tenet is, that all phænomena are *homogeneous*, whatever they may think upon the further question of their reality or unreality—it is actually such.

It is, indeed, therefore necessary "joindre la notion d'esprit" before these puzzles can be allowed to be altogether so difficult as they seem, at least to be such as they seem; and if mind or soul "have no name in the Chinese language," the reason of that at least is obvious; its existence is denied. Mind is only a peculiar modification of matter; et l'ordre unique de l'univers c'est l'ordre physique! Not fifty years since a man of genius in Europe declared that "the universal system does not consist of two principles so essentially different from one another as matter and spirit; but that the whole must be of some uniform composition; so that the material or immaterial part of the system is superfluous."*

This notion, unless I am mistaken, is to be found at the bottom of most Indian systems of philosophy, Brahmanical and Buddhist, connected with a rejection in some shape or other of phænomenal reality *in order to get rid of the difficulty of different properties existing in the cause (whether mind or matter) and in the effect*†.

The assertion that "material effects are subordinate to psychological" is no otherwise a difficulty than as two absolutely distinct classes of phænomena, are assumed to have a real existence; and I believe that there is scarcely one school of Bauddha philosophers which has not denied the one or the other assumption; and that the prevalent opinions include a denial of both. All known phænomena may be ascribed to mind or to matter without a palpable contradiction; nor, with the single exception of extent,‡ is there a physical phænomenon which does not seem to countenance the rejection of phænomenal reality. Hence the doctrines of Avidyá and of Máyá; and I would ask those whose musings are in an impartial strain, whether the Bauddha device be not as good a one as the Brahmanical, to stave off a difficulty which the unaided wit of man is utterly unable to cope with?§

* A writer in the Edinburgh Review for January 1852, p. 192, says that to make immortality dependant on immateriality is most illogical.

† Rémusat desired to know how the Buddhists reconcile multiplicity with unity, relative with absolute, imperfect with perfect, variable with eternal, nature with intelligence?

I answer; by the hypothesis of two modes—one of quiescence, the other of activity; one of development, the other of non-development. But when he joins "l'esprit et la matière" to the rest of his antitheses, I must beg leave to say the question is entirely altered, and must recommend the captious to a consideration of the extract given in the text from a *European* philosopher of eminence. Not that I have any sympathy with that extravagance, but that I wish merely to state the case fairly for the Buddhists.

‡ Time and Space; which however cannot, and are not classed among phænomena by Indian or European philosophers. Limited time and space are considered *quasi* phænomena by all.

§ See Ballantyne's *Vedanta*, p. 80: the very phrase "ignorance" or *Ajnána* is essentially the same and more precise than *Máyá*.

Questionless, it is not easy, if it be possible, to avoid the use of words equivalent to material and psychological; but the tenet obviously involved in the formal subordination of one to the other class of phænomena, when placed beside the tenet, that all phænomena are homogeneous, at once renders the former a mere trick of words, or creates an irreconcileable contradiction between the two doctrines, and in fact Rémusat has here again commingled tenets held exclusively by quite distinct schools of Buddhist philosophy.

If I have been held accountable for some of the notions above remarked on, I suspect that these my supposed opinions have been opposed by something more substantial than "des arguties mystiques." Rémusat expressly says, "M. Hodgson a eu parfaitement raison d' admettre, comme base du *système entier*, l'existence d'un seul être souverainement parfait et intelligent, de celui qu'il nomme Adi-Buddha." Now, I must crave leave to say that I never admitted anything of the sort; but, on the contrary, carefully pointed out that the "système entier' consists of *four* systems, all sufficiently different, and two of them, radically so—*viz.*, the Swábhávika and the Aiswarika. It is most apparent to me that Rémusat has made a melange out of the doctrines of all the four schools; and there are very sufficient indications in the course of this essay that his principal authority was of the Swábhávika sect.

In speaking of the two‡ bodies of Buddha he remarks, that "le véritable corps est identifié avec la science et la loi. La substance même est la science (Prajná)." He had previously made the same observation, "La loi même est son principe et sa nature." Now those who are aware that Prajná (most idly translated law, science, and so forth,) is the name of the *great material cause*,* can have no difficulty in reaching the conviction that the Buddhist authority from whence this assertion was borrowed,—'of Prajná being the very essence, nature, and principle of Buddha,'—belonged to the Swábhávika school, and would have laughed at the *co-ordinate* doctrine of his translator, that Buddha is the sovereign and sole cause, of whom nature (Prajná) is an effect.

The Swábhávika Buddhas, who derive their capacity of identifying themselves with the *first cause* from nature, which *is that cause*, are as all-accomplished as the Buddhas of the Aiswarikas, who derive the same capacity from Adi-Buddha, *who is that cause.*

In this express character of sovereign cause only, is the Adi-Buddha of the Aiswarikas distinguishable amid the crowd of Buddhas of all sorts; and such are the interminable subtleties of the 'système entier' that he who shall not carefully

‡ There are in fact five bodies named by me; see page 92.

* *Prakriteswari iti Prajná;* and again, *Dháranátmaka iti Dharma. Dharma* is a synonyme of *Prajná. Prajná* means Supreme Wisdom. Whose? Nature's — and nature's, as the sole, or only as the plastic, cause.

So, again, *Dharma* means mortality in the abstract, or the moral and religious code of these religionists, or material cause, in either of the two senses hinted at above; or, lastly, material effects, *viz.*, versatile worlds. These are points to be settled by the context and by the known tenets of the writer who uses the one or other word: and when it is known that the very texts of the Swábhávikas, differently interpreted, have served for the basis of the Aiswarika doctrine, I presume no further *caveto* can be required.

mark this cardinal point of primary causation, will find all others unavailing to guide him unconfusedly through the various labyrinths of the several schools.

Did Rémusat never meet with passages like the following?

"And as all other things and beings proceeded from Swabháva or nature, so did Vajra, Satwa, Buddha, thence called the *self-existent*." Even the Swábhávikas have their Dhyáni Buddhas, and their triad, including, of course, an Adi-Buddha. Names, therefore, are of little weight; and unmeasured epithets are so profusely scattered on every hand that the practised alone can avoid their snare. I did not admit a Theistic school, because I found a Buddha designated as Adi, or the first; nor yet because I found him yclept infinite, omniscient, eternal, and so forth; but because I found him explicitly contradistinguished from nature, and systematically expounded as the efficient cause of all. Nor should it be forgotten that when I announced the fact of a Theistic sect of Buddhists, I observed that this sect was, as compared with the Swábhávika, both recent and confined.†

If, in the course of this, and the three preceding letters, I have spoken harshly of Rémusat's researches, let it be remembered, that I conceive my labours to have been adopted without acknowledgment, as well as my opinions to have been miserably distorted. I have been most *courteously* told, that "the learned of Europe are indebted to *me* for the name of Adi-Buddha!" The inference is palpable that that is the extent of the obligation. Such insidious injustice compels me to avow in the face of the world my conviction that, whatever the Chinese and Mongolian works on Buddhism possessed by the French Savans may contain, no intelligible views were thence derived of the general subject before my essays appeared, or could have been afterwards, but for the lights those essays afforded.§ I had access to the original Sanskrit scriptures of the Buddhists, and they were interpreted to me by learned natives, whose hopes hereafter depended upon a just understanding of their contents. No wonder, therefore, and little merit, if I discovered very many things inscrutably hidden from those who were reduced to consult barbarian translations from the most refined and copious of languages upon the most subtle and interminable of topics, and who had no living oracle ever at hand to expound to them the dark signification of the written word—to guide their first steps through the most labyrinthine of human mazes.||

For the rest, and personally, there is *bienséance* for *bienséance*, and a sincere tear dropped over the untimely grave of the learned Rémusat.

† Burnouf seems to hold that the transcendentalists had very early an atheistic and a theistic section, the theistic being the Yogácháryas, whose founder was Arya Sangha, and that a sect apart from both held the middle path, and were therefore called Mádhyamikas.

§ The case is altered materially *now;* because my original authorities, which stand far less in need of living interpreters, are generally accessible.

|| I beg to propose, as an *experimentum crucis*, the celebrated text—*Ye Dharmánityá* of the *Sata Sáhasrika*. If the several theistic, atheistic, and sceptical meanings wrapped up in these few words, can be reached through Chinese or Mongolian translations uninterpreted by living authorities, I am content to consider my argument worthless.

NOTE ON THE INSCRIPTION FROM SARNATH.

I have just got the 39th Number of the Journal of the Asiatic Society and hasten to tell you, that your enigma requires no Œdipus for its solution at Kathmandu, where almost every man, woman, and child, of the Bauddha faith, can repeat the *confessio fidei* (for such it may be called), inscribed on the Sárnáth stone. Dr. Mill was perfectly right in denying the alleged necessary connexion between the inscription, and the complement to it produced by M. Csoma de Körös. No such complement is needed, nor is found in the great doctrinal authorities, wherein the passage occurs in numberless places sometimes containing but half of the complete dogma of the inscription;* thus:—"*Ye Dharmá hetu-prabhavá: hetus teshán Tathágato.*" Even thus curtailed, the sense is complete, without the "*Teshán cha yo nirodha, evam (vádi) Mahá* Sramana," as you may perceive by the following translation:—

"Of all things proceeding from cause, the cause is Tathágata;" or, with the additional word, "Of all things proceeding from cause, the cause (of their procession) hath the Tathágata explained." To complete the dogma, according to the inscription, we must add, "The great Sramana hath likewise declared the cause of the extinction of all things." With the help of the commentators, I render this passage thus, "The causes of all sentient existence in the versatile world, the Tathágata hath explained. The Great Sramana hath likewise explained the causes of the cessation of all such existence."§

Nothing can be more complete, or more fundamental, than this doctrine. It asserts that Buddha hath revealed the causes of (animate) mundane existence, as well as the causes of its complete cessation, implying, by the latter, translation to the eternal quiescence of Nirvritti, which is the grand object of all Bauddha vows. The addition to the inscription supplied by M. Csoma, is the *ritual application* merely of the general doctrine of the inscription. It explains especially the manner in which, according to the scriptures, a devout Buddhist may hope to attain cessation from mundane existence, *viz.*, by the practice of all virtues, avoidance of all vices, and by complete mental abstraction. More precise, and as usually interpreted here, more theistic too, than the first clause of the inscription is the terser sentence already given; which likewise is more familiar to the Nepalese, *viz.*, "Of all things proceeding from cause, the cause is the Tathágata:"—understanding by Tathágata, Adi-Buddha. And whenever, in playful mood, I used to reproach my old friend, Amirta Nanda, (now alas! no more) with the atheistic tendency of his creed, he would always silence me with, "*Ye Dharmá hetu-prabhavá hetus teshán Tathágato;*" insisting, that Tathágata referred to the supreme, self-existent *(Swayambhú)* Buddha.†

* This curtailed version is traditional not scriptural.

§ See pp. 79-80 for these causes, *viz.*, *Avidyá*, *Sanskára, etc.*

† The great temple of Swayambhu Nath is dedicated to this Buddha: whence its name. It stands about a mile west from Kathmandu, on a low, richly wooded, and detached hill, and consists of a hemisphere surmounted by a graduated cone.

The majestic size, and severe simplicity of outline, of this temple, with its burnished cone, set off by the dark garniture of woods, constitute the Chaitya of Swayambhu Nath, a very beauteous object.

Nor did I *often* care to rejoin, that he had taught me *so* to interpret that important word (Tathágata) as to strip the dogma of its necessarily theistic spirit! I have already remarked in your Journal,* that the Swábhávika texts, differently interpreted, form the groundwork of the Aiswarika tenets. It will not, however, *therefore*, follow, that the theistic school of Buddhism is not entitled to distinct recognition upon the ground of original authorities; for the oldest and highest authority of all—the aphorisms of the founder of the creed—are justly deemed, and proved, by the theistic school, to bear legitimately the construction put upon them by this school—proved in many ancient books, both Pauránika and Tántrika, the scriptural validity of which commands a necessary assent. As it seems to be supposed, that the theistic school has no other than Tántrika authorities for its support, I will just mention the *Swayambhú Purána* and the *Bhadra Kalpávadána*, as instances of the contrary. In a word, the theistic school of Buddhism, though not so ancient or prevalent as the atheistic and the sceptical schools, is as authentic and legitimate a scion of the original stock of oral dogmata whence this religion sprung, as any of the other schools. Nor is it to be confounded *altogether* with the vile obscenity and mystic mummery of the *Tantras*, though acknowledged to have considerable connexion with them. Far less is it to be considered peculiar to Nepaul and Tibet, proofs of the contrary being accessible to all; for instance, the *Pancha Buddha Dhyáni* are inshrined in the cave at *Bágh*, and in the minor temples surrounding the great edifice at *Gyá;* as to which see my old Bauddha Pandit's report further on. A. Cunningham of Bengal, Wilson of Bombay, and Chapman of Madras, have all recorded opinions substantially the same. And I have myself seen a fine image of Padma Páni, the æon of the Dhyáni Buddha Amitábha, at Karnagurh on the Ganges. As I was looking over your Journal, my Newári painter came into the room. I gave him the catch word, "Ye Dharmá,' and he immediately filled up the sentence, *finishing with Tathágata.* I then uttered "teshán cha," and he completed the doctrine according to the inscription. But it was to no purpose that I tried to carry him on through Csoma's ritual complement: he knew it not. After I had explained its meaning to him, he said, the substance of the passage was familiar to him, but that he had been taught to utter the sentiments in other words, which he gave, and in which, by the way, the *ordinary* Buddhist acceptation of *Kusal* and its opposite, or *Akusal*, came out. *Kusal* is good. *Akusal* is evil, in a moral or religious sense. Quod licitum vel mandatum: quod illicitum vel prohibitum.

I will presently send you a correct transcript of the words of the inscription, from some old and authentic copy of the *Rakshá Bhágavatí*, or *Prajná Páramitá*, as you seem to prefer calling it. So will I of Csoma's supplement so soon as I can lay my hands on the *Shurangamá Samádhi*, which I do not think I have by me. At all events, I do not at once recognise the name as that of a distinct Bauddha work. Meanwhile, you will notice, that as my draftsman, above spoken of, is no pandit, but a perfectly illiterate craftsman merely, his familiar acquaintance

* *i. e.*, J. A. S. B.

with your inscription may serve to show how perfectly familiar it is to *all* Buddhists. And here I would observe, by the way, that I have no doubt the inscription on the Dehli, Allahabad, and Behar pillars is *some* such cardinal dogma of *this* faith.

I am no competent critic of Sanskrit, but I have competent authority for the assertion, that Dharmá, as used in the inscription, means not *human actions* merely, but *all* sentient *existences* in the three versatile worlds (celestial, terrene, and infernal). Such is its meaning in the famous *Ye Dharmánitya* of the *Sata Sáhasrika*, where the sense is even larger, embracing the substance of all inanimate as well as animate entity, thus: "All things are imperishable," or, "The universe is eternal," (without maker or destroyer.) The passage just quoted from the *Sata Sáhasrika* serves likewise (I am assured) to prove that the signification of *ye* is not always strictly relative, but often expletive merely: but let that pass.

The points in question undoubtedly are,—*existence* in the *Právrittika* or versatile world, and *cessation* of such existence, by translation to the world of *Nirvritti;* and of such translation, animals generally, and not human beings solely, are capable. Witness the deer and the chakwa, which figure so much in Bauddha sculptures! The tales of their advancement to *Nirvritti* are popularly familiar. The word *nirodha* signifies, almost universally and exclusively, extinction, or total cessation of versatile existence; a meaning, by the way, which confirms and answers to the interpretation of *dharmá,* by general existences, entities, and not by merely human actions. The causes of versatile existence and of its extinction are given at pp. 79-80.

It is scarcely worth while to cumber the present question with the further remark that there *is* a sect of Bauddha philosophers holding opinions which confound onscious actions with universal entities throughout the versatile world, making the latter originate absolutely and *physically* from the former, (see my remarks on Rémusat in the Journal, No. 33, p. 431.)

It is not, however, admissible so to render generally received texts, as to make them correspondent to very peculiar dogmata. "*Dháranátmaka iti dharma,*" 'the holding, containing, or sustaining, essence (ens) is *dharma.*' The substratum of all form and quality in the versatile universe, the sustainer (in space) of versatile entity, mundane substances and existences, physical and moral, in a word, *all things.* Such is the *general* meaning of *dharma.* How many other meanings it has, may be seen by reference to a note at the foot of p. 502, No. 34, of your Journal.* The root of the word is *dhri,* 'to hold.' Wilson's dictionary gives *Nature* as Amara Sinha's explanation of *dharma.* This is essentially correct, as might be expected from a Bauddha lexicographer. The English word "substance" is the precise equivalent of *dharma,* which means that which supports qualities in space, and of the Brahmanic *mátrá,* meaning that which measures space or limits space, because space is only measurable by the substances it holds. I speak here merely of etymologies.

* See p. 109, in notes.

NOTE. If Mr. Hodgson's general interpretation of dharma is the true one, (which seems most probable, though its specification in the sense of *moral duties* is more agreeable to M. Csoma's supplement)—its implication, in the present reading, at least, appears manifestly atheistic. For that it cannot mean "Tathágata or the Adi-Buddha *is* the cause," is evident from the accusative hétún (which is also plural, *causas.*) Even if we were to strike out the word *avadat* or *áha*—the former of which is on the inscriptions, and the latter repeated in Ceylon—still some word of that meaning is plainly understood: and this may help to shew that the explication given by the Aiswarika Buddhists (as though the words were hétus téshám Tathágatas) is a more recent invention,—and that the Buddhist system properly recognizes no being superior to the sage expounder of physical and moral causes,—whose own exertions alone have raised him to the highest rank of existences,—the Epicurus of this great Oriental system,

qui potuit rerum cognoscere causas,
Atque metûs omnes et inexorabile fatum
Subjecit pedibus.

What is mere figure of speech in the Roman poet, to express the calm dignity of wisdom, becomes religious faith in the east; *viz.*, the elevation of a philosophical opponent of popular superstition and Brahmanical caste, to the character of a being supreme over all visible and invisible things, and the object of universal worship. —W. H. M.

———o———

Note on the Note of W. H. M.—My friendly and learned annotator is right as to the *comparative* recency of the Aiswarika school and may find that opinion long since expressed by myself. But he is wrong in supposing that that school has no old or unquestionable basis; for both Mr. Csoma and myself have produced genuine and ancient authorities in its support. So that it is hardly fair to revert to the fancies of Sir W. Jones' day, under cover of a Latin quotation! As to verbal criticism, it is surely scarce necessary to observe that the governing verb being removed, the noun will take the nominative case. I quoted popular words popularly and omitted the nice inflexions of case and number. That my terser text *is* familiar to the mouths of Buddhists, is an unquestionable fact; and I never said, either that this terser form was *that of the inscription*, or that I had seen scriptural authority for it, *ipsissimis verbis.*

The express causes of versatile existence, alluded to by Sákya, in the text graved at Sarnath, are Avidyá, Sanskára, etc., as enumerated in my "Quotations in Proof" under the head of the Kármika doctrine; and there, too, may be found the causes of the extinction of such existence. See pp. 79-80 of this vol. This passage is the true complement or exponent of the *ye dharmá*, and leaves no possible doubt as to its meaning.

NOTICE OF ADI-BUDDHA AND OF THE SEVEN MORTAL BUDDHAS,*

(With reference to Nepaul chiefly)

FROM THE SWAYAMBHU PURANA.

The *Swayambhú Purána* relates in substance as follows: That formerly the valley of Nepaul was of circular form, and full of very deep water, and that the mountains confining it were clothed with the densest forests, giving shelter to numberless birds and beasts. Countless waterfowl rejoiced in the waters. The name of the lake was Nâga Vâsa;§ it was beautiful as the lake of Indra; south of the Hemáchal, the residence of Karkotaka, prince of the Nâgas; seven *cos* long, and as many broad. In the lake were many sorts of water-plants; but not the lotos. After a time, Vipasyî Buddha arrived, with very many disciples and Bhikshus, from Vindumatí Nagar, in Madhya Desa, at the Lake of Nâga Vâsa, in the course of his customary religious peregrinations. Vipasyî, having thrice circumambulated the lake, seated himself in the N. W. (Váyukona) side of it, and, having repeated several mantras over the root of a lotos, he threw it into the water, exclaiming, "What time this root shall produce a flower, then, from out of the flower, Swayambhú, the Lord of Agnishtha Bhuvana, shall be revealed in the form of flame; and then shall the lake become a cultivated and populous country." Having repeated these words, Vipasyî departed. Long after the date of this prophecy, it was fulfilled according to the letter.

After Vipasyî Buddha, came Sikhí Buddha to Nâga Vâsa with a great company of respectful followers, composed of rajas and persons of the four castes (chatur varna). Sikhí, so soon as he beheld Jyotí-rúpa-Swayambhú, offered to him many laudatory forms of prayer: then rising, he thrice walked round Nâga Vâsa, and, having done so, thus addressed his disciples. "This place shall hereafter, by the blessing of Swayambhú, become a delightful abode to those who shall resort to it from all quarters to dwell in it, and a sweet place of sojourn for the pilgrim and passenger: my apotheosis is now near at hand, do you all take your leave of me and depart to your own country." So saying Sikhí threw himself into the waters of Nâga Vâsa, grasping in his hands the stalk of the lotos, and his soul was absorbed into the essence of Swayambhú. Many of his disciples, following their master, threw themselves into the lake, and were absorbed into Swayambhú, *i. e.,* the self-existent; the rest returned home. Viswabhú was the third Buddha who visited Nága Vâsa. Viswabhú was born in Anupama-puri-nagar, of Madhya Desa; his life was devoted to benefitting his fellow-creatures. His visit to Nepaul was long after that of Sikhí, and, like Sikhí, he brought with him a great many disciples and Bhikshus, Rajas and cultivators, natives of his own land. Having repeated the praises of Swayambhú-jyotí-rúpa, he observed; "In this lake

* Printed from the Bengal Asiatic Journal, No. 29, A. D. 1834.

§ When the lake was desiccated (by the sword of Manjusrí says the myth—probably earthquake) Karkotaka had a fine tank built for him to dwell in; and there he is still worshipped, also in the cave-temple appendant to the great Buddhist shrine of Swayambhú Náth.

Prajná-surúpa-Guhyeswarí* will be produced. A Bodhisatwa will, in time, make her manifest out of the waters: and this place, through the blessing of Swayambhú, will become replete with villages, towns, and tirthas, and inhabitants of various and diverse tribes." Having thus prophesied he thrice circumambulated the lake, and returned to his native country. The Bodhisatwa above alluded to is Manju Srí,‡ whose native place is very far off, towards the north, and is called Pancha Sírsha Parvata, [which is situated in Mahá Chína Des.§] After the coming of Viswabhú Buddha to Nâga Vâsa, Manju Srí, meditating upon what was passing in the world, discovered by means of his divine science that Swayambhú-jyotí-rúpa, that is, the self-existent, in the form of flame, was revealed out of a lotos in the lake of Nâga Vâsa. Again, he reflected within himself: "Let me behold that sacred spot, and my name will long be celebrated in the world;" and on the instant, collecting together his disciples, comprising a multitude of the peasantry of the land, and a Raja named Dharmákar, he assumed the form of Viswakarma, and with his two Devís (wives,) and the persons above-mentioned, set out upon the long journey from Sírsha Parvata to Nâga Vâsa. There having arrived, and having made púja to the self-existent, he began to circumambulate the lake, beseeching all the while the aid of Swayambhú in prayer. In the second circuit, when he had reached the central barrier mountain to the south, he became satisfied that that was the best place whereat to draw off the waters of the lake. Immediately he struck the mountain with his scimitar, when the sundered rock gave passage to the waters, and the bottom of the lake became dry. He then descended from the mountain, and began to walk about the valley in all directions. As he approached Guhyeswarí,|| he beheld the water bubbling up violently from the spot, and betook himself with pious zeal to the task of stopping it. No sooner had he commenced than the ebullition of the water became less violent, when, leaving bare only the flower of the lotos, the root of which is the abode of Guhyeswarí, he erected a protecting structure of stone and brick over the recum-

* That is the mystic form of Prajná, who is the same with Dharmá and the Sakti of Swayambhú or Adi-Buddha, according to the Triadists. The type of Adi-Buddha in Nepaul is fire—that of Adi-Dharmá or Prajná or Guhyeswarí is water—or she has no type, is of a secret form, *i. e.*, Guhyeswarí, or lastly, according to the Tantras, her type is the Yoni, which, as well as the whole ritual belonging to it, is Guhya or esoteric and concealed.

‡ The Tibetans identify Manjusrí with Thu mi Sam bho ta, minister of King Srong-tsan, who lived in the seventh century, and was the great introducer of Buddhism into Tibet. Manjusrí's Tibetan name is Jam yang; Thumi is an incarnation of him.

§ The bracketed portions are from the commentators.

|| The site of the temple is near the centre of the valley, on the skirts of the lovely grove of Pasupati; and above 2½ or 3 miles east from Mount Sambhú. The fable says, that the root of the lotos of Guhyeswarí is at the former place, and the flower at the latter; the recumbent stalk being extended throughout the interval between them. Swayambhú or Adi-Buddha is supposed to reside in the flower, in the form of flame; Prajná Páramitá or Guhyeswarí, in or at the root, in the form of water. The temple of Guhyeswarí has been appropriated by the Brahmans, who worship this goddess as the Sakti of Pasupati Náth, whose symbol is the four-faced Lingam. But it may be that the Buddhists are wrong in identifying Guhyeswarí with Prajná, and that Guhyeswarí, the Sakti of Pasupati Náth, is really one of the deities of Náthism—a half orthodox (Goraksha náth) and half heterodox (Matsyendra náth) divinity.

bent stalk, and called the structure, which rose into a considerable elevation as it neared the flower of the lotos, *Satya Giri.* This work completed, Manju Srí began to look about him in search of a fit place of residence, and at length constructed for that purpose a small hill, to which he gave the name of Manju Srí Parvata, (the western half of the little hill of Sambhú Náth,) and called the desiccated valley, *Népálá—Né* signifying 'the sender' (to paradise,) who is Swayambhú, and *pálá* 'cherished'; implying that the protecting genius of the valley was Swayambhú or Adi-Buddha. Thus the valley got the name of Népálá: and, since very many persons had come from Mount Sírsha (or China) with Manju Srí, for the residence of Dharmákar Raja and his suite, Manju constructed a large place of abode [half way between Mount Swayambhú and Guhyeswarí,] and named it after himself, *Manju Pattana*, and established therein Dharmákar [of Mahá Chína] as Raja, subjecting the whole of the inferior sort of people who came from Sírsha Parvata to Dharmákar's rule, and providing abodes for them in the city of Manju Pattana.

[Thus was Nepaul peopled, the first inhabitants of which came all from Mount Sírsha, which is in Mahá Chína, and thus the valley got the name of Népálá, and its inhabitants, that of Népálí, whose primitive language was Chinese.* This language in course of time came to be much altered by the immigration of people from Madhya Desa, and by the necessary progress of corruption and change in a new country, till a new language arose in Nepaul by the natural course of things. The primitive inhabitants of Nepaul were all of one caste, or had no caste. But their descendants, in the course of time became divided into many castes, according to the trades and professions which they followed; and of these, such as abandoned the world and shaved their heads became Bhikshu, Sramana, Chailaka, and Arhana, and took up their abode in forests or in monasteries. These four orders all monastic; and in strictness absolutely excluded from all worldly commerce. But should any of them, still retaining the custom of tonsure, become worldly men, such are called Srávaka, etc. to a great extent of diverse names.] Manju Sri, having by such deeds as these acquired the highest celebrity in Nepaul, [ostensibly, and for the instruction of the people] relinquished his mortal form and became *nirván;* [but in truth departed for Mount Sirsha with his two Devis, and in due course arrived at Pancha Sírsha Parvata.] Some time after the disappearance of Manju Srí, Karkut Sand Buddha came to Nepaul, with some Bhikshukas, Dharmapála Raja, and a multitude of the common people, from Kshêmâvatinagar, of Madhya Desa. The beauty of the country delighted

* Manju Srí or Manju Ghosha (sweet voice) and Dharmákar are pure Sanskrit words, which fact makes against the alleged location of Mount Sírsha (also Sanskrit) in China, and there are grounds for supposing that mount Sîrsha was in Assam.

In the Nepaulese Vansavalis the first race of kings are apparently Gwallas and Saivas, or rather Pasupatas, who worshipped Pasupati and received the throne from a Rishi called Neyam. But this dynasty is open to doubt in all ways. The next dynasty is clearly barbarian and utterly alien to Sanskrit and India. It is of the Kiranti tribe now located in all the eastern part of Nepaul. This evidence is indecisive. What says the *Skand Purana*, and what is its age compared with that of the *Sambhu Purana?* Physiognomy and speech decisively refer the *Newars* to the Tibetan stock.

him, and he remarked that in such a land the cultivator must be sure to reap as he sowed. He paid his devotions to Swayambhú, and then launched out in praise of the merits of Manju Srí, the Nepaulese patriarch. Afterwards he performed *pújá* to Guhyeswarí, and then ascended Sankhocha mountain (Siva Púra:) the prospect of that valley from that mount filled him with fresh delight, and he again celebrated the excellence of the country. Gunadhvaja, a brahman, and Abhayandada, a kshetriya, and others of the four castes (chatúr varna,) respectful followers of Kurkut Sand, here solicited at his hands the favour of being made Bhikshukas, in order that they might remain in this happy land, and by the worship of Swayambhú attain to high merit and honour. Kurkut cheerfully complied, and agreed to make a great many of the company Bhikshukas; and since the mountain top afforded no water for that ceremony, he by his divine power caused a spring to issue from the rock, and with its waters gave to his followers the requisite Abhishêka or baptism. He called the river that originated with this spring Vángmati;* and then related to his followers both the past and future history of the valley watered by the Vángmati. Then, having left behind him in Nepaul, Raja Dharmapál and some Bhikshus and common folks, who had come with him, and desired to stay, Kurkut Sand departed with the rest of them to his native city of Kshemávati. [These companions of Kurkut Sand, or Krakucchand, were the first natives of the plains of India (Madhya Desa) who remained in Nepaul. Many of them, addicting themselves to the business of the world, became householders, and the founders of several towns and villages in Nepaul; whilst others, who adopted the ascetical profession, dwelt in the forests and Vihárs. When these Madhyadésiyas had become numerous in Nepaul, they and their descendants were confounded with the former or northern colonists under the common appellation of Népáli and Néwárí; being only separated and contradistinguished by the several trades and professions which they hereditarily practised. Thus, in the early ages, Nepaul had four classes of secular people, as Brahman, Kshetriya, Vaisya, and Sudra, and four ascetical classes, namely, Bhikshu, Sramana,† Chailaka, and Arhanta, dwelling in forests and monasteries, and all were *Buddha-márgi.*]

ACCOUNT OF DHARMAKAR RAJA AND DHARMAPAL RAJA.

Dharmákar, the before noted [Chinese] prince of Nepaul, being disgusted with the world, abandoned his sovereign power, and placed *Dharmapál*, the Raja of Gaur-des, already mentioned, upon his throne. *Dharmapál* governed his subjects with perfect justice and clemency, and made pújá at the Chaitya erected by Dharmákar, and regarded with equal favour his subjects that came from Mount Sírsha [or Mahá Chína,] and those who immigrated from Madhya-desa.

ACCOUNT OF PRACHANDA DEVA.—Prachanda Deva, a Raja of Gaur-des, which is adjacent to Madhya-des, and of the Kshatriya tribe, was the wise man of his age and country. At length, being inspired with the ambition of becoming

* From *Vach*, 'speech.'
† *Srávaka* and *Sramana* are equivalent.

nirvâna, he abandoned his princely sway; and taking with him a few sages, he began to wander over various countries, visiting all the shrines and pilgrimages, and in the course of his peregrinations arrived at Nepaul. He was delighted with the beauty of the country, and having visited every *tírtha*, and *píth*, and *devatá*, and having made *pújá* to the *Tri Ratna*, or triad, he went to the temple of Swayambhú, and there performed his devotions. He then ascended Manju Srí Parbat, and offered his prayers to Manju Srí, and finished by becoming a disciple of Gunákar Bhikshu, a follower of Manju Srí. One day Prachanda Deva so delighted Gunákar with the display of his excellent qualities, that Gunákar made him a Bhikshuka; and the said Raja Prachanda after becoming a Bhikshu, obtained the titular appellation of Sánta Srí. [A great many Brahmans and others who accompanied Prachanda to Nepaul received the tonsure, and became Bhikshus at the same time with Prachanda, and took up their abode in the monasteries of Nepaul. Some others of those that came with Prachanda to Nepaul preferring the pursuits of the world, continued to exercise them in Nepaul, where they also remained and became Buddhists. A third portion of Prachanda's companions returned to Gaur-des.] After a time, Sánta Srí represented to his Guru Gunákar his desire to protect the sacred flame of Swayambhú with a covering structure. Gunákar was charmed with the proposition and proposer, and having purified him with thirteen sprinklings of sacred water (*trayodasábhisheka*,) gave him the title of Díkshita Sántikar Vajra Achárya. [From these transactions is dated the arrival of the people of Gaur-des in Nepaul, and their becoming Buddhists.]

ACCOUNT OF KANAKA MUNI.—Once on a time, from Subhávatínagar of Madhyades, Kanaka Muni Buddha, with many disciples, some illustrious persons, and a countless multitude of common people, arrived at Nepaul, in the course of his religious peregrinations, and spent some months in the worship of Swayambhú, and the Tri Ratna, and then departed with most of his attendants. A few remained in Nepaul, who became Buddha-márgí and worshippers of Swayambhú; [and these too, like all the preceding, soon lost their name and character as Madhyadesiyas, and were blended with the Népálí or Néwárí race.]

ACCOUNT OF KASYAPA BUDDHA.—Once on a time in Mrigadâba-vana, near Benares, Kásyapa Buddha was born. He visited Nepaul in pilgrimage, and made his devotions to Sambhunáth. [Most of the people who came with him staid in Nepaul, and soon became confounded with the aborigines.]

ACCOUNT OF SAKYA SINHA BUDDHA.—Some time after Kásyapa's visit at Gangá Ságara,* in the sthan of Kapila Muni, and city of Kapila-vastu, and reign of Suddhódana Raja, of the Sâka-vansa, was born (as the son of that Raja) Sarvártha Siddha, who afterwards became a Buddha with the name of Sákya Sinha. Sákya, with 1,350 Bhikshukas, and the Raja of Benares, several counsellors of state,

* Gangâ Ságara, says Wilson, has no necessary connection with the ocean. For the site of Kapila-pur see Laidlay's *Fahian*. But I doubt if the site were so far from the hills. Ṭimur, in his annals, says that he took it and speaks of it as though it were actually in the hills, a mountain fastness in fact.

and a crowd of peasantry of that kingdom, set out on the pilgrimage to Nepaul. Having paid his devotions to the self-existent, in the form of flame, he went to the Chaitya or Puchhágra Hill,† and repeated to his disciples the past history of Nepaul, as well as its whole future history, with many praises of Manju Srí Bodhisatwa: he then observed, "In all the world are twenty-four *Píths*, and of all these that of Nepaul is the best." Having so said, he departed. His companions, who were of the Chatur varna, or four castes, [Bráhman, Kshatriya, Vaisya, and Súdra,] and belonged to the four orders, [Bhikshu, and Sramana, and Chailaka, and Arhant,] being much pleased with Népál-des continued to dwell in it; [and in course of time were blended with the aboriginal Népálís, and became divided into several castes, according to the avocations which they hereditarily pursued.] Some time after the date of the above transaction, Raja Gunakáma Deva, prince of Káthmándú, [a principal city of Nepaul,] became the disciple of the above-mentioned Sántikar Vajra Achárya. Guna Káma Deva, with the aid derived from the divine merits of Sántikar, brought the Nág Raja Karkotaka‡ out of the lake or tank of Adhâr, and conveyed him to Sántipúr with much ceremony and many religious rites. The *cause* of this act was that for many previous years there had been a deficiency of rain, whereby the people had been grievously distressed with famine; and its *consequence* was an ample supply of rain, and the return of the usual fertility of the earth and plenty of food.§

Subsequently, Srí Narendra Deva became Raja of Bhagatpattan, (or Bhatgaon;) he was the disciple of Bandudatta Achárya, and brought Aryávalokiteswara* (Padma Páni) from Pútalakáparvat (in Assam) to the city of Lalita pattan in Nepaul. The reason of inviting this divinity to Nepaul was a drought of twelve years duration, and of the greatest severity. The measure was attended with like happy results, as in the case of conveying the Nág Raja with so much honour to Sántipúr.

NOTE ON THE PRIMARY LANGUAGE OF THE BUDDHIST WRITINGS.||

I have read article II. of the 66th No. of your Journal with great interest. With regard to the language in which the religion of Sákya, 'was preached and spread among the people,' I perceive nothing opposed to my own opinions in the fact that that language was the vernacular.

There is merely in your case, as priorly in that of Mr. Turnour, some misapprehension of the sense in which I spoke to that point.

† Part of Mount Sambhu, west of the great Chaitya; also called Go-pucch.

‡ Karkotaka is named in the *Sanhitá*. And in the annals of Cashmir he figures as conspicuously as in Nepaul. The Nágas and Indra maintain still in Nepaul a deal of their pristine authority, and in connection one with the other: for the Nágas are invoked for rain.

§ The *Nágas* are still worshipped in Sántipur whenever the rains are deficient, in conformity with this legend and with the original one of the lake as being the *Nágvása*.

* Is Avalokeswara the same as Matsyendra Náth, whose arrival in Nepaul is referred to the fifth century of Christ by well known memorial verses? The identification with Padma Páni rests on Sástras of Nepaul and of China. See J.R.A.S., new series, vol. ii., part i., p. 137.

|| Printed from the Bengal Asiatic Journal, No. 68, A. D. 1837.

The preaching and spreading of the religion is a very different thing from the elaboration of those speculative principles from which the religion was deduced. In the one case, the appeal would be to the many; in the other, to the few. And whilst I am satisfied that the Buddhists as practical reformers addressed themselves to the people, and as propagandists used the vulgar tongue, I think that those philosophical dogmata which formed the basis of the popular creed, were enounced, defended and systematised in Sanskrit. I never alleged that the Buddhists had eschewed the Prákrits: I only denied the allegation that they had eschewed the Sanskrit; and I endeavoured, at the same time, to reconcile their use of *both*, by drawing a distinction between the means employed by their philosophers to establish the principles of this religion, and the means employed by their missionaries to propagate the religion itself.

Joinville had argued that Buddhism was an original creed, older than Brahmanism, because of the grossness of its leading tenets which savour so much of 'flat atheism.'

I answered that Buddhism was an innovation on the existing creed, and that all the peculiarities of the religion of Sákya could be best and only explained by advertence to shameful *prior abuse* of the *religious sanction*, whence arose the characteristic *Bauddha* aversion to gods and priests, and that enthusiastic self-reliance taught by Buddhism in express opposition to the servile extant reference of all things to heavenly and earthly mediation. Jones, again, had argued that the Buddhists used only the Prákrit, *i. e.*, Páli, because the books of *Ceylon* and *Ava*, (the only ones then forthcoming,*) were solely in that language or dialect. I answered by producing a whole library of Sanskrit works in which the principles of Buddhism are more fully expounded than in all the legendary tomes of *Ceylon* and *Ava;* I answered, further, by pointing to the abstruse philosophy of Buddhism, to the admitted preeminence, as *scholars*, of its expounders; and to their location in the most central and literary part of India (*Behar* and *Audh*). With the Sanskrit at command, I asked and ask again, *why* men so placed and gifted, and having to defend their principles in the schools against ripe *scholars* from all parts of India (for those were days of high debate and of perpetual formal disputation in palaces and in cloisters) should be supposed to have resorted to a limited and feebler organ when they had the universal and more powerful one equally available? The presumption that they did *not* thus postpone Sanskrit to Prákrit is, in my judgment, worth a score of any inferences deduceable from monumental slabs, backed as this presumption is by the Sanskrit records of Buddhism discovered here. Those records came direct from the proximate head quarters of Buddhism. And, if the principles of this creed were not expounded and systematised in the schools of India in Sanskrit, what are we to make of the Nepaulese Sanskrit originals and of the avowed Tibetan translations? In my judgment the *extent* and *character* of these works settle the question that the philosophic founders of Buddhism used Sanskrit and Sanskrit only, to expound, defend

* Sir W. Jones had, however, in his possession a Sanskrit copy of the *Lalita Vistara*, and had noticed the personification of *Diva Natura* under the style of Arya Tárá.

and record the speculative principles of their system, principles without which the vulgar creed would be (for us,) mere leather and prunella! Nor is this opinion in the least opposed to your notion (mine too) that the *practical system of belief*, deduced from those principles, was spread among the people of the spot, as well as propagated to remoter spots, by means of the vernacular.

It is admitted that Buddhism was long taught in *Ceylon* without the aid of Books: and that the first book reached that island nearly 300 years after the introduction of the creed.

Here is a distinct admission of what I long since inferred from the general character of the religion of Sákya in that island, *viz.*, the protracted total want, and ultimate imperfect supply, of those standard written authorities of the sect which regulated belief and practice in *Magadha*, in *Kosala* and *Rájagriha*,—in a word, in the *Metropolis* of Buddhism. From this metropolis the authorities in question were transferred directly and *immediately* to the *proximate hills* of *Nepaul*, where and where only, I believe, they are now to be found. If not translations, the books of *Ceylon* have all the appearance of being ritual collectanea, legendary hearsays, and loose comments on received texts—all which would naturally be written in the vulgar tongue.* To these, however, we must add some very important historical annals, detailing the spread and diffusion of Buddhism. Similar annals are yet found in Tibet, but, as far as I know, not in *Nepaul*, for what reason it is difficult to divine.

But these annals, however valuable to us, for historical uses, are not the original written standard of faith; and until I see the *Prajná Páramitá* and the nine *Dharmas*† produced from *Ceylon*, I must continue of the opinion that the Buddhists of that island drew their faith from secondary, not primary sources; and that whilst the former were in *Ceylon* as elsewhere, vernacular; the latter were in *Magadha* and *Kosala*, as they are still in *Nepaul*, classical or Sanskrit!

Certainly Buddhism, considered in the practical view of a religious system, always appealed to the common sense and interest of the many, inscribing its most sacred texts (Sanskrit and Prákrit) on temple walls and on pillars, placed in market, highroad and cross-road.

This material fact (so opposite to the genius of Brahmanism,) I long since called attention to; and thence argued that the inscriptions on the láts would be probably found to be of scriptural character.

The tendency of your researches to prove that the elaborate forms of the Devanágarí were constructed from simpler elements, more or less appropriated to the popular Bháshás, is very curious; and seems to strengthen the opinion of those who hold Hindí to be indigenous, older than Sanskrit in India, and not (as Colebrooke supposed) deduced from Sanskrit. If Buddhism used these primitive letters before the Devanágarí existed, the date of this creed would seem to be

* *Such* works written in the vulgar tongue are common in *Nepaul* and frequently we have a Sanskrit text with a vernacular running commentary.

† They *have* one of the nine, *viz.*, the *Lalita Vistara;* but M. Burnouf assures me, in a miserably corrupted state. Now, as this work is forthcoming in a faultless state in Sanskrit, I say the Pali version *must* be a translation.

thrown back to a remote æra, or, the Sanskrit letters and language must be comparatively recent.

I can trace something *very like* Buddhism into far ages and realms: but I am sure that that Buddhism which has come down to us in the Sanskrit, Páli and Tibetan books of the sect, and which alone therefore we do or can *know*, is neither old nor exotic. *That* Buddhism (the doctrines of the so called *seventh* Buddha) arose in the middle of India, in comparatively recent times, and expressly out of those prior abominations which had long held the people of India in cruel vassalage to a bloated priesthood.

The race of *Sáka*, or progenitors of *Sákya Sinha* (by the way, the *Sinha* proves that the princely style was given to him until he assumed the ascetic habit) may have been Scythians or Northmen, in one sense; and so probably were the Brahmans in that same sense, *viz.*, with reference to their original seat. (*Brachmanes nomen gentis diffusissimæ, cujus maxima pars in montibus degit; reliqui circa Gangem.*)

If one's purpose and object were to search backwards to the original hive of nations, one might, as in consistency one should, draw Brahmanism and Buddhism, Vyása and Sákya, from Tartary.* All I say is, that *quoad* the known and recorded man and thing—Sákya Sinha and his tenets—they are indisputably Indian and recent.†

I incline to the opinion that Hindí may be older in India than Sanskrit, and independent, originally, of Sanskrit. But were this so, and were it also true that the Buddhists used the best dialect of Hindí (*that* however is saturated with Sanskrit, whatever its primal independence), such admissions would rather strengthen than weaken the argument from language against the exotic origin of Buddhism.‡

According to this hypothesis, Hindí is not less, but more, Indian than Sanskrit: and, *à fortiori*, so is the religion assumed to have committed its records to Hindí.

But, in very truth, the extant records of Buddhism, whether Sanskrit or Prákrit, exhibit both languages in a high state of refinement; and though one or both tongues came originally from Tartary, they received that refinement in India, where, certainly, what *we know* as Buddhism, (by means of these records) had its origin, long after Brahmanism had flourished there in all its mischievous might.

P. S. You will, I hope, excuse my having adverted to some other controverted topics besides that which your paper immediately suggested. These questions are a good deal linked together: for instance, if Buddhism furnishes *internal* evidence throughout its most authentic records that it is the express antithesis of

* That is from a country to the north-west of Hindostan—a country beyond the Indus—and no doubt the country called Ariana or Irán, in the widest sense, but not Túrán or Tartary as we call it, for none of the Tartar races were literary, and even to this hour the Turks only have some poor and borrowed pretensions to literature. The Uighours got their alphabet from the Nestorians, and the Mongols theirs from the Uighours.

† According to all Bauddha authorities the lineage of the whole seven mortal Buddhas is expressly stated to be Brahmanical or Kshatriya! What is the answer to this?

‡ Our own distinguished Wilson has too easily followed the continental European writers in identifying the *Saka vansa* with the classical Sacæ or Scythians, and Buddhism with Samanism. The Tartars of our day avow that they got all their knowledge from India; *teste Kahgyur et Stangyur.*

Brahmanism, its posteriority of date to the latter is decided, *as well as* its *jealousy* of *priestly pretensions. Nec clericis infinita aut libera potestas*, is a deduction which only very precise and weighty evidence will suffice to set aside: I have seen none such yet from *Ceylon* or from *Ava*. And be it observed, I here advert to authentic scriptural tenets, and not to popular corruptions resulting from the facile confusion of the monastic with the clerical character.

———o———

NOTE. We are by no means prepared to enter into a controversy on a subject on which we profess but a slight and accidental acquaintance: nor will we arrogate to ourselves the distinction of having entered the lists already occupied by such champions as Mr. Hodgson and Mr. Turnour, who have both very strong arguments to bring forward, in support of their opposite views. As far as the *Dharmalipi* could be taken as evidence the vernacularists had the right to it; but on the other hand there can be no doubt, as Mr. Hodgson says, that all scholastic disputation with the existing Brahmanical schools which Sákya personally visited and overcame, must have been conducted in the classical language. The only question is, whether any of these early disquisitions have been preserved, and whether, for example, the Life of Sákya, called the *Lalita Vistara*, found by Professor Wilson to agree verbatim with the Tibetan translation examined simultaneously by Mr. Csoma, has a greater antiquity than the *Pitakattayan* of *Ceylon?* We happen fortuitously to have received at this moment two letters bearing upon the point in dispute from which we gladly avail ourselves of an extract or two:—Mr. Turnour, alluding to the notice of the life of Sákya from the Tibetan authorities by Mr. Csoma in the As. Res. vol. xx. writes—"The Tibetan life is apparently a very meagre performance, containing scarcely anything valuable in the department of history; whereas had the materials whence it was taken been genuine, the translator would have been able to bring forward and illustrate much valuable information on the pilgrimages and the acts of Sákya in various parts of India during the forty-five years he was *Buddha*. Even the superstitious facts recorded are much more absurd than they are represented in the *Pitakattayan*. Thus the *dream* of Máyá Deví of having been rubbed by a *Chhadanta* elephant, during her pregnancy,—is converted into a matter of fact, of Sákya, 'in the form of an elephant having entered by the right side into the womb or cavity of the body of Máyá Deví!' '*Chhadanta*' is taken literally as a *six-tusked* elephant, whereas by our books *Chhadanta* is the name of a lake beyond the *Himálaya* mountains where the elephants are of a superior breed.* It is mentioned twice in the *Maháwanso;* chaps. v. and xxii."

If the rationality of a story be a fair test of its genuineness, which few will deny, the *Páli* record will here bear away the palm:—but it is much to be regretted that we have not a complete translation of the Sanskrit and of the Ceylonese "life" to place side by side. It is impossible that instruction should not be gained

* Let zoologists say what they think of the rationality of this story. I would add that refining of the sense of old legends is a common practise of later times.

by such an impartial examination.|| But to return to the subject under discussion; my friend Mr. Csoma writes from *Titalya* in the *Purniya* district:—

"In reference to your and Mr. Turnour's opinion that the original records of the Buddhists in ancient India, were written in the *Mágadhí* dialect, I beg leave to add in support of it, that in the index or register, (d*kar-chhag*) of the *Kahgyur*, it is stated that the *Sútras* in general—*i. e.*, all the works in the *Kahgyur*, except the twenty-one volumes of the *Sher-chhin** and the twenty-two volumes of the r*Gyud* class, after the death of Sákya, were first written in the *Sindhu* language, and the *Sher-chhin* and r*Gyud* in the Sanskrit: but part of the r*Gyud* also in several other corrupt dialects. It is probable that in the seventh century and afterwards, the ancient Buddhistic religion was remodelled and generally written in Sanskrit,§ before the Tibetans commenced its introduction by translation into their own country."

This explanation, so simple and so authentic, ought to set the matter at rest, and that in the manner that the advocates of either view should most desire, for it shews that both are right!—It is generally allowed that the *Páli* and the *Zend* are derivatives of nearly the same grade from the Sanskrit stock; and the modern dialect of Sindh as well as the *Bháshá* of upper and western India present more striking analogies to the *Páli*, in the removal particularly of the *r*, and the modification of the auxiliary verbs, than any of the dialects of *Bengal*, *Behar*, or *Ceylon*.|| Plausible grounds for the existence of this western dialect in the heart of *Magadha*, and the preference given it in writings of the period, may be found in the origin of the ruling dynasty of that province, which had confessedly proceeded from the north-west. At any rate those of the *Sákya* race, which had emigrated from *Sinde* to *Kapilavastu* (somewhere in the *Gangetic* valley) may

|| As an example of the information already obtained from Mr. Csoma's translated sketch, we may adduce the origin of the custom seemingly so universal among the Buddhists of preserving pictorial or sculptured representations of the facts of his life.—After his death the priests and minister at *Rájagriha* are afraid of telling the king Ajáta Satru thereof lest he should faint from the shock, and it is suggested by Mahákásyapa by way of breaking the intelligence to him, that the *Mahámantra* or chief priest should "go speedily into the king's garden, and cause to be represented in painting, how Chomdandas *(Bhagaván)* was in *Tushita:* how in the shape of an elephant he entered his mother's womb: how at the foot of the holy fig-tree he attained supreme perfection: how at *Váránasi* he turned the wheel of the law of twelve kinds, (taught his doctrines:)—how he at *Srávasti* displayed great miracles; —how at the city of *Ghachen* he descended from the *Trayastrinsa* heaven, whither he had gone to instruct his mother: —and lastly, how having accomplished his acts in civilizing and instructing men in his doctrine at several places, he went to his last repose in the city of *Kusha* in *Assam*." Now whether the book in question was written sooner or later, it explains the practice equally and teaches us how we may successfully analyze the events depicted in the drawings of *Ajanta*, perchance, or the sculptures of *Bhilsa*, with a full volume of the life of *Sákya* in our hand. Similar paintings are common in *Ava*, and an amusing, but rather aprocryphal, series may be seen in Upham's folio history of Buddhism.

* This exception embraces the whole speculative tenets or philosophy of Buddhism.

§ This is a daring hypothesis, contrary, I think, to all legitimate presumptions. Where were the books remodelled, and why in Sanskrit if their prototypes were Sanskrit.

|| See the Rev. Dr. Mill's note on this subject in the Jour. B. As. Soc. vol. v., p. 30; also Professor Wilson's remarks, vol. i. p. 8.

have preserved the idiom of this native province and have caused it to prevail along with the religion which was promulgated through its means.*

We are by no means of opinion that the *Hindí*, *Sindhí*, or *Páli* had an independent origin prior to the *Sanskrit.* The more the first of these, which is the most modern form and the farthest removed from the classical language, is examined and analyzed the more evidently is its modification and corruption from the ancient stock found to follow systematic rules, and to evince rather provincial dialectism (if I may use the word) than the mere engraftment of foreign words upon a pre-existent and written language. The aboriginal terms of Indian speech must be rather sought in the hills and in the peninsula; in the plains and populous districts of the north the evidences of their existence are necessarily smothered by the predominance of the refined and durable languages of the court, of religion, and of the educated classes. A writer in the Foreign Quarterly has lately been bold enough to revive the theory of Sanskrit being merely a derivative from the Greek through the intervention of the Zend, and subsequent to the Macedonian invasion! The Agathocles' coin ought to answer all such speculations. The *Páli* of that day alone with its appropriate symbols is proved to have held the same precise derivative relation to the Sanskrit as it does now—for the records on which we argue are not modern, but of that very period. All we still want is to find some graven Brahmanical record of the same period to shew the character then in use for writing Sanskrit; and to add ocular demonstration to the proofs afforded by the profound researches of philologists as to the genuine antiquity of the venerable depository of the Vedas.§

——o——

A DISPUTATION RESPECTING CASTE BY A BUDDHIST.

One day my learned old *Bauddha* friend brought me a little tract in Sanskrit, with such an evident air of pride and pleasure, that I immediately asked him what it contained. "Oh, my friend!" was his reply, "I have long been trying to procure for you this work, in the assurance that you must highly approve the wit and wisdom contained in it; and, after many applications to the owner, I have at length obtained the loan of it for three or four days. But I cannot let you have it, nor even a copy of it, such being the conditions on which I procured you a sight of it." These words of my old friend stimulated my curiosity, and with a few fair words I engaged the old gentleman to lend me and my *pandit* his aid in making a translation of it; a task which we accomplished within the limited period of my possession of the original, although my *pandit* (a Brahman of Benares) soon declined co-operation with us, full of indignation at the author and his work! Notwith-

* This is Csoma in No. 14 of Jour. Bengal As. Soc. But Wilson in the Hindu Drama (*Notes on the Mrichhakati*) derives the Bihar dynasty from Andhra or Telingana.

§ If the Sanskrit literature be so old as alleged (tenth to fourteenth century B. C.) it is most strange that we have no Brahmanical monument or inscription nearly so old as the Buddhist Pali ones. The Rigveda Sanhitá suggests at once that this cannot be referred to ignorance, and may be referred to the Sabæan genius of primitive Hinduism, which was averse to idols and temples.

standing, however, the loss of the *pandit's* aid, I think I may venture to say that the translation gives a fair representation of the *matter* of the original, and is not altogether without some traces of its *manner*.

It consists of a shrewd and argumentative attack, by a *Bauddha*, upon the Brahmanical doctrine of caste: and what adds to its pungency is, that, throughout, the truth of the Brahmanical writings is assumed, and that the author's proofs of the erroneousness of the doctrine of *caste* are all drawn from those writings. He possesses himself of the enemy's battery, and turns their own guns against them. To an English reader this circumstance gives a puerile character to a large portion of the treatise, owing to the enormous absurdity of the data from which the author argues. His inferences, however, are almost always shrewdly drawn, and we must remember that not he but his antagonists must be answerable for the character of the data. To judge by the effect produced upon my Brahman *pandit*—a wise man in his generation, and accustomed for the last four years to the examination of *Bauddha* literature—by this little treatise, it would seem that there is no method of assailing Brahmanism comparable to that of "judging it out of its own mouth:" and the resolution of the Committee of the Serampore College to make a thorough knowledge of Hindu learning the basis of the education of their destined young apostles of Christianity in India, would thence appear to be most wise and politic. But to return to my little treatise.

We all know that the Brahmans scorn to consider the Súdras as of the same nature with themselves, in this respect resembling the bigoted Christians of the dark ages, who deemed in like manner of the Jews. The manner in which our author treats this part of his subject is, in my judgment, admirable, and altogether worthy of an European mind. Indeed it bears the closest resemblance to the style of argument used by Shakespeare in covertly assailing the analogous European prejudice already adverted to. I need not point more particularly to the glorious passage in the Merchant of Venice: "Hath not a Jew eyes, hands, organs, dimensions, senses, passions; fed with the same food, hurt by the same diseases?" etc.

The *Bauddha* treatise commences in the sober manner of a title page to a book; but immediately after the author has announced himself with due pomp, he rushes "*in medias res*," and to the end of his work maintains the animated style of *vivá voce* disputation. Who Ashu Ghosha, the author, was, *when* he flourished, and *where*, I cannot ascertain. All that is known of him in Nepaul is, that he was a *Mahá pandit*, or great sage, and wrote, besides the little treatise now translated, two larger *Bauddha* works of high repute, the names of which are mentioned in a note.*

——o——

I, Ashu Ghosha, first invoking Manju Ghosha, the *Guru* of the world, with all my soul and all my strength, proceed to compose the book called *Vajra Súchi*,† in

* The *Buddha Charitra Kávya*, and the *Nandi-Mukhasughosha Avadána*, and other works.

† Burnouf has said that the very term *Vajra* proves this to be a very recent work.

accordance with the *Shastras* (Hindu or Brahmanical *Sástras*).

Allow then that your *Vedas* and *Smritis*, and works involving both *Dharma* and *Artha*, are good and valid, and that discourses at variance with them are invalid, still what you say, that the Brahman is the highest of the four castes, cannot be proved from those books.

Tell me, first of all, what is Brahmanhood? Is it life, or parentage, or body, or wisdom, or the ritual (*áchára*), or acts, *i.e.*, morality (*karma*) or the *Vedas*?

If you say it is life (*jíva*), such an assertion cannot be reconciled with the *Vedas*: for it is written in the *Vedas* that "the sun and the moon, Indra, and other deities, were at first quadrupeds; and some other deities were first animals and afterwards became gods; even the vilest of the vile (*Swápaka*) have become gods." From these words it is clear that Brahmanhood is not life (*jíva*), a position which is further proved from these words of the *Mahábhárata*: "Seven hunters and ten deer, of the hill of Kalinjal, a goose of the lake Manasarovara, and a *chakwa* of Saradwípa, all these were born as Brahmans, in the *Kurukshetra* (near Dehli), and became very learned in the *Vedas*." It is also said by Manu, in his *Dharma Sástra*, "Whatever Brahman learned in the four *Vedas*, with their *anga* and *upanga*, shall take charity from a Súdra, shall for twelve births be an ass, and for sixty births a hog, and seventy births a dog." From these words it is clear that Brahmanhood is not life; for if it were, how could such things be?

If, again, you say that Brahmanhood depends on parentage or birth (*játi*); that is, that to be a Brahman one must be born of Brahman parents,—this notion is at variance with the known passage of the *Smriti*, that Achala Muni was born of an elephant, and Kesa Pingala of an owl, and Agastya Muni from the *Agasti* flower, and Kausika Muni from the *Kusa* grass; and Kapila from a monkey, and Gotama Rishi from creeper that entwined a saul-tree, and Drona Achárya from an earthern pot, and Taittiri Rishi from a partridge, and Parasu Ráma from dust, and Sringa Rishi from a deer, and Vyása Muni from a fisherwoman, and Kausika Muni from a female Súdra, and Viswámitra from a *Chándálini*, and Vasishtha Muni from a strumpet. Not one of them had a Brahman mother, and yet all were notoriously called Brahmans; whence I infer, that the title is a distinction of popular origin, and cannot be traced to parentage from written authorities.

Should you again say, that whoever is born of a Brahman father *or* mother is a Brahman, then the child of a slave even may become a Brahman; a consequence to which I have no objection, but which will not consort with your notions, I fancy.

Do you say, that he who is sprung of Brahman parents is a Brahman? Still I object that, since you must mean pure and true Brahmans, in such case the

But Weber in his new printed edition of it (original and translation) has shewn that the *Vajra Súchi* is at least a thousand years old, for in a work of Sankara áchárya not only is the term *Vajra* used, but strange to say, the first paragraph of his work is identical with one in the work before us, though of course differently intended as to scope and purpose, Sankara only proposing to exalt his ideal of Brahmanhood by contrasting it with the ordinary and actual types. But this shews what I have elsewhere remarked, *viz.*, that Saintism by its very genius and character (above ordinances) tends to obliterate the distinctive marks of Brahmanism and Buddhism.

breed of Brahmans must be at an end; since the fathers of the present race of Brahmans are not, any of them, free from the suspicion of having wives, who notoriously commit adultery with Súdras. Now, if the real father be a Súdra, the son cannot be a Brahman, notwithstanding the Brahmanhood of his mother. From all which I infer, that Brahmanhood is not truly derivable from birth; and I draw fresh proofs of this from the *Mánava Dharma*, which affirms that the Brahman who eats flesh loses instantly his rank; and also, that by selling wax, or salt, or milk, he becomes a Súdra in three days; and further, that even such a Brahman as can fly like a bird, directly ceases to be a Brahman by meddling with the fleshpots.

From all this is it not clear that Brahmanhood is not the same with birth? since, if that were the case, it could not be lost by any acts however degrading. Knew you ever of a flying horse that by alighting on earth was turned into a pig?—'Tis impossible.

Say you that body (*Sarira*) is the Brahman? this too is false; for, if body be the Brahman, then fire, when the Brahman's corpse is consumed by it, will be the murderer of a Brahman; and such also will be every one of the Brahman's relatives who consigned his body to the flames. Nor less will this other absurdity follow, that every one born of a Brahman, though his mother were a *Kshatriya* or *Vaisya*, would be a Brahman—being bone of the bone, and flesh of the flesh of his father: a monstrosity, you will allow, that was never heard of. Again, are not performing sacrifice, and causing others to perform it, reading and causing to read, receiving and giving charity, and other holy acts, sprung from the body of the Brahman?

Is then the virtue of all these destroyed by the destruction of the body of a Brahman? Surely not, according to your own principles; and, if not, then Brahmanhood cannot consist in body.

Say you that wisdom* constitutes the Brahman? This too is incorrect. Why? Because, if it were true, many Súdras must have become Brahmans from the great wisdom they acquired. I myself know many Súdras who are masters of the four *Vedas*, and of philology, and of the *Mimánsá*, and *Sánkhya*, and *Vaiseshika* and *Jyotishika* philosophies; yet not one of them is or ever was called a Brahman. It is clearly proved, then, that Brahmanhood consists not in wisdom or learning. Then do you affirm that the *Achára* is Brahmanhood? This too is false; for if it were true, many Súdras would become Brahmans; since many *Nats* and *Bhats*, and *Kaivartas*, and *Bhands*, and others, are everywhere to be seen performing the severest and most laborious acts of piety. Yet not one of these, who are all so pre-eminent in their *Achára*, is ever called a Brahman: from which it is clear that *Achára* does not constitute the Brahman.

Say you that *Karma* makes the Brahman? I answer, no; for the argument used above applies here with even greater force, altogether annihilating the notion that acts constitute the Brahman. Do you declare that by reading the *Vedas* a

* Perhaps it should rather be translated *learning*. The word in the original is *jnána*.

man becomes a Brahman? This is palpably false; for it is notorious that the *Rákshasa* Ravan was deeply versed in all the four *Vedas:* and that, indeed, all the *Rákshasas* studied the *Vedas* in Rávan's time: yet you do not say that one of them thereby became a Brahman. It is therefore proved that no one becomes a Brahman by reading the *Vedas.*

What then is this creature called a Brahman? If neither reading the *Vedas,* nor *Sanskára,* nor parentage, nor race *(Kula),* nor acts *(Karma),* confers Brahmanhood, what does or can? To my mind Brahmanhood is merely an immaculate quality, like the snowy whiteness of the *Kund* flower. That which removes sin is Brahmanhood. It consists of *Vrata,* and *Tapas,* and *Niyama,* and *Uparása,* and *Dána,* and *Dama,* and *Shama,* and *Sanyama.* It is written in the *Vedas* that the gods hold that man to be a Brahman who is free from intemperance and egotism; and from *Sanga,* and *Parigraha,* and *Rága,* and *Dwesha.* Moreover, it is written in all the *Sástras* that the signs of a Brahman are these, truth, penance, the command of the organs of sense, and mercy; as those of a *Chandála* are the vices opposed to those virtues. Another mark of the Brahman is a scrupulous abstinence from sexual commerce, whether he be born a god, or a man, or a beast. Yet further, Sukra Achárya has said, that the gods take no heed of caste, but deem him to be the Brahman who is a good man, although he belong to the vilest class. From all which I infer, that birth, and life, and body, and wisdom, and observance of religious rites *(áchára),* and acts *(karma)* are all of no avail towards becoming a Brahman.

Then again, that opinion of your sect, that *Pravrajyá* is prohibited to the Súdra; and that for him service and obedience paid to Brahmans are instead of *pravrajyá,* —*because,* forsooth, in speaking of the four castes, the Súdra is mentioned last, and is therefore the vilest,—is absurd; for if were correct, Indra would be made out to be the lowest and meanest of beings, Indra being mentioned in the *Páni Sútra* after the dog, thus—"*Shva, Yuva, Maghava.*" In truth, the order in which they are mentioned or written, cannot affect the relative rank and dignity of the beings spoken of.

What! is Parvati greater than Mahesa? or are the teeth superior in dignity to the lips, because we find the latter postponed to the former, for the mere sake of euphony, in some grammar sentence? Are the teeth older than the lips; or does your creed teach you to postpone Siva to his spouse? No; nor any more is it true that the Súdra is vile, and the Brahman high and mighty, because we are used to repeat the *Chatur Varna* in a particular order. And if this proposition be untenable, your deduction from it, *viz.,* that the vile Súdra must be content to regard his service and obedience to Brahmans as his only *pravrajyá,* falls likewise to the ground.

Know further, that it is written in the *Dharma Sástra* of Manu, that the Brahman who has drank the milk of a *Súdrání,* or has been even breathed upon by a *Súdrání,* or has been born of such a female, is not restored to his rank by *práyaschitta.* In the same work it is further asserted, that if any Brahman eat and drink from the hands of a *Súdrání,* he becomes in life a Súdra, and after death a dog. Manu further says, that a Brahman who associates with female Súdras, or keeps a Súdra concubine, shall be rejected by gods and ancestors, and after death shall go to hell. From all these assertions of the *Mánava Dharma*, it is clear that Brahmanhood is nothing indefeasibly attached to any race or breed, but is merely a quality of good men. Further, it is written in the *Sástra* of Manu, that many Súdras became Brahmans by force of their piety; for example, Kathina Muni, who was born of the sacrificial flame produced by the friction of wood, became a Brahman by dint of *Tapas;* and Vasishtha Muni, born of the courtezan Urvasí; and Vyása Muni, born of a female of the fisherman's caste; and Rishyasringa Muni, born of a doe; and Visvamitra, born of a *Chándální;* and Nárada Muni, born of a female spirit-seller; all these became Brahmans by virtue of their *Tapas.* Is it not clear then that Brahmanhood depends not on birth? It is also notorious that he who has conquered himself is a *Yati;* that he who performs penance is a *Tapasyi;* and that he who observes the *Brahma charya* is a Brahman. It is clear then that he whose life is pure, and his temper cheerful, is the true Brahman; and that lineage *(Kula)* has nothing to do with the matter. There are these *slokas* in the *Mánava Dharma*, "Goodness of disposition and purity are the best of all things; lineage is not alone deserving of respect. If the race be royal and virtue be wanting to it, it is contemptible and useless." Kathina Muni and Vyása Muni, and other sages, though born of Súdras, are famous among men as Brahmans; and many persons born in the lowest ranks have attained heaven by the practice of uniform good conduct *(síla).* To say therefore that the Brahman is of one particular race is idle and false.

Your doctrine, that the Brahman was produced from the mouth, the Kshatriya from the arms, the Vaisya from the thighs, and the Súdra from the feet, cannot be supported. Brahmans are not of one particular race. Many persons have lived who belonged to the *Kaivarta Kul*, and the *Rajaka Kul*, and the *Chandála Kul*, and yet, while they existed in this world, performed the *Chúdá Karan*, and *Munja-bandhan*, and *Dant-káshtha*, and other acts appropriated to Brahmans, and after their deaths became, and still are, famous under the name of Brahmans.

All that I have said about Brahmans you must know is equally applicable to Kshatriyas; and that the doctrine of the four castes is altogether false. All men are of one caste.

Wonderful! You affirm that all men proceeded from one, *i.e.*, Brahma; how then can there be a fourfold insuperable diversity among them? If I have four sons by one wife, the four sons, having one father and mother, must be all essentially alike. Know too that distinctions of race among beings are broadly marked by differences of conformation and organization: thus, the foot of the elephant is very different from that of the horse; that of the tiger unlike that of the deer; and so of the rest: and by that single diagnosis we learn those animals belong to very different races. But I never heard that the foot of a Kshatriya was different from that of a Brahman, or that of a Súdra. All men are formed alike, and are clearly of one race. Further, the generative organs, the colour, the figure, the ordure, the urine, the odour, and utterance, of the ox, the buffalo, the horse, the elephant, the ass, the monkey, the goat, the sheep, etc., furnish clear diagnostics whereby to separate these various races of animals: but in all those respects the Brahman resembles the Kshatriya, and is therefore of the same race or species with him. I have instanced among quadrupeds the diversities which separate diverse genera. I now proceed to give some more instances from among birds. Thus, the goose, the dove, the parrot, the peacock, etc., are known to be different by their diversities of figure, and colour, and plumage, and beak: but the Brahman, Kshatriya, Vaisya, and Súdra are alike without and within. How then can we say they are essentially distinct? Again, among trees the *Bata*, and *Bakula*, and *Palás*, and *Asoka*, and *Tamál*, and *Nagkesar*, and *Shirísh*, and *Champa*, and others, are clearly contradistinguished by their stems, and leaves, and flowers, and fruits, and barks, and timber, and seeds, and juices, and odours; but Brahmans, and Kshatriyas, and the rest, are alike in flesh, and skin, and blood, and bones, and figure, and excrements, and mode of birth. It is surely then clear that they are of one species or race.

Again, tell me, is a Brahman's sense of pleasure and pain different from that of a Kshatriya? Does not the one sustain life in the same way, and find death from the same causes as the other? Do they differ in intellectual faculties, in their actions, or the objects of those actions; in the manner of their birth, or in their subjection to fear and hope? Not a whit. It is therefore clear that they are essentially the same. In the *Udumbara* and *Panasa* trees the fruit is produced from the branches, the stem, the joints, and the roots. Is one fruit therefore different from another, so that we may call that produced from the top of the stem the Brahman fruit, and that from the roots the Súdra fruit? Surely not. Nor can men be of four distinct races, because they sprang from four different parts of one body. You say that the Brahman was produced from the mouth; whence was the Brahmani produced? From the mouth likewise? Grant it—and then you must marry the brother to the sister! a pretty business indeed! if such incest is to have place in this world of ours, all distinctions of right and wrong must be obliterated.

This consequence, flowing inevitably from your doctrine that the Brahman proceeded from the mouth, proves the falsity of that doctrine. The distinctions between Brahmans, Kshatriyas, Vaisyas, and Súdras, are founded merely on the observance of divers rites, and the practice of different professions; as it clearly proved by the conversation of Vaishampáyana, 'Whom do you call a Brahman; and what are the signs of Brahmanhood?' Vaisham answered, 'The first sign of a Brahman is, that he possesses long-suffering and the rest of the virtues, and never is guilty of violence and wrong doing; that he never eats flesh; and never hurts a sentient thing. The second sign is, that he never takes that which belongs to another without the owner's consent, even though he find it in the road. The third sign, that he masters all worldly affections and desires, and is absolutely indifferent to earthly considerations. The fourth, that whether he is born a man, or a god, or a beast, he never yields to sexual desires. The fifth, that he possesses the following five pure qualities, truth, mercy, command of the senses, universal benevolence, and penance.* Whoever possesses these five signs of Brahmanhood I acknowledge to be a Brahman; and, if he possess them not, he is a Súdra. Brahmanhood depends not on race *(Kula)*, or birth *(Játi)*, nor on the performance of certain ceremonies. If a *Chándál* is virtuous, and possesses the signs above noted, he is a Brahman. Oh! Yudhisthira, formerly in this world of ours there was but one caste. The division into four castes originated with diversity of rites and of avocations. All men were born of women in like manner. All are subject to the same physical necessities, and have the same organs and senses. But he whose conduct is uniformly good is a Brahman; and if it be otherwise he is a Súdra; aye, lower than a Súdra. The Súdra who, on the other hand, possesses these virtues is a Brahman.'

'Oh, Yudhisthira! If a Súdra be superior to the allurements of the five senses, to give him charity is a virtue that will be rewarded in heaven. Heed not his caste; but only mark his qualities. Whoever in this life ever does well, and is ever ready to benefit others, spending his days and nights in good acts, such an one is a Brahman; and whoever, relinquishing worldly ways, employs himself solely in the acquisition of *Moksha*, such an one also is a Brahman; and whoever refrains from destruction of life, and from worldly affections, and evil acts and is free from passion and backbiting, such an one also is a Brahman; and whoso possesses *Kshema*, and *Dayá*, and *Dama*, and *Dán*, and *Satya*, and *Sauchana*, and *Smriti*, and *Ghriná*, and *Vidyá*, and *Vijnán*, etc., is a Brahman. Oh, Yudhisthira! if a person perform the *Brahmacharya* for one night, the merit of it is greater than that of a thousand sacrifices *(yajna)*. And whoso has read all the *Vedas*, and performed all the *Tírthas*, and observed all the commands and prohibitions of the *Sástra*, such an one is a Brahman! and whoso has never injured a sentient thing by act, word, or thought, such a person shall instantly be absorbed (at his death) in Brahma.' Such were the words of Vaishampáyana. Oh, my friend, my design in the above discourse is, that all ignorant Brahmans and others should acquire wisdom by studying it, and take to the right way. Let them, if they approve it, heed it; and if they approve it not, let them neglect its admonitions.

——o——

ON THE EXTREME RESEMBLANCE THAT PREVAILS BETWEEN MANY OF THE SYMBOLS OF BUDDHISM AND SAIVISM.

It is the purpose of the following paper to furnish to those who have means and inclination to follow them out, a few hints relative to the extreme resemblance that prevails between many of the symbols of Buddhism and Saivism. Having resided myself some few years in a Bauddha country, I have had ample opportunities of noting this resemblance, and a perusal of the works of Crawfurd, of Raffles, and of the Bombay Literary Society, has satisfied me that this curious similitude

* The word in the original is *Tapas*, which we are accustomed to translate "penance," and I have followed the usage, though "ascetism" would be a better word. The proud *Tapasyi*, whom the very gods regard with dread, never dreams of contrition and repentance.

is not peculiar to the country wherein I abide. I observe that my countrymen, to whom *any degree* of identity between faiths in general so opposite to each other as Saivism and Buddhism, never seems to have occurred, have in their examinations of the monuments of India and its Islands, proceeded upon the assumption of an absolute incommunity between the types of the two religions as well as between the things typified. This assumption has puzzled them not a little so often as the evidence of their eyes has forced upon them the observation of images in the closest juxta-position which their previous ideas nevertheless obliged them to sunder as far apart as Brahmanism and Buddhism!

When in the country in which I reside, I observed images the most apparently Saiva placed in the precincts of Saugata temples, I was at first inclined to consider the circumstance as an incongruity arising out of an ignorant confusion of the two creeds by the people of this country: but upon multiplying my observations such a resolution gave me no satisfaction; these images often occupied the very penetralia of Saugata temples; and in the sequel I obtained sufficient access to the conversation, and books of the Bauddhas to convince me that the cause of the difficulty lay deeper than I had supposed.* The best informed of the Bauddhas contemptuously rejected the notion of the images in question being Saiva, and in the books of their own faith they pointed out the Bauddha legends justifying and explaining their use of such, to me, doubtful symbols. Besides, my access to the European works of which I have already spoken exhibited to me the very same apparent anomaly existing in regions the most remote from one another, and from that wherein I dwell. Indeed, whencesoever Bauddha monuments, sculptural or architectural, had been drawn by European curiosity the same dubious symbols were exhibited; nor could my curiosity be at all appeased by the assumption which I found employed to explain them. I shewed these monuments to a well informed old Bauddha, and asked him what he thought of them, particularly of the famous Tri-Múrti image of the Cave temple of the West. He recognised it as a genuine Bauddha image! As he did many many others declared by our writers to be Saiva! Of these matters you may perchance hear hereafter, suffice it at present to say that I continued to interrogate my friend as to whether he had ever visited the plains of India, and had there found any remains of his faith. Yes, was the prompt reply, I made a pilgrimage to Gayah, in my youth: I then asked him if he remembered what he had seen, and could tell me. He replied that he had, at the time, put a few remarks on paper which he had preserved, and would give me a copy of, if I desired it. I bade him do so, and was presented with a paper of which the enclosed is a translation. Let me add that never having visited Gayah, I cannot say anything relative to the accuracy of my friend's details, and that in regard to the topographical ones, there are probably a few slight mistakes. I am aware that an accurate explanation

* Causes are not at present my game: but consider the easy temper of superstition: the common origin of Buddhism and Brahmanism in India; the common tendency of both Saivaism and Buddhism to asceticism, etc. Even Christianity adopted many of the rites and emblems of classic paganism.

from the Bauddha books of the drawings that accompany my paper, would be of more value than that paper. But, Sir, *non omnia possumus omnes*, and I hope that a Bauddha comment on Brahmanical ignorance will be found to possess some value, as a curiosity; and some utility, for the hints it furnishes relative to the topic adverted to in this letter.

P.S.—Captain Dangerfield's five images in the cave at Bág, and which the Brahmans told him were the five Pándús, are doubtless the "Pancha Buddha Dhyáni;" as is the Captain's "Charan," said to be that of Vishnu, the Charan of Sákya Sinha; or that of Manju Ghosha. If it be the latter, it has *an eye* engraved in the *centre of each foot;* if the former, it has the *ashtmangal* and *sahasra chakra.*

——o——

Buddh Gayah, according to a Nepaulese Bauddha who visited it.

In Buddh Gayah there is a temple* of Mahá Buddha in the interior of which is enshrined the image of Sákya Sinha: before the image is a Chaitya of stone, close to which are the images of three Lokeswaras, *viz.*, Halá halá Lokeswara, Hari hari váhana Lokeswara, and Amogha pása Lokeswara.† This temple of Mahá Buddha, the Brahmans call the temple of Jagat Nátha, and the image of Sákya Sinha they denominate Mahá Muni;‡ of the three Lok Náths, one they call Mahá Deva, one Párvatí, and the third their son. On the south side of the temple of Mahá Buddha is a small stone temple in which are the images of the seven Buddhas:§ and near to them on the left three other images, of Halá halá Lokeswara, Maitreya Bodhisatwa, and Dípankara Buddha. The Brahmans call six of the seven Buddhas, the Pándús and their bride, but know not what to make of the seventh Buddha, or of the remaining three images.

* The word in the original is Kútágár, and I understand that the temple of Mahá Buddha in the city of Patan, in this valley, is built after the model of the Gayah temple. If so, the latter is of the same general form with the Orissan Jagannath. The Patan temple is divided in the interior into five stories. Sákya Sinha, the genius *loci*, is enshrined in the centre of the first story; Amitábha, the fourth Dhyáni Buddha, occupies the second story; a small stone Chaitya, the third; the Dharma Dhátu *mandal*, the fourth; and the Vajra Dhátu *mandal*, the fifth and highest story, and the whole structure is crowned, on the outside, by a Chúrá Mani Chaitya.

† Halá halá Lokeswara, a form of Padma Páni, the fourth Dhyáni Bodhisatwa, and active creator and governor of the *present* system of nature. Three Dhyáni Bodhisatwas preceded him in that office, and one remains to follow him.

‡ This name is equivocal: the Brahmans mean I suppose, to designate by it the chief of their own Munis. The Bauddhas recognise it as just, since the Tri-Kánd Sesh, and many of their scriptures give this name to Sákya Sinha.

§ The Bauddha scriptures say that one form is common to all the seven great Mánushi Buddhas. The figure I have given of Sákya has the Bhúmisparsa Mudrá, or right hand touching the earth. The Gayah image of him is said to have the Dhyán Mudrá for the position of the hands. That is, the two hands open and laid one on the other and both resting on the doubled thighs, the figure sitting tailor-wise. There is nothing improper in giving that Mudrá to Sákya or other Mánushi Buddhas, but *usually* it is appropriated to Amitábha; and almost all the images of Sákya that I have seen are characterised by the Bhúmi-sparsa Mudrá. Sákya's image is generally supported by lions, sometimes however by elephants, Sákya's appropriate *colour* is yellow or golden, which colour, like the other characteristics, belongs also to the remaining six great Mánushis.

Upon the wall of the small temple containing the Sapta Buddha, and immediately above their images is an image of Vajra Satwa,† one head, two hands, in the right hand a Vajra, and in the left a bell, with the lock on the crown of the head, twisted into a turban: the Brahmans call this image of Vajra Satwa Mahá Brahmá. At the distance of fifteen yards, perhaps, east of the great temple of Mahá Buddha is another small temple in which is placed a circular slab having the print of the feet of Sákya Sinha graven on it. The feet are known to be those of Sákya, because the stone has the eight mangals,§ and the thousand-fold chakra upon it. The Brahmans of Gayah call this Charan, the Charan of Vishnu, but they are silent when the mangals and chakras are pointed out to them as decisive proofs of their error.

Somewhat further (perhaps 150 yards) from the great temple of Mahá Buddha towards the east, is a Kund called Páni Hata, and at the eastern corner of the well is the image of Maitreya Bodhisatwa.

The Kund is called Páni Hata because Sákya produced the spring of water by striking his hand on the ground there. That water has eight peculiar qualities. The Brahmans say that the Kund is Saraswatí's, and insist that Maitreya's image is the image of Saraswatí. At a little distance to the north of the great Mahá Buddha temple are many small Chaityas,‖ which the Brahmans call Siva Lingas, and as such worship them, having broken off the Chúrá Mani from each.* Much astonished was I to find the great temple of *my* religion consecrated to Brahman worship, and Brahmans ignorantly falling down before the Gods of *my* fathers.

——o——

The purpose of my paper is to show that very many symbols, the most apparently Saiva, are notwithstanding strictly and purely Bauddha; and that, therefore, in the examination of the antiquities of India and its islands, we need not vex ourselves, because on the sites of old Saugata temples we find the very genius *loci* arrayed with many of the apparent attributes of a Saiva God; far less need we infer from the presence, on *such* sites, of seemingly Saiva images and types, the presence of actual Sivaism.

† Vajra Satwa is a Dhyáni or celestial Buddha. There is a series of five celestial Buddhas, to whom are assigned the five elements of matter, the five organs of human sense, and the five respective objects of sensation. There is also a series of six Dhyáni Buddhas, which is composed of the above five, with the addition of Vajra Satwa, and to him are ascribed intellectual force and the discrimination of good and evil.

§ These are symbols of the Vítarágas, which are portions of the eight Bodhisatwas. See Naipálya Kalyána, in *Jour. Ben. As. Society.*

‖ The Chaitya is the only proper temple of Buddhism, though many other temples have been adopted by the Saugatas for enshrining their Dii Minores. In Nepaul, the Chaitya is exclusively appropriated to five Dhyáni Buddhas, whose images are placed in niches around the base of the solid hemisphere which forms the most essential part of the Chaitya. Almost every Nepaul Chaitya has its hemisphère surmounted by a cone or pyramid called Chúrá Mani. The small and unadorned Chaitya might easily be taken for a Linga. It was so mistaken by Mr. Crawfurd, etc.

* The like metamorphosis of the Chaitya into a Lingam and its worship as the latter, may now be seen in numerous instances in Nepaul, *e.g.*, at Kali's temple on the road side near Tundi Khél.

Crawfurd, standing in the midst of hundreds of images of Buddhas, on the platform of a temple, the general form and structure of which irresistibly demonstrated that it was consecrated to Saugatism, could yet allow certain *appearances* of Sivaism to conduct him to the conclusion, that the presiding Deity of the place was Hara himself! Nay, further, though he was persuaded that the ancient religion of the Javanese was Buddhism, yet having always found what he conceived to be the unequivocal indices of the presidency of the Hindu destroyer, in all the great Saugata temples, he came to the *general* conclusion, that "genuine Buddhism" is no other than Sivaism. I thought when I had shewn no reliance could be placed upon the inference from seemingly Saiva symbols to actual Sivaism, I had smoothed the way for the admission that those cave-temples of the west of India, as well as those fine edifices at Java, whereat the majority of indications both for number and weight prove Buddhism, are *Bauddha* and *exclusively Bauddha;* notwithstanding the presence of symbols and images occupying the post of honour, which, strongly to the eye, but in fact, erroneously in these cases, seem to imply Sivaism, or at least a coalition of the two faiths. For such a coalition at any time and in any place, I have not seen one plausible argument adduced; and as for the one ordinarily derived from the existence of supposed Saiva images and emblems in and around Bauddha temples, it is both erroneous in fact, and insufficient were it true. However probably *borrowed* from Sivaism, these images and symbols became genuinely Bauddha by their adoption into Buddhism—just as the statue of a Capitoline Jupiter became the very orthodox effigy of St. Paul, because the Romanists chose to adopt the Pagan idol in an orthodox sense. And were this explanation of the existence of *seeming* Sivaism in sites which were beyond doubt consecrated to Buddhism, far less satisfactory than it is, I would still say it is a thousand times more reasonable than the supposition of an identity or coalition* between two creeds, the speculative tenets of which are wide asunder as heaven and earth, and the followers of which are pretty well known to have been, so soon as Buddhism became important, furiously opposed to each other.

Upon the whole, therefore, I deem it certain, as well that the types of Sivaism and Buddhism are very frequently the same, *as that the things typified are, always more or less, and generally radically, different.*

Of the aptness of our writers to infer Sivaism from apparently Saiva images and emblems, I shall adduce a few striking instances from Crawfurd's second volume, chap. i., on the ancient religion of the Islanders; and to save time and avoid odium, I speak rather to his engravings, than to his text; and shall merely state matters, without arguing them.

Let me add, too, that Crawfurd's mistakes could not well have been avoided. He had no access to the dead or living oracles of Buddhism, and reasoning only

* In regard to those cave-temples of the Western Continent of India, called mixed Saiva and Bauddha, the best suggested solution is *successive possession*—but I believe them to have been wholly Buddhist.

from what he saw, reasonably inferred that images, the most apparently Saiva, were really what they seemed to be; and that Saiva images and emblems proved a Saiva place of worship.

In his chapter already alluded to, there are several engravings. No. 27 is said to be "a figure of Mahá Deva as a devotee." It is, in fact, Sinha-Nátha-Lokeswara. Plate 28 is called "a representation of Siva." It is, in fact, Lokeswara Bhagawán or Padma Páni,§ in his character of creator and ruler of the *present* system of nature. How Mr. Crawfurd could take it for Siva, I do not know, since in the forehead is placed a tiny image of Amitábha Buddha, whose son Padma Páni is feigned by the Bauddha mythologists, to be. Again, the principal personage in plate 21 is said to be "Siva in his car." It is, in truth, Namuchi Mára, (the Bauddha personification of the evil principle,) proceeding to interrupt the Dhyán of Sákya Sinha; and plate 22 gives a continuation of this exploit, exhibiting Sákya meditating, and the frustration of Namuchi's attempt by the opposition of force to force.|| The whole legend is to be found in the Sambhu purána.

The same work contains likewise the elucidation of plate 24, of which Mr. C. could make nothing.

Of the remaining plates, and of the text of this chapter of Mr. C.'s, on other subjects, very able work, it would be easy, but it would to me be wearisome, to furnish the true explanation from the books or oral communications of the Bauddhas of Nepaul, to the more learned of whom the subjects of the plates in Mr. C's book are perfectly familiar. One quotation from Mr. C.'s *text*, and I have done. At p. 209, vol. ii., he observes: "The fact most worthy of attention, in respect to the images of Buddha, is that they never appear *in* any of the great central temples as the primary objects of worship, but in the smaller surrounding ones, seeming themselves to represent votaries. They are not found as single images, but always in numbers together,* seeming, in a word, to represent, not Deities themselves, but sages worshipping Siva."

The whole secret of this marvel is, that the temples seen by Mr. C. were not genuine Chaityas, but either composite Chaityas, or structures still less exclusively appropriated to the Dii majores of Buddhism. The genuine Chaitya is a *solid* structure exclusively appropriated to the Dhyáni Buddhas, whose images are placed in niches round the base of its hemisphere. Mánushi Buddhas and Dhyáni and Mánushi Bodhisatwas and Lokeswaras, with their Saktis, are placed *in* and *around* various *hollow* temples, less sacred than the Chaityas.‡ These Bodhisatwas and

§ At Kurnagush (the ruins near Bhagulpur) there is a fine and perfect image of Padma Páni, with Amitábha in the forehead. The Pujári to me called it a Krishna, and was astounded when he heard my explanation and whence derived.

|| See *Jour. Amer. Ori. Soc.*, vol. ii., part ii, pp. 31-35, for another version of this story.

* And why not? for Buddha is a mere title: and though there are but six Dhyáni Buddhas, there are hundreds of Mánushis, which latter are constantly placed about temples in vast numbers; always as objects, though not, when so placed, special ones, of worship.

Lokeswaras never have the peculiar hair of the Buddhas, but, instead thereof, long-braided locks like Siva; often also the sacred thread and other indications apt to be set down as proofs, "strong as holy writ," of their being Brahmanical Deities. Such indications, however, are delusive, and the instances of plates 27 and 28, shew how Mr. C. was misled by them.

By the way, Mr. C. is biassed by his theory to discover Sivaism, where it did not and could not exist, of which propensity we have an odd instance (unless it be an oversight or misprint) in p. 219: for no one needs be told that Hari is Vishnu, not Siva,§ and I may add that in adopting as Dii minores the Gods of the Hindoo Pantheon, the Bauddhas have not, by any means, entirely confined themselves to the Sectarian Deities of the Saivas.

——o——

P.S.—A *theistic* sect of Buddhas having been announced as discovered in Nepaul, it is presently inferred that this is a local peculiarity. Let us not be in too great haste: Mr. Crawfurd's book *(loco citato)* affords a very fine engraving of an image of Akshobhya, the first Dhyáni, or Celestial Buddha, (see plate 29,) and I have remarked generally that our engravings of Bauddha architecture and sculpture, drawn from the Indian cave temples, from Java, etc., conform, in the minutest particulars, to the existing Saugata monuments of Nepaul—which monuments prove here, (as at Java,) the *Foreign* and *Indian origin of Buddhism*, animals, implements, vehicles, dresses, being alien to Nepaul, and proper to India.

THE PRAVRAJYA VRATA OR INITIATORY RITES OF THE BUDDHISTS ACCORDING TO THE PUJA KAND.

If any one desires to become a Bandya (monastic or proper Buddhist) he must give notice thereof, not more than a month or less than four days, to his Guru, to whom he must present *paun*, *supári*, *dakshiná*, and *akshat*, requesting the Guru to give him the Pravrajya Vrata. The Guru, if he assent, must accept the offerings and perform the *Kalasí pújá*, which is as follows: The Guru takes a *kalasí* or vessel full of water and puts into it a lotos made of gold or other precious metal, and five confections, and five flowers, and five trees (small branches), and five drugs, and five fragrant things, and five Bríhi, and five Amrita, and five Ratna, and five threads of as many diverse colours. Above the vessel he places rice, and then makes *pújá* to it. He next seats the aspirant before the vessel in the *Vajra ásan* fashion and draws on the ground before the aspirant four *mandals* or circular diagrams, three of which are devoted to the Tri Ratna, and the fourth to the officiating Guru. Then the aspirant repeats the following text: 'I salute Buddhanáth, Dharma, and Sangha, and entreat them to bestow the *Pravrajya Vrata* on me, wherefore I perform this rite to them and to my Guru, and present

‡ As for example, Sákya Sinha in the great temple of Gya, which is a Kútágár, and wherein Sákya appears as the genius *loci*.

§ See also pp. 221-2, for a singular error into which apparently Mr. C's pursuit of his theory could alone have led him. Flowers not offered by Hindoos to their Gods, and *therefore* Buddha was a sage merely, and not a God!!

these offerings.' Reciting this text and holding five *supáris* in each hand, the aspirant, with joined hands, begs the Guru to make him a Bandya. The offerings above mentioned he gives to the Guru, and *dakshiná* proportioned to his means. This ceremony is called *Gwál Dán.* On the next day the ceremony above related is repeated, with the under-mentioned variations only. As in the *Gwál Dán* the *Kalasí pújá* and *Deva pújá* are performed, so here again: but the aspirant on the former occasion is seated in the *Vajra ásan* manner, in this day's ceremony in the *Sustaka ásan.* The *Sustaka ásan* is thus: first of all *kús* is spread on the ground, and above it, two unbaked bricks, and above them the *Sustaka* is inscribed, thus –

upon which the aspirant is seated. Then the aspirant is made *Niranjana*, that is, a light is kindled and shown to him, and some *mantras* repeated to him. Then the *Vajra Rákshá* is performed, that is, upon the aspirant's head a *Vajra* is placed and the Guru reads some *mantras.* Next comes the ceremony of the Loha Rakshá, that is, the Guru takes three iron padlocks, and places one on the belly and the two others on the shoulders of the neophyte, repeating some more *mantras,* the purport of which is an invocation of divine protection from ill, on the head of the aspirant. This rite is followed by the *Agni Rakshá,* that is, the Guru puts a cup of wine *(surá-pátra)* on the head of the Chela and utters some prayers over him.

Next is performed the *Kalasí-Abhisheka* or baptism; that is, holy water from the *Kalasí* is sprinkled by the Guru on the Chela's head and prayers repeated over him; after which, the *Náyaka Bandya* or head of the Vihár (Abbot or Prior,) comes and puts a silver ring on the finger of the aspirant. The Náyaka, or superior aforesaid, then takes four seers of rice and milk mixed with flowers, and sprinkles the whole, at three times, on the aspirant's head. Next the Náyaka performs the *Vajra Rakshá,* and then makes *pújá* to the Guru Mandal before mentioned, which ceremony completed, he rings a bell, and then sprinkles rice on the aspirant and on the images of the Gods.

Then the aspirant, rising, pays his devotions to his Guru, and having presented a small present and a plate of rice to him, and having received his blessing, departs. This second day's ceremony is called Dùsala.

The third day's is denominated Pravra Vrata,* and is as follows:—

Early in the morning the following things, *viz.*, the image of a Chaitya, those of the Tri Ratna or Triad, the Prajná Páramitá scripture, and other sacred scriptures, a kalas, or water-pot filled with the articles before enumerated, a platter of curds, four other water-pots filled with water only, a Chivara and Newás, a

* The monastic vows properly so called.

Pinda pátra and a Khikshari, a pair of wooden sandals,‡ a small mixed metal plate spread over with pounded sandal wood, in which the image of the moon is inscribed, a golden razor and a silver one, and lastly, a plate of dressed rice, are collected, and the aspirant is seated in the Sústak Asan and made to perform worship to the Guru Mandala, and the Chaitya, and the Tri Ratna and the Prajná Páramitá Sástra. Then the aspirant, kneeling with one knee on the ground with joined hands, entreats the Guru to make him a Bandya, and to teach him whatsoever it is needful for him to know. The Guru answers, 'O! disciple! if you desire to perform the Pravrajjya Vrata, first of all devote yourself to the worship of the Chaitya and of the Tri Ratna; you must observe the five precepts or Pancha Sikshá, the fastings and the vows prescribed; you must hurt no living thing; nor amass property of any kind; nor go near women; nor speak or think evilly; nor touch any intoxicating liquors or drugs; nor be proud of heart in consequence of your observance of your religious and moral duties.'

Then the aspirant pledges himself thrice to observe the whole of the above precepts; upon which the Guru tells him, 'If while you live you will keep the above rules, then will I make you a Bandya.' He assents, when the Guru, having again given the three Rakshás above mentioned to the Chela, delivers a cloth for the loins to him to put on. Then the Guru brings the aspirant out into the court yard, and having seated him, touches his hair with rice and oil, and gives those articles to a barber. The Guru next puts on the ground a little pulse and desires the Chela to apply it to his own feet. Then the Guru gives the Chela a cloth of four fingers' breadth and one cubit in length, woven with threads of five colours, and which is especially manufactured for this purpose, to bind round his head. Then he causes the aspirant to perform his ablutions; after which he makes *pújá* to the hands of the barber in the name of Viswakarma, and then causes the barber to shave all the hair, save the forelock, off the aspirant's head. Then the paternal or maternal aunt of the aspirant takes the vessel of mixed metal above noted and collects the hair into it. The aspirant is now bathed again and his nails pared; when the above party puts the parings into the pot with the hair. Another ablution of the aspirant follows, after which the aspirant is taken again within, and seated. Then the Guru causes him to eat, and also sprinkles upon him the Pancha Garbha, and says to him, 'Heretofore you have lived a householder; have you a real desire to abandon that state and assume the state of a monk? The aspirant answers in the affirmative, when the Guru or Náyaka,* or maternal uncle, cuts off with his own hand, the aspirant's forelock. Then the Guru puts a *tiara* adorned with the images of the five Buddhas on his own head, and taking

‡ These, with the water-pot or Gahdhar and an umbrella constitute the equipments of a Bauddha ascetic. The *chívar* and *nivás* are the upper and lower garments. The *pinda pátra* is the begging platter: *khikshari*, the appropriate baton or distinctive staff (carried in the hand and surmounted by a model of a Chaitya). The *Mani* or prayer-cylinder, which is so universally in the hands of the Tibetan monks, is not in use in Nepaul. The *chívar* and *nivás* are of a deep red color.

* Náyaka is Abbot, that is, head of the Religious House into which the neophyte purposes to enter.

the kalas or water-pot, sprinkles the aspirant with holy water, repeating prayers at the same time over him.

The neophyte is then again brought below, when four Náyakas or superiors of proximate Viháras and the aspirant's Guru perform the Pancha Abhisheka, *i.e.*, the Guru takes water from the kalas and pours it into a conch; and then, ringing a bell and repeating prayers, sprinkles the water from the conch on the aspirant's head; whilst the four Náyakas, taking water from the other four water-pots named above, severally baptize the aspirant. The musicians present then strike up, when the Náyakas and Guru invoke the following blessing on the neophyte: 'May you be happy as he who dwells in the hearts of all, who is the universal Atman, the lord of all, the Buddha called Ratna Sambhava.' The aspirant is next led by the Náyakas and Guru above stairs, and seated as before. He is then made to perform *pújá* to the Guru Mandal and to sprinkle rice on the images of the Deities. The Guru next gives him the Chívara, and Nivása, and golden earrings, when the aspirant thrice says to the Guru, 'O Guru, I, who am such an one, have abandoned the state of a householder for this whole birth, and have become a monk.' Upon which the aspirant's former name is relinquished and a new one given him, such as Ananda Sháli Putra, Kásyapa, Dharma Srí Mitra, Páramitá Ságar. Then the Guru causes him to perform *pújá* to the Tri Ratna, after having given him a golden *tiara*, and repeated some prayers over him. The Guru then repeats the following praises of the Tri Ratna: 'I salute that Buddha who is the lord of the three worlds, whom Gods and men alike worship, who is apart from the world, long-suffering, profound as the ocean, the quintessence of all good, the Dharma Rája and Muníndra, the destroyer of desire and affection, and vice and darkness; who is void of avarice and lust, who is the ikon of wisdom. I ever invoke him, placing my head on his feet.'

'I salute that Dharma, who is the Prajná Páramitá, pointing out the way of perfect tranquillity to all mortals, leading them into the paths of perfect wisdom; who, by the testimony of all the sages, produced or created all things; who is the mother of all the Bodhisatwas and Srávakas. I salute that Sangha, who is Avalokiteswara and Maitreya, and Gagan Ganja, and Samanta Bhadra, and Vajra Páni, and Manju Ghósha, and Sarvani varana Viskambhi, and Kshiti Garbha and Kha Garbha.'† The aspirant then says to the Guru, 'I will devote my whole life to the Tri Ratna, nor ever desert them.' Then the Guru gives him the Das Sikshá or ten precepts observed by all the Buddhas and Bhikshukas; and commands his observance of them. They are: 1. Thou shalt not destroy life; 2. Thou shalt not steal; 3. Thou shalt not follow strange faiths; 4. Thou shalt not lie;

† These are nine Bodhisatwas, whereof the first, or Padma Páni, is now lord of the ascendant, and as such constitutes the Sangha of the present cycle, and is therefore associated to Buddha and Dharma of the triad as the third member of it. But there is confusion of celestial and mortal Bodhisatwas, and so also in the general enumeration. (See and compare pp. 95 and 96.) The Padma Páni here spoken of is probably Avalokiteswara, who seems to be the same with Matsyendra Náth—a mortal clearly, and therefore improperly identified with Padma Páni, a celestial. Of the rest all but four or five are mortal Sanghas.

5. Thou shalt not touch intoxicating liquors or drugs; 6. Thou shalt not be proud of heart; 7. Thou shalt avoid music, dancing, and all such idle toys; 8. Thou shalt not dress in fine clothes, nor use perfumes or ornaments; 9. Thou shalt sit and sleep in lowly places; 10. Thou shalt not eat out of the prescribed hours.

The Guru then says, 'All these things the Buddhas avoided. You are now become a Bhikshu and you must avoid them too;' which said, the Guru obliterates the Tri Ratna Mandala. Next, the aspirant asks from the Guru the Chívara and Nivása, the Pinda Pátra and Khikshari and Gandhar, equipments of a Bauddha ascetic: they are an upper and lower garb of special form, a begging platter, a short staff surmounted by a Chaitya and a waterpot. Add thereto an umbrella and sandals to complete it. The aspirant proceeds to make a Mandala and places in it five flowers, and five Druba-Kund, and some *khil*, and some rice, and assuming the Utkútak Asan, and joining his hands, he repeats the praises of the Tri Ratna above cited, and then again requests his Guru to give him three suits of the Chívara and the like number of the Nivása—one for occasions of ceremony, as attending the palace, another for wearing at meals, and the third for ordinary wear. He also requests from his Guru the like number of Gandhár or drinking cups, of Pinda Pátra, and of Khikshari. One entire suit of these the aspirant then assumes, receiving them from the hands of the Guru, who, previously to giving them, consecrates them by prayers. The aspirant then says, 'Now I have received the Pravrajya Vrata, I will religiously observe the *Síla-skandha* the *Samádhi-skandha*, the *Prajná-skandha*, and the *Vimuktiskandha*.'

Then the Guru gives him four sprinklings of holy water and presents him with an umbrella having thirty-two radii. Next he sprinkles him once again and gives him a pair of wooden sandals—after which the Guru draws on the ground linearly, and near to each other, seven images of the lotos flower, upon each of which he puts a *supári*, and then commands the aspirant to traverse them, placing a foot on each as he proceeds. When the Chela has done so, the Guru placing the Pancha Rakshá Sástra on his head, sends him into the sanctum, where stands the image of Sákya Sinha, to offer to it *pán*, and *supári*, and *dakshiná*. All this the Chela does, and likewise performs the Pancha Upachárya pújá; when, having circumambulated the image, he returns to the Guru.

Then the Guru performs the ceremony called Shik Adhivásan, which is thus: The ball of five-coloured thread mentioned in the first day's proceedings as being deposited in the kalas, is taken out of the kalas and one end of it twisted thrice round the neck of the kalas; it is then unrolled and carried on to the Chela and twined in like manner round the Khikshari he holds in his hands, whence it is continued unbroken to the Guru and delivered into his hands. The Guru holding the clue in his hands, repeats prayers and then rolls up the thread and then redeposits it in the kalas. He next performs the Pancha Upachárya pújá to the kalas and to the Khikshari; next he gives flowers and a blessing to the aspirant; next he gives him the Abhisheka, invests his neck with a cord composed of a piece of the thread just adverted to; places the Pancha Rakshá Sástra on his head, and repeats over him some prayers. The Mandal is then obliterated, when the aspirant is made to perform the Mahá Bali ceremony, which is thus:—

In a large earthen vessel four seers of dressed rice, and a quarter of the quantity of Bhatmas, and a noose and a mask faced like Bhairava,* having a small quantity of flesh in the mouth of it, are placed; and the aspirant makes *pújá* to Bhairava, presenting to the mask the Naived and a light, and pouring out water from a conch he holds in his hands so that it shall fall into the vessel. The Guru repeats *mantras*, and invoking the Devatas and Nágas, and Yakshas, and Rákshasas, and Gandharvas, and Mahoragas, and mortals, and immortals (Amánushas), and Pretas, and Pisáchas, and Dákas, and Dákinis, and Mátrika Grahas, and Apas Márgas, and all motionless and moving things, he says, 'Accept this Bali and be propitious to this aspirant, since the sacrifice has been performed according to the directions of *Vajra Satwa*.' Such is the Sarva Bhúta Bali. In like manner the Balis of Mahá Kála, and of the Graha, and of the Pancha Rakshá, and of the Graha Mátrika, and of Chand Mahá Rakshana, and of the guardians of the four quarters, and of Ekavinsati, and of Basundhará, and of the Chaitya, and of Pindi Karma, and of Amoghpása, and of Sarak Dhára, and of Tárá, and of Hevajra, and of Kurkulla, and of Vajra Krodha, and of Maríchi, and of Ushnísha, and of Háríti, are performed. Next the Balis denominated the Tyága Bali, and the Sankha Bali, are thus performed. In the conch are put flesh, and blood and spirits, which are poured as before, into the great vessel, whilst the Deities of all the six quarters are invoked with prayers. Then the Pancha Upachára pújá is made in the vessel, after which the aspirant is commanded to perform the Chakra pújá, which completed, he returns to his seat. The Chakra pújá is that which is made to all the images in the Vihára by going round to them all. The Guru then causes the aspirant to perform the Guru Mandal pújá and afterwards to sprinkle rice on all the images, which done, the aspirant gives Dakshiná to the Guru, and the Guru, in return, gives the aspirant a small quantity of rice and a trifle of money. Then the Guru causes him to perform the Des-Bali-Yátra, which is, the aspirant removes the great earthen vessel with its contents, by means of carriers, and distributes the contents in small quantities to all the shrines of Daityas, and Pisáchas, and other evil spirits throughout the city; and having distributed them, returns with the empty vessel.

Then the Guru and ten Náyakas take the aspirant to make the circuit of all the shrines in the neighbourhood and to present at each, offerings of rice, and *pán*, and *supári*, and flowers; after which they go to the Chela's home, when his relatives come out and give him four seers of rice, and then conduct the aspirant and the rest within and feed them with *khìl* or rice and milk. The Guru then returns to the Vihára, and the Chela remains at home.§ Then the aspi-

* Thus far all is conducted according to the Pauranik exoteric and purely Buddhist ritual: what follows is derived from the Tantrik esoteric, and not purely Buddhist ritual.

§ Here end the scriptural injunctions: what follows rests on customary authority only, and has reference to the fact that in Nepaul the Buddhists have long since abandoned the monastic restraints. Tonsure is the only mark of the old monastic habits still re-

rant must, at all events, practise mendicity and the other rules of his order, for four days: but if at the end of that time, he feel no serious call to the monastic profession he must go to his Guru at the Vihára and to his Upádhyáya, (the latter is his instructor in the forms of *pújá*, according to the Pújá Kánd) and addressing the Guru, must say, 'O Guru! I cannot remain an ascetic, pray take back the Chívara and other ensigns of monachism; and, having delivered me from the Srávaka Charya, teach me the Mahá Yán Charya.' The Guru replies, 'Truly, in these degenerate days to keep the Pravrajya Vrata is hard; adopt then the Mahá Yán Charya. But if you abandon the Pravrajya, still you cannot be relieved from observing the following commandments:—Not to destroy life. Not to steal. Not to commit adultery. Not to speak evilly. Not to take spirituous liquors and drugs. To be clement to all living beings. The observance of the above rules shall be a pravrajya to you, and if you obey them, you shall attain to Mukti.' The aspirant then washes the Guru's feet, and having done so, returns to his seat, when the Guru having prepared the materials of *pújá* noted in the first day's ceremonies, makes *pújá* to the Kalas, after which he makes *pújá* to the vessel, holding the aspirant's shorn locks. He then draws Mandals for the Tri Ratna and for himself, and makes the aspirant offer *pújá* to all four; when he obliterates the whole and says, 'You have abandoned the Bhikshu Charya and adopted the Mahá Yán Charya; attend to the obligations to the latter, as just explained to you.'

The badges of monachism are then taken from the aspirant by the Guru, who gives him the Pancha Rakshá as before related, and then sends him to make the Chakra pújá, which done, he causes him to perform the Guru Mandal pújá, and then to sprinkle rice on the Deities. Then the Guru Mandal is erased, the aspirant makes an offering to the Guru, and the Guru gives him his blessing. The Guru then sends the aspirant to throw into the river the hair shaven from his head and on his return makes the Agam pújá and Kumárí pújá; when the whole is concluded by a feast.*

P.S.—Since the above papers were written, I have perused Mr. Turnour's essays in the Bengal Asiatic Journal, and I fully admit (as anticipated by Mr. Prinsep) that the honours of Ceylonese literature and of the Páli language are no longer disputable. I may add in regard to the latter point, that recent research has established the following very curious fact, *viz.*, that the Sanskrit Buddhist works discovered by me in Nepaul, are now found to be copiously interspersed with passages in various Prakrits—Pali among the rest—pretty much in the manner of the Hindoo Drama wherein this mixture of less finished dialects with the Sanskrit is of common occurrence.

tained by the Nepalese Bandyas, who are now divided into Vajra Acháryas, Bhikshukas, Sákyavansikas, and Chivaha Bares.

* In the above Srávakcharya and Bhikshucharya are made equivalents, equally representing the strict rule opposed to Mahá Yán charya as the designation of the lax rule or that of the non-monastic many. This sense of the latter term is contrary to some authorities. The Triyána are elsewhere specified as Pratyeka, Srávaka and Mahá, but in another sense a scripture of the highest class or that treats of transcendental topics is called a Maháyána Sútra.

Comparative Vocabulary of the several Languages or Dialects of the Eastern Sub-Himálayas, from the Káli or Ghógra, to the Dhansri, with the written and spoken Tibetan for comparison, by B. H. Hodgson, *Esq.*

English.	*Tibetan, written.*	*Tibetan, spoken.*	*Sérpa.*	*Bhútáni or Lhópa.*	*Lepcha.*	*Limbú.*	*Kiránti.*	*Múrmi.*	*Néwár.*	*Gúrung.*	*Magar.*	*Súnwár.*
Air	r Lungma	Lhakpá	Lúngbo	Lúng	Sagmat	Sammit	Hak	Lhábá	Phai	Náng mro	Namsú	Phasé
Ant	Grogma	Thómá	Rhúnmá	Kyómá	Takphyúl	Sikchémba	Sáchakáva	Syouri	Imo	Chiji	Mhár	Ragmachi
Arrow	mDah	Dá	Dá	Dá	Chóng	Tong	Mé	Myá	Bálá	Myá	Myá	Blá
Bird	Byú	Chyá	Jhá	Bhyá	Phó	Bú	Chongwá	Námyá	Jhango. Jhá-ngo	Némyá	Gwájá. g-wá-já	Chívá. Chí-vá
Blood	Khrag	Thák	Thák	Thyak	Ví	Makhi	Hau	Ká	Hí.	Koh	Hyú	U'si
Boat	Grú	Kóá. Syén	Thú	Drú. Tú	Navar	Khombé	Náva	Dúngá	Dónga	Plava	Dúngá	Dúngá
Bone	Rúspá	Rúkó	Rúbá	Rutok	Arhet	Sayet	Saiba	Nákhú	Kwé	Núgri	Misyá ros	Rúshé
Buffalo	Mahi, S.	Máhé	Méshi, S.	Mahi	Mahí	Sáwét	Sánwá	Mahi	Mé	Mái	Bhainsa, H.	Méryé, S.
Cat	Byila S.	Simi	Bérmo	Pilli[b]	Aleu	Myóng	Myong	Táwar	Bhoa	Nawár	Súthú	Bérmó
Cow	Bá	Phá chúk	Chú ma	Gnó[c]	Bik	Bit	Pit	Mhé	Sá[g]	Myau	Nhet	Bi
Crow	Kháta	Ablak	Ká lak	O'lá	Alok	A'hwá	Káhwá	Káwá	Kó	Mlóngyá	Kág	Khad
Day	Nyin mo	Nyi'mo	Ni'mo	Nyi'm	Sakni'	Léndik	Lén	Dini'	Nhi'	Dini'	Na msin	Náthi
Dog	Khyí	U'yó	Khí	Khi'	Kazéu	Khía	Kochú	Nángi	Khichá	Nagyú	Chhyú	Kúchúng
Ear	r Na	Amchó	Am chúk	Navo	Anyór	Nekho	Nábá	Nápé	Nhai pong	Nábé	Ná kyep	Nophá
Earth	Sa	Sá	Sá	Sáh	Phat	Kham	Bákhá	Sá	Chá	Sa. Nhé	Jhá	Kha pi
Egg	s Gonga	Góng ná	Góngná	Gong do	Atí. ä-ti	Thín. thí-n	U'di'ng[l] U-di'-ng = di'-mba Denwar, etc.	Phúm	Khyén. Khé-n	Phúng	Rhú	Bá-phú. bá = fowl, bird
Elephant	g Langchhen	Lámboché	Lángbó	Lángchen	Tyánmó	Hettí, H.	Háthi	Háthi	Kísí	Háthi	Háthi	So dá
Eye	Mig	Mik	Mik	Mido	Amik	Mik	Mak	Mí	Mí khá	Mí	Mik	Mí chí
Father	Phá	Pála	A'bá	Appá	Abó	Amba	Upa. Opa	A'pá	Abú	Abo	Bai	Bávé
Fire	Mé	Mé	Mé	Mi	Mí	Mé	Mí	Mé	Mí	Mí	Mhé	Mí
Fowl	..	Pya. Chya	Jha	..	Hí-k[d]	Wá	Wá	Ná-gá	Khá[h]	Na-ka	Gwá	Pä á
Fish	Nyá	Gna	Gná	Gnyá	Gnó	Gná	Gná	Tár nyá	Nyá	Tángná	Díshé	Gnan
Flower	Metog	Méntok	Mendok	Mentog	Ríp	Phúng	Búngwai	Méndú	Swong	Táh	Sár	Phú
Foot	r Kangpa	Kángó	Kángó	Kánglep	Diángliok	Lángdáphé[f]	U'khúro	Baló	Páli'	Bhaló	Mihil	Khwét
Goat	Rá	Rá	Rá	Ráh	Saar	Ménda	Chhengár	Rá	Choló	Rá	Rhá	Chársyé
Hair	s Krá. sPú	Tá. Krá	Tá	Ryá	Achom. A-chóm	Thágí	Moá	Krá	Song	Moí	Chham	Chúng
Hand	Lagpa	Lángó	Lángó	Láppa	Kaliók	Húktáphé[f]	Chúkúphémá[f]	Yá	Pá láhá[g]	Laptá	Hút piak	Tablé. Gwí
Head	mGo	Gó	Gó	Gútoh	Áthiak. A-thiák	Thagék	Táng	Thóbó	Chhon	Krá	Mí tálú	Píyá
Hog	Phag	Phak-pá[j]	Phak	Phagpo[j]	Mon	Phag	Bhag	Dhwá	Phá	Tílí	Wak	Pó
Horn	Rá	Rajo	Arkyok	Rou	Aróng. A-ron'g	Táng	Usángá	Rhú	Nékú	Rú	Mírháng	Gúró
Horse	r Tá	Tá	Tá	Táh	O'n[k]	O'n[k]	Ghora, H.	Tá	Sala	Ghora, H.[j]	Ghora	Ghora
House	Khyim	Náng	Khángbá	Khyim	Lí	Him	Khim	Dhim	Chhén	Tín	Yúm	Khí
Iron	l Chags	Chhyá	Chhyá	Chyá	Panjing	Phenjé	Phalám	Phái	Na'	Pái	Phalám	Wá ak'li
Leaf	Lómá	Hyómá	Hyómáp	Syóma. Dáma	Lop	Pellá	U'báva	Lápté	Lapté. Hau	Lau	Lhá	Saphá
Light	Hod	Hwé. Eu	Rhip. Eu	Eu. Dam	A'óm. Achúr	O't. Thorú	U'láwa chámi	U'jálo	Jala	Bhla	Tyáwongcho. Ráp	Hángo
Man	Mi	Mi	Mi	Mi	Maro. Tagri[e]	Yapmi. Yembecha	Mana. Dúwachha	Mí	Mano.[i] Mi jang	Mhi	Bharmi	Múrú
Monkey	s Prébú	Tyú	Rhú	Pyá	Sahéu	Sobá	Héláwá	Máng	Máko	Timyú	Báner, P.	Moro
Moon	z Láva	Dáwá	Oulá	Dau	Lavó	Láva	Lá dima	Lhá ni	Twó mila. Túyú mila	Lau-gni	Gyá hút	Lá to sí
Mother	Amá	A'má	A'má	A'i'	Amo. A-mó	Amma	Oma. Uma	Amma	Mang	A'mo	Má	Amai
Mountain	Ri	Rí	Rí	Róng	Rok	Toksóng	Bhar	Gang	Gún	Kwón	Dándá, P.	Dándá
Mouth	Khá	Khá	Khá	Khá	Abóng	Múrá	Dóh	Súng	Mhútú	Súng	Gner	So
Moschito	Srinbú. mChurings	Syé-dongma	Dángma	Zéndóng	Mang kóng	Lámjonkhi	Lámkhútia. P.	Lám khútia. P.	Patí	Chwé	Lám khútia, P.	Lám khútia. P.
Name	Ming	Ming	Min	Ming	Abriáng	Ming	Núng	Min	Náng	Ming	Ming	Né
Night	mTshanmo[a]	Chénmó	Chémó	Phírú. Nammo	Sana'p	Kúsén. Sendik	Khákwé	Món	Chá	Mhóis	Námbik	Nádo
Oil	hBrumar	Núm	Núm	Mákhú	Nam	Ningé	A'wá	Chi gú	Chikang	Chúgú	Sidí	Gyo

[a] *Tshan, Chén, Ché, San, Sén, Nam, Nap,* the radical words of the six first columns. In the ninth we have *Chá*, void of suffix. The others have the suffixes or formative particles.

[b] *Pho* and *mo*, as post-fixes, for mas. et fœm.

[c] *Bhá* the cow; *Láng*, the bull. *Gno* = *Bos*, both sexes.

[d] *Hik bong*, hen; *Hik beu*, cock; *Hik len*, chick.

[e] *Tagri*, mas. *Tayú*, fœm.

[f] *Chukuphema* = arm flat; so also *Huktaphe* of Limbu, and *Langdaphe* for foot is leg flat. There are no proper words for hand or foot. The words for arm and leg are used with the sign of flat things (*phe*) suffixed. So also in Limbu *Huktaphe*, foot; and *Langtaphe*, hand.

[g] *Ba* and *Ma*, used prefixually, are the sex signs, and unchangeable in Néwári. Here *Ba-sa* equals bull, *Ma-sa* cow. The occasional omission or insertion of these signs, and, still more, of the generic or segregative signs, as in *Pá-lá-ha* = hand, from the root *lá*, with *pá*, the mark of flat things, and *há* the sign of long things, create many differences more apparent than real, since the use or neglect of these additions is to a great extent optional, *e.g.*, *pali* 'foot' has the *pá* sign only; *lilá* 'hands and feet' has neither.

[h] *Khá* equals fowl in Takpa and Uraon, *Kai* in Chinese.

[i] *Mi-jang*, vir.; *Mi-sá*, mulier; *Mano*, like *Maro*, mankind; and so *Yapmi* in Limbu, whilst *Yem bi cha*, *Menchima*, are man and woman in that tongue.

[j] *Pa*, *Ma*, merely sexual adjuncts, mas. et fœm., identical with the *Pho mo* and *Po mo* elsewhere occurring.

[k] Compare *On-hya*, the Dhimali word.

[l] *U* is the pronominal definitive, as in *Upa, Uma, Ukhura, Uhokwa*, etc., of the sequel.

*** H, Hindi.—P, Parbatya.—S, Sanskrit.

English.	Tibetan, written.	Tibetan, spoken.	Sérpa.	Bhútáni or Lhópa.	Lepcha.	Limbú.	Kiránti.	Múrmi.	Néwár.	Gúrung.	Magar.	Súnwár.
Plantain	căret	Gnálá	Lánga sí	Gnálá	Kardúng	Lá seh'	Gnák si	Moché	Mwai	Kala	Mocha	Múhji
River	*g* Tsang po	Chángpó	Hyúng	Ohhu kyong	Ongkyong	Chua. Wo-hong	Hong-kú	Syóng	Khúsi	Khwóng	Kholá	Lí kú
Road	Lam	Lam	Lam	Lám	Laum	Lam	Lam	Ghyám	Lón	Kyán	Lam	Lá
Salt	Tshá	Chhá	Chhá	Chhá	Vom	Yúm	Yúm	Chåchá	Chhi	Chåchá	Ohá	Yú sí
Skin	Pagspa	Pág-pa	Kop-pa	Páko. Kompo	Athún. Kómbó	Horik. Saho	U'hók bá	Dí bhí	Syú. Chégú	Dhí	Chåla	Kúsyúl
Sky	Nam khah	Nam	Nam	Nam	Ta liáng	Témsákpá	Nam cho	Mú	Sarag	Túndi. Mún	Sarang	Sarángi
Snake	*s* Brúl	Deu	Drúl	Beu	Beu	Osék	Pachám	Púkúri	Bí	Bhúgúri	Búl	Bú sa
Star	*s* Karma	Karma	Karma	Kám	Sahór	Khéséвá	Sángyen	Karchin	Nagú	Pírá. Tárgya	Bhúga	Sorú
Stone	*r* Do	Dó	Doh	Doh	Lóng	Lúng	Lúng tá	Yúmbá	Lohong	Yúma	Lhúng	Phúnglú
Sun	Nyimá	Nyí ma	Nimo	Nyim	Sachak	Nam	Nam	Dini	Sújá, S.[g]	Dhini	Nám khán	Ná
Tiger	*s* Tag	Ták	Jik	Táh	Sathóng	Keh'va	Kíwá	Chyan	Dhun	Chén	Ránghú	Gúpsa
Tooth	So	So	So	Soh	Aphó	Hébo	Kang	Swá	Wá	Sak	Syak	Kryú
Tree	*l* Jonshing	Shingdóng	Dóngó	Shing	Kúng	Sing	Sangtáng	Dhong	Símá	Sindú	Sing	Rawa
Village	Yúl tsho	Thóng	Yúl	Kyong	Kyong	Bángphé	Téng	Namso	Dé. Gang, S.	Nása	Lángbá	Gáún
Water	Chhú	Chhú	Chhú	Chhú	Ong	Chúá	Cháwa	Kwí	Lau. Lá	Kyú	Dí	Pánkhú
Yam	Dova	Thómá	Dhóá	Kyú	Búk	Khé	Sá kí	Témé	Hí	Taya	Námi	Rébé
I	Na	Gnyá	Gná	Gná	Go	Ingá	A'nka	Gná	Ji	Gná	Gná	Go
Thou	Khyod	Khé	Khyo	Chhú	Hau	Khéné	Khaná	Ai	Chha	Kén	Náng	Gai
He, she, it	Kho	Khú	Khwo	Khó	Heu	Khúné	Moko	Thé	Wo	Thi	Hos	Hari
We	Na chag	Gnánjo	Nírang	Gná chá	Kayú. Ká	Anigé	Ankan	Gnáni	Ji ping	Gni mo	Kan kúrik	Gov ki
Ye	Khyod chag	Khenjo	Khyoráng	Khá chá	Háyú	Khénih'	Khananin	Aini	Chha ping	Kén mo	Náng kúrik	Gaiv ki
They	Khochag	Khónjo	Khworáng	Khóng	Hóyú	Khúnchi	Moko chi	Théni	Wo ping	Thi mo	Hos kúrik	Harév ki
Mine	Nahi. Nayi	Gná yi	Gná ti	Gné gi	Kaseusa	Ingá in	Ang ko	Gná lá	Ji[h] gú mha	Gná lá	Gnou	A'ké
Thine	Khyodkyi	Khé yi	Khyo ti	Chhé gi	Hadosa	Khene in	Am ko	Ai lá	Chhang gú	Kén lá	Núwo	I'ké
His, etc.	Khóyi. Khóhi	Khó yi	Khwoti	Kheu gi	Heusa	Khúne in	Mósó	Thé lá	Waya gú	Thí lá	Ho chú	Hareá ké. Méró ké
Our's	Nachaggi	Gnánjo yi	Nirátí	Gná chégi	Kayú pongsa	Anigen in	A'in ko	In ná	Jiping gú	Gni molo	Kan kúrikúm	Go ain ké
Your's	Khyod chaggi	Khenjo yi	Khyérátí	Kheu chégi	Hayú pongsa	Khenih' in	A'mno	An ná	Chhaping gú	Keme molo	Nang kúrikúm	Gai ain ké. In ké
Their's	Khochaggi	Khonjo yi	Khworátí	Khong gi	Hoyú pongsa	Khúnchi in	Myaucho. Moyo so	Thenná	Waping gú	Thaméla	A kúrikúm	Hari ain ké
One	*g* Chig	Chik	Chik	Chi	Kát[b]	Thit. Thí-t[b]	Ektai. *k*Tai'	Ghrik. *gh*Ri-k	Chhi' (All take hma gu, suffixed like other qualitives.)	Kri. *k*Rí	Kát[b]	Ká[b]
Two	*g* Nyis	Nyi	Nyi	Nyi	Nyet	Nyetsh	Hasat	Gni. *g*Ni	Ni'	Ni	Nis	Níshi
Three	*g* Súm	Súm	Súm	Súm	Sam	Syúm sh	Súmyá	Sóm	Són	Song	Song	Sáng
Four	*b* Zhi	Zhyi	Zhyi	Zhi	Pha lí[c]	Lí sh	Láyá	Blí[c]	Pi	Plí[c]	Buli. Bu-li[c]	Lé
Five	Hna	Gná	Gná	Gná	Pha gnón[c]	Gná sh	Gnáyá	Gná	Gná	Gná	Bangá. Ba-gná	Gnó
Six	Drúk. *d*Ru-k	Thú	Túk	Dhú	Tarok'	Túk sh	Túkyá	Dhú	Khú	Tú	caret	Rúk
Seven	*b* Dún	Dún	Dyún	Dún	Ka kyok'	Nú sh	Bhágyá	Nis	Nhé	Nis. Limbu, S.	caret	Cha ni
Eight	*br* Gyud	Gyé	Gyé	Gyé	Kakeú	Yet sh[e]	Réyá = *r* Tibet	Pré = *br* Tib.	Chyá	Pré. *p*Ré	caret	Yoh
Nine	*d* Gú	Gúh	Gúh	Gú	Ka kyót	Pháng sh	Phángyá	Kúh	Gún	Kúh	caret	Gúh
Ten	*b* Chú. Thámbá	Chúh	Chúh	Cha-tham	Ka tí	Bong. Thi bong. Gip[f]	Kip	Chíwai	Sá nhó	Chúh	caret	Sa shi
Twenty	Nyi shú	Nyi shú	Nyi shú	Nyi sho. Khéchik	Khakát	Ní bong	caret	Nhi shú	Sang sánho. Níé	Kúti	caret	Khalká
Thirty	Súmchú	Súm chú	Súm chú	Khé phédáni	Khakátsa katí	Súm bong, 3 tens	caret	Bokal ché shú	Ní sánho. Súyé	caret	caret	Sasi sán
Forty	*b* Zhibchú	Hip chú	Hip chú	Khé ní	Kha nyét	Lí gip, 4 tens	caret	Bokal nhi	Sú sánho. Pí-yé	caret	caret	Khák néshi
Fifty	Hna bchú[a]	Gnap chú	Gnap chú	Khè phedángsúm	Kha nyét sa ká ti	Gná gip, 5 tens	caret	Bokal ni shú chú	Pí sánho. Gniáye	caret	caret	Khák nishisasika
Hundred	*br* Gya thámbá	Gyá. Gya thámbá	Gyá	Khé gná	Khá pha gnón	Thi bong gip, one [10 + 10	caret	Bokal gná	Gún sánho. Sat chi	caret	caret	Swaiká
Of	Kyi. Gi. Hi. Yi.	Gi	Tí	Gi. Yé	Sa	Le. In	Wó	Lá	Yá	Yé. Lá. Bó	Yó. Wó	Kwé. Kyé
To	La. Tú. Dú. Ra. Sú.	Lá	Lá	Ló	Ká. Rem	Mo. Nin	caret	Dá	Yáta	Dé	Kí	Kali
From	Nas. Las.	Né. Diné	Nébá	Nálá. Lé chálé	Nan. Liang[d]	Nú. Manú	Dánká	Yanché	Ni shyang. Ang. Yákén	Wájé	In	Gná
By, *instr.*	Kyis. Gis. S: His. Yis.	I'	Elongation of terminal vowel	Ki. Dá	Nan. Sa	Illé. Nú	Yá	Syé	Ang[i]	Jí	Ye. I'	Mi

[a] *hNab Chu* = 5 tens, and so of 20, 30, 40, which also give the radicals of the decimal scale, and show how serviles are always dropt in compounds. See and compare all.

[b] Compare *Tá-ku*, Burmese *Ta*, with the neuter sign *ku* = Néwári *gu*, *G vel K*, final of Tibetan; Sérpa and Múrmi is the same. So also is initial *ka* of Lepcha 7 to 10. In Néwári the numeral adjuncts are invariable. *Hic et hæc*, *hma*; *hoc*, *gu*, suffixed as in the other qualitives. The varied position and optional use of these addenda create much false semblance of diversity.

[c] *Pha* pre-fixed here, like *sh* post-fixed in the Limbú columns, is not radical. *Pha* is equal to the silent *b* of written Tibetan. *Bú*, *Bá*, in the Magar column, are equivalent to the Lepcha *Pha*, that is, pre-fixed only, and so *B* in the Múrmi and *P* in the Gúrung columns. The mutation throughout is very instructive: compare *h na*, *b zhi*, of the written Tibetan.

[d] *Liang* added, forms the locative form. Its meaning is place, locus. *Nan* alone is instrumental.

[e] The terminal *sh vel si* is omissible, though usual; so also the Kiránti final *ya*.

[f] Limbú *gip* = Kiránti *kip*, got from 40, just as Gyarung *pli* for four got from *plisi* 40. *Bong* equals *Gip*.

[g] Corruption merely of Súryá.

[h] *Hma* mas. et fœm.; *Gu* neuter. *Jihma*, myself; *Jigu*, my goods. *Hma* and *gú* are affixed to every qualitive whatever. See note [i] *voce long*.

[i] *Anuswar* merely, and for instr. and abl. alike: also *yakén*, which likewise expresses with, or *sáth* of Urdu; the Latin *cum*.

[j] Chinese 到 *tau*, equals first, a verb to arrive at, second a prep. to (going to.)

Comparative Vocabulary of the languages of Hôr Sôkyeul and Sifán, by B. H. HODGSON, *Esq.*

English.	*Thóchú.*	*Sókpa.*	*Gyámi.*	*Gyárúng.*	*Hórpa.*	*Tákpa.*	*Mányak.*
Air	mozyú	sálki	sphún	talí	púryú	rhót	mérdah̲
Ant	tú-khrá	khoró-khwé	•mai-thún	ko-rok	ˢkhró	rhok-pó	ba-rah̲
Arrow	jáh̲*	sé li mé	chen	ki-pi	ˡdá	mlá	má (rili, bow)
Bird	marwó	thá-kól	sphúï-chher	pyé-pyé	gyó	pyá	há
Blood	sáh̲	khoro-gwé	syé	tá-shí	syé, séh	khrá	sháh̲
Boat	phyá	sákersú	sí-thú,-thú	brú { tabrú, small shabru, great	grá	grú	gú
Bone	ripat	yá sô	kú-thó	syá-rhú	ré-râ	róspá	rúkhú
Buffaloe	caret	caret	swi-nyú	caret	caret	caret	dingmi
Cat	lô-chi	si-mí	mau, myau	ta-rhú	chúlah̲	syimbú	macheu
Cow	gwah̲ (bull zyáh)	sá-lo	neu, nyeu	nyé-nyé	gnaumeh̲	báh̲	womi (gnázi, bull)
Crow	nyágwo	khéré	láwa	ta-brok	kalé	ákpo	kali
Day	styákló	wúndúr	póth-yan	pish-né, nyé	nyé-lé	nyénti	nashchah̲
Dog	khwah̲	nhókhwé	kou	khí	katáh̲	khi	kehah̲
Ear	núkh	khikhé	airto	tir-né	nyó	nebláp	nápi
Earth	zip	wonnish	ti, thou	séh̲	kcha	sáh̲	malí, mlí
Egg	kiwóst	caret	chitun	kí-tan	sgangá	khálúm	rácha
Elephant	caret	lhábochê	syáng	láng-chhen	lámochhén	láng-chhén	„
Eye	kan	nútû	yen-chin	tai-myék, tam-myek	mó	mélóng	mni
Father	ai	I'chiki	dhá-dá	ta-pé	ápá	ápá	ápá
Fire	méh̲	kwál	ak-khá	ti-mi	u-mah̲	meh̲	sa-méh̲
Fish	izháh̲	khôlé	yúé	chú-ngyó	hyá	gná, nyá	yú
Flower	lámpáh̲	chichúk	khwá	tau-den	métô	ménto	ménto
Foot	jákó	khóil	chya a	tá-mi	kó	lémi	lipchhéh̲
Goat	tsáh	yá má	chúlyú	kús-só	chhé	rá	tsáh
Hair	hompá, grong { kachû of head	kéchigé	thou phwá	tár-ní	spú†	pú (krá of head)	múi (tsi of head)
Hand	jipah̲	kar	syú, syeu	ta-yak	lhá	lá	láp-chéh̲
Head	kapat	tholá-gwé	thau	ta-kó	ghó (hard)	gók-ti	wúlli
Hog	pi	khá-khai	dhú	kí	váh	phá	wáh
Horn	rak	yé-bour	tiko	ta-rú	k-rúm-bo	rú-ba	rú-bu
Horse	róh̲	má-ri	má	bo-róh	rhí, ryi	téh̲	bó-roh̲, bróh
House	kih̲	pá-syáng	shhangcha	chhém	hyó	khém	nyéh̲
Iron	sormo	thúmár	thé	shóm	chú	lékh	shi
Leaf	thrompi	náí	yé-cha	tai-mek	báláh̲	blap	nipchéh̲
Light	uik	caret	réyai	caret	sphó	wot	wúh̲
Man	náh̲	khún	rin	tir-mí	vzih	mih̲	chhoh̲
Monkey	wáisal	móchi	khouch	shé-prí, ti	zumdeh̲	prá	miyaháh
Moon	chháh̲	sárá	yoliáng	tsi-le, chileh̲	slikno	leh̲	lheh̲
Mother	ou	ákhi, yekhi	má	to-mó	ama	ama	amá
Mountain	spyáh̲	tává	sán, syan	ta-vet'	rih̲rap	ri	ᵐbi
Mouth	dsúkh	ámá	chwé	ti-khé	ya	khá	yebá
Moschito	boup	khó-khwé	wocha	caret	ˡvasa	pholi	bimo
Name	ʳmáh̲	nér	minn	tir-min'g	amen	myéng	ming
Night	áshá	sú	khólo	tó-di, tong-mor	spha	senti	kwakah̲
Oil	chingyú	má-chin-thóso	eué, yú	chin-swí, { mustard seed juice	marnak	kyamar	íchírá, ítira
Plantain	sarmi	caret	máchouker	„	„	lamrep	„
River	chabráh̲	wassú, úsú	shúi	ti-chí	hráh̲	chhi	dyáh̲
Road	grih̲	chám	lú	tri	chéh̲	lémdaug	ráh̲
Salt	cheh̲	távósó	yan	chhé	chháh̲	tsá	cheh̲
Skin	rápí	sárú	phi-cha	tidri	gla	phyekh (hard)	grah̲
Sky	mahto	théng-gré	khen	tú-món, teumeun	koh̲	namdúng	mah̲
Snake	brigi	thólé	shré	kha-bri	phri	mrúi	brú
Star	ghada	„	singh-syú	tsi-ni	sgré	karma	krah̲

English.	*Thóchu.*	*Sókpa.*	*Gyámi.*	*Gyárúng.*	*Hórpa.*	*Tákpa.*	*Mányak.*
Stone	gholopi { hard arabic gh	chhilo	hri-thou	rú-gú	ʳgámé	gôrr	wobi
Sun	mún	nára	ré-thou	ki-ni	gna	pláng	nyi-ma
Tiger	khoh̲ (hard)	pár	khú	kông	sták	tê ê	léphê
Tooth	swéh̲	syú-chi	yá	ti-swè	syô	wáh̲	phwih̲
Tree	gwozósi	mótó	hrú	shi	nah	shêng dong	sápoh̲
Village	wékhá	hótó	twáng-cha	wo-khyú, túk-	rhava	yú	hú
Water	chah	wassú, u'sú	shúi	tichí [hyú	hráh̲	chhi	dyáh̲
Yam	jyah	caret	yángsú	sé-ten	zó	khé	zgwáh̲
I	chi, ká	mi, bi, a'bú	gnô	gná, gná-yo	gná	gnê, nyê	á
Thou	kwá, kwé	chhá	ni	sán-ré	ni	í	nô
He, she, it	kwán-tá-cha	thá‡	thá	gná-pos, watú	ʳja, ʳjya	pê, bê	thi
We	chúklar chiki, pl. cheun, dual	mini‡	gnômé	yô	gnáni, gnáriggi or rigya	gna-rá	plural, ádúr dual, ajú
Ye	kwéniko, kwa-nik-lar	chhini‡	nimé	nyô	nini, ni-riggi	i-rá	nóndúr
They	táho, thák-lar	tháni‡	thámé	ya pos [prefix‖	ʳjini, ʳji-riggi	pe-rá	thídúr
Mine	kák chi	caret	gnóti	gná, conjunct	gná-á (elonga-	gnokú	aí
Thine	kwék chi	caret	niti	ní, the same	níí [tion)	i'kú	nöö
His, hers, its	thákchi, kwanák-chi	caret	tháti	wá, the same	ʳjyá a	pêkú	thíô
Our's	chikúk	caret	gnométi	caret	gnáárígya	gná rá kú	ádurí
Yours	kwánikúk	caret	nimeti	caret	nii rigya	í'rákú	nônduri
Their's	thakúk	caret	thámeti	caret	ʳjaa rigya	pêrákú	thidúri
One	a'ri	négé	í-ku, í	katí	rá	thi	tábí
Two	gnári	hóyúr	liáng-ku, ar'	ka-nés**	gnó	nai	nabi
Three	ksbíri	kórbá	sang'-ku, san	ka-sám	sú	súm	síbi
Four	gzháré	tírbá	sí-ku, sí	ka-dí	hla	pli	rêbi
Five	wáré	thábá	wú-ku, wu	kung-gnó	gwé	liágnô	gnábi
Six	khatáré	chórka	leuku, leu	kú-tók	chhô	kró	trúbi
Seven	stáré	tóló	chhí-ku, chhi	kúsh-nés	znó	nis	skwibí
Eight	khráré	némá	pá-ku, pá	or-yét	rhiéó	gyet	zibi
Nine	ʳgúre	yésó	chyú-ku,	kúng-gú	gô	dúgú	gúbi
Ten	hadúré§	árbá	ish'-sa [chyú	sih̲	sgá	ᴾchi	chéchibi
Twenty	gninaso	hóré	air'-sa	kinis-si	naská	khali	náchábi
Thirty	kshyáso	kóchhen	san'-sa	ka-sam-si	súská	caret	sá chá bi
Forty	ghyiso	téché	syú-sa	kaplis-si	léska	caret	zyizabi
Fifty	wásso	sháché	wú-sa	kúngnôsi††	gwêská	caret	gná zábi
Hundred	akshi	chóvó	ípé	paryé	ʳhyá	caret	tôjô
Of	k	na, né	ti	caret.§§ um?	i dang?	kú	í
To	shil	tú	khá	caret	gi? da	syá, lá	wô
From	k, to, gé	gásá	li	sbis. s	lháno, gha	í	tha, ni
By, instru.	í	rá	lá	gi	khá, wú	gi	lô
With, cum	ong	thángdi	khang chhen	kri, khyás	áchô	núm láng	pháô
Without, sine	marúk	úg gwé	mómá, meyú	kameí	máchú	ma nôna	májú
In, on	kúkú, tik, ti	thú, tú	lá [má	s. pri	ná, no, chá	ná	khú, choh̲
Now	patino	ótô, wótô	chhá yé	púz-dúi	habdeu	dá	millê
Then	stáka, hatús	caret	lá khún	tis-dúí [tra	tabdeu	têné	thilê
When?	thisni	khech ché	ná khún	this-dúi, kwús-	sa deu	kashú	ninkhê
To-day	pashi	in dúr	chin thé	pish-nyi	pas-ni	tashi	tanyúr
To-morrow	sózyú	mágás	min thé	sôs-nyi	khasi	nogor	sôrúh̲
Yesterday	narr	nokhor	hou thé	púsyúr	naga, a'wesni	dáng	yáhá
Here	cho, kúzga, chaksi	in dé	thi mé	chidú	údú	wo cho	khopú, dait
There	háto, thúzga,	yá bú	lá mé	hadú	oúthú	wo tho	thúngá pu, kwa-
Where?	tano [thaksi.	thyerthar	lá li	katú	lóré	gá, gáhá	khadê [nait
Above	tíkh	téré	syáng thou	úrkyé	chhá	gáng	chú

* h, underscored thus, h̲, marks the abrupt accent.

† Ghórmé, hair of head. Yâ-spú, hair of mouth or moustache. Pá-spú, hair of body.

‡ But for the analogy of the Hórpa plural in 'ni,' I should say these were genitives and possessives, and that the plurals were wanting.

§ Ri, re, final, is a servile. So ba and ka of Sókpa and ku of Gyámi.

‖ A disjunct and complete series of possessives formed by adding the suffix 'young' to the personals has been alleged to me, but it is so rarely used, I doubt its genuineness. Here it is Gnayong vel Gnong, Nayong vel Nong, Gnaposyong, Yoyong, Nyoyong, Yaposyong.

** In composition these names of the numerals are liable to variation, as tirmi-targé, one man; tirmi-tage, two men; but three men is tirmi-kasam, unchanged.

†† Ka, prefix, varying to ku, and taking a nasal or other euphonic appendage, ku-ng, ku-sh, is servile. It is the common and almost inseparable adjunct of nouns, verbs, etc., and it interchanges with ta. Sometimes both are used. Compare ta-pé, a father, with ka-pá, Kassia, and Ta-gappen, Tamil, etc.

§§ No declensional signs as the general rule; but úm has been obtained as an anomalous exception of very special and narrow use, as Láma-úm-boroh, the Láma's horse.

English.	Thóchú.	Sókpa.	Gyámi.	Gyárúng.	Hórpa.	Tákpa.	Mányak.
Below	kól	tóro	ti syá	wáki	wó	wá	zyé
Between	tigú	toung dú	túng jen	úle. tilé	kyúkú	‡ bút ká, képá	onglhé
Without, outside	khanyis	gáchá	wai thú	wónpo	pheu-so	phit ka	nwá
Within, inside	kúkú	tótar	lithú	úgú, wógú	náng	néngá	khú
Far	grikho	khóló	ywén	ka sri	chéchi	ringbú	rassá
Near	grin, grinista	nangni	jhin	kaching	tháne	thúngbú	rini
Little	khwini	bágá	syóti (small)	kúh ché	ámché	chúti	tameh
Much	brobo	elvik	tá-ti (great)	kak-ti	kagaré	shibo	tabrá
How much?	nikal	in chhin yûbi	tó syó	this-ti	haisyi	gó	trimni
As, rel.	tek	caret	ah men-ti	caret	naya	dantang	mi
So, correl.	stáká	caret	lá men-ti	caret	nyú	dantarang	thûzyó [moh
Thus, poss.	cheu [chan	yénichhin	thi men-ti	caret	wodé	ustúm	thûsû, thûsû,
How?	nikanjú nika-	caret	thi má	thígúpso, this-pé	áchibi	katin gyá	hanus moh
Why?	niblin, nishi	tharichhin	syá chú	thús-pé	áchú gnó	sagyak	hámilé
Yes	gnówá, gno	bi	syó	do-mos	gnór*	in*	zyi
No	mángwá, mang	bi si	púsitiéyó	di-mek	nyér	men	má zyi
(Do) not	chí	puthi ké	púsyo	met	má, di†	má,** magyá	thá**
And, also	tah, dah	pichhé	orcha	caret	ré	”	”
Or	gnóá	”	tháng	kó. wóvé	ná	na, iná	lé
This	chá	ani. yéni	thikou	chidi	údé	wochû	thú
That	thá	théni	lákou	hadi	outhá, yé	wotho	quathú
Which, who, jon	”	”	hi mé	caret	caret	caret	”
Which, who, ton	”	”	lá mé	caret	caret	caret	”
Which, who, kon?	”	”	syá, himé	sú	sú, ló	sû	sú
What? kyá	sú	”	syácha, hima	thú	achin	si	hánó
Anything	ning [wan	”	hiong	tenzi, tizzé	áke	sirang	táká
Anybody	ningwan sóng-	”	ohki, hiong	sú	súyó	sirang	súyé
Good	náí	chháng béné	houkhou. houti	kasné	gáyé gnor	lihûni	deundah
Bad	ghái. ghé. mari	má béné	hou ti myú	ma-kasné	gáyé nyér	lihûmani	mánda
Cold	styú	khou thún	sidi	kavandró, kamishta	kúrkú	krang-mo	phemphé
Hot	si	há lon	ré-di	kassí. kavassi	ché ché	gromo	chéché
Ripe	an. min	bál chhén	phú-ti	ka-smán	núlúmsi	choso	demi
Raw	a-min	chhik thé	myúphû	ma-ka-smán	númálúmsi	machoso	demámi
Sweet	jam	am thé thé	syángdi	kam-gnar	thú-thú	nyok-pa	debi
Sour	chak	ammahálon	lá-ti	kúch-chúr	'gús-go	kyûr-pû	da-chá
Bitter	khák	” [(good)	khû-ti	kúch-chék	'nésmé	khák-bó	dá-khá
Handsome	'kwi	cháng béné	houti (good)	kúm-chúr	kam-syúr	lihúmi. gnómánó	phyún phú
Ugly	márkwi	má béne (bad)	houti myú (bad)	ma-kumchhúr	mem-syúr	lihúmáni. gnómá	mám phyu
Straight	kasth [gwê	”	ting-di	ka-kas'to	kathông	tráng bó [-mano	chú chú
Crooked	jaggra. jablá	”	ting-di myû	ma-kas'to	gúngú	kyok po	kho kho
Black	nyik	”	khidi	ka-nak'	nyá nyá	nak po	dáná
White	phyokh	chhágán	pi-di	kaprom'	phrú phrú	khérû	dallú

English.	Thóchu.	Sókpa.	Gyámi.	Gyárúng.	Hórpa.	Tákpa.	Mányak.
Red	shidzi	ulán	khóng-di	kaver' ni	gingi	leu	dani
Green	zyángkú	khó khó	lig-di	karmyák.	jhángú	chángú	chúgindo
Long	drithu	úr thú	tháng-ti	kasri	kachi	ringbo	shásha
Short	wóngchithá	”	thóng-ti	kachan'	kalgó	thongpo	dridra
Tall	bráthá	ún dúr	kou-ti	kasri'	gakhyé	zúgring	hra hra
Short	'thátha	”	ti-ti	kachin'	gádé	zúg thung	dridrá
Small	bratsi tha	bágá	syou-ti	kachhai'	kamma	chúngbo. prû	yú
Great	pwí tha	íkbî	tá-ti	kahti'	kamthú	thénbo	kah kah
Round	ashyara	”	eang-di. yáng	kálárlar ‡‡	lóló	birhi	wáh wah
Square	ghzírú	”	pyáng-di [-di	zhirdo	súr zhi	túp-zhi	drazo
Fat	charwá	yokhwé thé	hou-ti (good)	kwipan	kalbo. galvo	gyák pa	dachúh
Thin	charghé	ókhú ná é	syou-ti (small)	kwichem	chú chú	kámrháng	kárí
Weariness	darvatch	yá tava	sphwá leu	disdúk	nerthá	”	ná brída
Thirst	tírpitch	úléso	kháng-ti	taskom'	nasyá	”	depsyá
Hunger	ashpitch	wolúso	ó-ti. wó-ti	tomos	namjóngsi	”	vitengné
Eat	adz	éthé	thyé, khyé	ta-zô	nangi	zó	gnajeu
Drink	áthí	wúó	khwá	ta-mot	wathi	thong	gnachhóh
Sleep	ánan	wúm tha	swikyór	korman'	gúrgyún	nyet	khaiyah
Wake	toron (get up)	pós	khilé	tar-was	taryén	láng (get up)	dougwáh
Laugh	daran	enna	syó	ka-náré	khá khé	gyé	narir
Weep	arzan [kochin	wún na	shúhrin	da-ka-krú	nakabrá	gun	dangwá
Be silent	'sgástan. dzúk	áh má hóp-	quápótho	nák-chún	yá-gúzi	thómá	thathadyú
Speak	kwor, kúrr	caret [chhi	caret	ta-chén	nap-shéh, tayin	syát	thadyu
Come	hai	iré	lé	ka-pún. papún‖	kwi-lhen	syô	lemo
Go, depart	dákan	yá bú? ††	chhi	yeyen, da-chin	ta-shin. wa-shin	gai	yú
Stand up	toron	posth	chhilé	tar-yúp ya-chin	zúryén	lang	khanjéh
Sit down	ajon	só	chó	ná-nen	únzun. wanzún	zúk	naijeu
Move, walk	dákan	áhyar yábo	chú. chhi	ye-yen. ya-chin	tashin	gai	yú
Run	dádran	thûr keng	théwo	danar-gyúk	tamgyo	pshet	tachimoyú
Give	dagsh (cuivis)§ kwúgsh (mihi)	wûg. euk	kí. yoho	davo (cuivis) da-vong (mihi)	tú-khyé (cuivis) tú-không (mihi)	bé. bin	wa-khi. ta-
Take	jádjh	caret	rákwô	da-ven	gwonkhé. túshthú	ya. §§ lóngá	dangó [khi
Strike	da-gatch	chhok ka	tá	ta-túp	nazbi	dúngé	dan-thá
Kill	ta-séh	”	sá	ná-sé	ta-shé	sòtá	na-sya
Bring	dzi-la	áhba-táhira	lá-le	ko-pet	wú-khyé	rotá	trúlhé
Take away	doukwa	áhba-chhi	lá-chhé	di-cháng	wúm-bé	khor	túyú
Lift up	ta-chi	wúra	máyú	ta-yok	rang-ké. rházi	longna	da-chi
Put down	kwaksh	caret	caret	na-tok	rale	nina	wúchi
Hear	kokshustan	súnú	thyén	kar-nyou	wul min	nyan	khabé ní
Understand	ákhchan	hériya	syá	ti-sen	sam tenchú	séin	najinjó
Tell, relate	kúrr	khala	shró	ta-chen	ta-yin. nap-shé	syat	thui-dyú

* These are the positive and negative forms of the substantive verb = the Persian hast, nést, exactly.

† Di, an in-fix, medial; ma, pre-fix.

‡ Horizontal and perpendicular betweenity. ** Initial and medial.

§ In all these tongues there is a special and general term, indicated by the Latin appendage.

†† Quære! Iré bú, come not, in Kalmak.

‡‡ Ka pre-fix of all the above words is the same as that added to the numerals. (See note at "ten.") To the verbs the analogous prefix ta vel da is usually added. But ka is also used with verbs, *e.g.*, dong = it is, he is, in Bodo (Du of Newari and Tibetan du-g) is ka-m-dong in Gyárúng.

‖ Ka prefix becomes pa, according to that alliterative principle which prevails so greatly, though irregularly.

§§ Bé, Yá, have a special sense. Give to *me:* take from me. Bin, Láng, a general sense. One solicits; the other commands.

English.	*Tibetan, written.*	*Tibetan, spoken.*	*Sérpa.*	*Bhutáni or Lhópa*	*Lepcha.*	*Limbú.*	*Kiránti.*	*Múrni.*	*Néwár.*	*Gúrung.*	*Magar.*	*Súnwár.*
With, cum. latin Sath in Hindí and Urdú	Lhanchig	Lá, Dá	Táng	Chá. Dá. Cháro	Sá. Tyol	Nú. Téng	Dá	Ta	Yákêṉ. Nápo	Dê. Dêyê	Lê tháng. Khńta	Núh
Without, sine	......	Tháng	Mé tála	Té ló	Ti	Menné	Mádang	A'dá ná	Madaya kang	Ar esyá	Mámúlê	Mabáthú
In, on	La. Na	Lá	Lá	Ná	Ká. Plóng	Mo. khep mo. kúthung tho	Dá	Ri	E'. Tê[d]	Rí	Yáng. Ang	Mi
Now	Dengtsé. Da. Deng	Thándá	Tángdá	Dháto	Along	Alo	Handê	Dande	A'	Tasso	Chamlán	I'chí
Then	Dé tsé	Thi dwi	Támá	O'dé. O'di gáng	Othá	Khem pha lé	Khwomlo	Jámá	Wala	Chok lênê	Arnan	Méná
When?	Gang tsé. Nam.	Kha dwi	Tanam	Nam	Sathá	A'phá lé	Dêmkhê	Ká ï ma	Gola	Khaï mo	Syen	Géná
To-day	Déring	Thíring	Táring	Dharing	Sarong	Ain	Ai	Tíní	Thawoṉ	Tíní	Chini	Múṉ láti
To-morrow	Sang. Thoré	Sáng	Thoráng	Nábah	Lúk	Tándik	Mang koleng	Nangar	Ka nhai	Nhá gá	Pyúngúra	Dís
Yesterday	mDang	Dáng	Dáng	Kháchá	Tasso	Meh ma	Akhománg	Tilá	Mhigo	Tela	Tisyêngmi	Sináti
Here	Hadina	Dicho	Dirú	Dí té	Alim. Aba	Kót ná	Wadá	Jyásê	Thúkê	Chúri	Ilak	Watha
There	Héna	Hácho	Chúrú	Phá té	Pil. Woba	Ná	Míyánú	Kersyê	Ukê	Kyúri	Alak	Yéré
Where?	Gangna	Khácho	Káni	Káná té	Saba. Sabi	Atáng	Khádánú	Kháin	Gúkhê	Khaníri	Kúlak	Dôtha. Gétha
Above	sTengna	Téng. Ché. Yégi	Tyáng	Tén khá. Téng	Atún. Tal. Aplóng	Tháng	Madhani	Toyáng	Cho	Túṉri	Dhênam	Ri
Below	Hog na	Wó. Syú. Magi	Wag	Wóh	Achúm Cheul. Sadam	Mó	Móyúni	Moyang	Ko	Muṉri	Mháka	Yû
Between	Bar, du	Bhar	Par	Páná	Abik. Achúk	Kúlúm	Lúndi	Gúngari	Dathú	Khíonri	Mi khiáng	Daté
Without, outside	Phyi, rohna	Chi	Yáng	Phí	Póng	Báhar	Udúng yá	Mang gyêr	Pine	Hú jêri	Báhar	Báhir
Within	Nang, na	Náng	Náng	Náng	Sagong	Kúsi gang. Hong	Ukúng yá	Túng gyêr	Dúnô	Nhóri	Bhitar	A'gá
Far	Né. Nyé	Tháring	Rimbo	Thá ring	Maram	Mánkhó	Mángsá	Tháring	Tápá	Rhêmô	Lôs	Gnani
Near	Ring	Tháni	Thak nimbo	Thá ni	Athol	Neng dáng	Nêktá	Jyat na	Satti	Kêṉdo	Khwep	Néthá
Little	Nyúng	Nigúva	Cháyak chik	Nyúng bo	Amán	Mí sa	Chichi	U'dit	Bhati	Chigi dê	Chêk já	Iská
Much	Mang. Tumo.	Má gúá	A'lá	Máng bo	Ag yáp	Yorik	Badho	Lháná	Apá	Lhêyô	Dher	Itch ká
How much?	Tsam. Tsoma.	Khá chwé	Kajó	Kájéu	Satet	A'khen	Dê móyê	Gádê	Guli	Ká tê	Kúdit	Gísí
As, rel.	Hadétsúg	Khándá	Kándé	Katé	Salom	Aphá dong bá	Khaïn súko	Khájú	Gathing	Khaga líyon	Kúdáng cha	Dôdiv
So, corr.	Détsúg	Théndá	Phándé	O'té	Olom	Khem phá dong ba	Khöïn súko	Waspa	Athing	Húchúga líyon	Adáng cha	Môdiv
Thus, poz.	Jitsúg	Dindá	Dindé	O'dé. Dé	A'lom	Kon phá dong ba	Woïn súko	Chúspá	Tha thing	Chúga líyon	I'dáng cha	Akko
How?	Tsúg. Chitsúg.	Kháché. Khánda	Kánde. Kándá	Káte bé	Salom	A'pha	A'ínsúko	Khátpá	Gathé. Gé.	Khaga líyon	Kúdáng cha	Dódiv
Why?	......	Khá ïn	Káng	Kám bé	Shú mat	Thé áng Thé jokma Thé yambókle	Kháïnsê	Tik	Chhá	Ta	Kútta	Mara
Yes	......	I'ṉ	In	Túp. In	Ak. Euk	O'k	Angá	Minná. Yá	Khau. Da Ang.	Wôï	Hô. Le Au	Mai
No	Ma. Mi	Meṉ	Men	Mé túp. Men	Má né	Men. Ná	Máng	A'nin	Ma Khú. Mai Ahang	A'ní	Mále	Ma mai
(Do) not	......	Má	Má	Má	Má-nan[a]	Ma-ne[b]	Man	Thá	Matê	A'. Wachê	Má	Mo
Also, and	Yang	Yáng	Dáng. Ang	Dá	Lá	Ang	Ning	Yen. Den.	Ang. Nang	Yê	Ra	Nû
Or	......	Mo	Nam. Inam	Yáng. Mo	Yáng. Eu	Bi	Hê	Wá	Lá	Bani. Gi	Kí	Dó
This	Hadé	Dí	Diráng	Di. Didi	Aré	Kon	Wó	Chún	Thô	Chúṉ yo	Ise ná	Yékwó
That	Dé	Phi-dí	Phi diráng	Phé. Phedi	Oré	Khen	Mô	Hô chan	Wô	Hô chun yo	O'sé ná	Makwé
Which, rel.	......	Thinda?[e]	Swín?	Kádi?	Sare?	A'ti?	Sá?	Khá chúá?	Gú. Sú?	Sú?	Kós?	Tékwé
Which, corr.	......	Thé	Thi dáng	U'di	Waré	Khen	Khó	Ho chúá	Wo	Thí	Hôs	Me kwé
Which?	Gang	Khangi	Swín	Ká di	Saré	Thé	Khá. Ko	Khá chúá?	Gú. Sú	Sú	Kôs	Té kwe
What?	Chi	Kháng	Kháng	Kang chi. Kan	Shú	Thé	Dí. Dê	Tigi	Chhú	Tá	Hí	Maro
Who?	Sú. Kha.	Khangi. Sú	Sú	Ká	To	Hát	Dí	Khá	Sú	Sú	Súra. Hira	Sûká
Any thing	Chizhig	Khá ïn	Khai náng	Kándochi	Shúri. Tham	Thé ré	Dimin	Tigi	Chhúng	Tayáng	Hihi ko	Márká
Any body	Súzhig. Khachig	Sú ín	Súi náng	Káyé. Ka imchi	Tólá	Hát ló	Aktai	Khá láï	Súṉg	Súyáng	Súr	Sûká
Eat!	Zó	Só	Sé. Só	Sah	Zó. Thá	Ohé	Chô	Chou	Na	Chad	Jeú	Jau
Drink!	hThúng	Thúng	Thúng	Thóng	Thóng	Thúng nê	Dúng	Thúng	Tôṉ	Thú nú	Gau	Tûng
Sleep	Nyah	Nyé	Nyól	Nyé	Dá	Ipsê[c]	Imsá[c]	Gnúng	Dyôn	Rod	Mís	I'p
Wake	......	*caret*	*caret*	Lhóng	Sí	Phókê	*caret*. Pôka.	*caret*	Don	Rêd	Swôn	Bôk
Laugh	bGad	Gá	Gwet	Gá	Then	Yêrê	I'yá. Isa	Nyet	Nhyú	Nyêd	Ret	Ris[c]

[a] *Ma* is a pre-fix and *nan* a conjunct post-fix, thus *ma mat nan*, 'do not.'

[b] In Lepcha and Limbú the double negative is used in composition precisely in the same way. Thus, from the Lepcha verbal root *ryu*, 'it is good,' is formed *ma-ryún-né*, 'bad,' and from the Limbú root *noh*, 'it is good,' *men-noh-mina*, 'bad,' an extra-harmonic consonant being added in both tongues to the root.

[c] Of the verbs, the final silibant is the neuter sign: the final nasal, almost neutral, referring rather to root than grammar: the final dentals (*t*, *d*) and gutturals (*k*, *g*) and labials (*p*, *b*, *v*, *m*), with or without a consonant, transitive signs; but all more or less passing into oblivion, as well as all sense of that suffixed pronoun, which, in the more complex tongues, helps to difference the transitive and reflex forms, as in Hayu, Gyarung, etc. See on, to weep and give.

[d] Elongation of terminal vowel merely often expresses in *eu* in declension. Also the abl. and instr. sign *kén*, *yáken*.

[e] Though the list exhibits relative and correlative terms, pronominal, and others, it may be doubted if there be any such, or any conjunction equivalent to English "and." The correlative pronoun and the conjunction "also" may indeed be had.

to me'; *Pire* 'give to him'; *cuivis præter me*. One root, *Bá vel Pá vel Phá*, runs throughout. In Lepcha *Bo* and *Bi* are similarly partitive, and in Néwári *Byú* and *Ti*. The paucity of grammatical forms is compensated by minute distinction in vocables. *Pá-áng* in Kiránti is 'give to me,' and therefore exactly equivalent to the Limbúan *Piráng*: but for comparison both should have been given in the non-reflective or non-pronominalized form.

[g] *Pi* is the root, and *ré*, *ángne* addenda. The whole series of equivalents for *give* is very demonstrative of the unity of this group of tongues; so also the series for *kill*, of which the root is *sa*, and the added *t* the widely diffused transitive sign; *e. g.*, *ha-to* = give, in Khamti and Hayu.

[h] *Ak*, *A*, Lepcha, is a formative pre-fix like *ke*, *ku*, in Limbu, and the root is verbal, as *kri*, *khi*, 'it is bitter'; *chí*, *chu*, 'it is small'; *li*, *lim*, 'it is sweet.'

[i] All Néwár adjectives are crude roots or verbal roots. It is hot. It is cold. To make them attributive, they take the generic signs, *hic et hæc*, *hma*, and *dhi*; *hoc*, *gu*, and *dhang*. *Ta* and *chi*, great and small, or plus and minus, are used as general qualifyers of all the rest. Of things each sort has class mark; thus, *há* is the sign of long things; and *ta ha* long, and *chi ha* short are literally, great in length and small in length. *Ta-dhi*, *Chiki-dhi* are *magnus et parvus*: *ta-dhang* and *chiki-dhang* are *magnum et parvum*.

* Specialization is the essence of these tongues, and if I say 'give' or 'take,' I shall be asked to whom or from whom before I can get an answer, the sense being so vague and large in the English words. The pronominalized tongue, distinguished by the personal suffix, as in Limbú and Kiránti; the non-pronominalized by the use of separate roots or words, as in Lepcha, *Bo* or *Bi*, and in Newári, *Byu* or *Ti*, *voce* give.

† Kiránti *pá-äng* equals 'give to me'; *kháta* equals 'take.' Read *bátuki* for bring and *bátakikhara*, take and go, for 'take away.'

‡ *ú* prefix is the pronominal, and *yang pang* suffix, the definitive appendage of the qualitives. ‖ Double negative, like Lepcha.

§ *Khâng-nuh'va* equals 'to see good,' *khâng-uva* equals 'to see bad.' So also *voce* 'tell,' *khang metta*, the causal of *khang* 'to see,' supplies the lack of a primitive term.

English.	*Tibetan, written.*	*Tibetan, spoken.*	*Sérpa.*	*Bhutáni or Lhópa*	*Lepcha.*	*Limbú.*	*Kiránti.*	*Múrmi.*	*Néwár.*	*Gúrung.*	*Magar.*	*Súnwár.*
Weep	Ngú. Shum	Gnó	Gnúni	Gnú	Rhiop	Hábé[f]	Khává	Krájh	Khwo	Krod	Ráp	Gnák
Be silent	Khrog	Chúm	Khárá	Khá chúm	Sakmá	Swáté	Man chebda	Kú dyú	Súmú khá chòn	Táya pun	Má chák	Pálo ma pau
Speak	*br*Jod. Smrás	*caret*	*caret*	Lap	Lí	Páré	Chéwa	*caret*	Nava	*caret*	*caret*	Bák
Come	Hóng. *s*Byon	Syó	Syok	Syó	Dí=move: aba dí, come, or here move	Phéré	Báná	Khou. Jyan [gou	Wá	Khó	Rá ní	Pyû
Go	Sóng. Gro. Gyu.	Gyó. Song	Gyok	Sóng	Nón	Bégé	Khára	Nyú. Syégo	Hon	Yád	Nú ni	Lau
Stand up	*h*Chhar	Lóng	Lóng	Lóng	Lúk. Ding	Pôgé = wake	Yéwá lúnta	Ráb	Dong	Rádh	Swóná	Bok
Sit down	*h*Dúg	Deh	Det	Deu	Gnán	Yúng né	Yúnga	Tyú	Phé tú	Tídh	Nú ná	Bák
Move, walk	*h*Gro	Gyó	Dóng	Dyú	Nón. Dí	Bégê	Bíyá	Brou	Nyá. Hún	Yáy	Wha ni	Gák
Run	*r*Gyúg	Gyúgé. Chóng	Chóng	Pán kyap	Deung	Lòkté. Lôk-te	Lòyá	Yár	Bwá	Dhíd	Yani	Plok
Give to me[e] / to any	*h*Búh. Phul. Tong	Phin. Bá. Bak	Bin	Náng. Bá. Bak	Bô. Bí	Piráng né[g] Pi-re	Páäng Pái	Pin	Byú. Tí	Pín	Láni	Gi
Take from me / from any	*b*Lan. Júng. Hen	Lén. Yá	Ling	Lén. Nang	Lyô	Lé	Bátu. Khátu	Thob	Ká. Ná	Kínú	Léó	Hyê
Strike	*b*Dún. *r*Deg	Dúng	Dúng	Dúng	Bak	Hipté. Hip-té	Mòú	Rob	Dá	Tan	Dûng	Tûp
Kill	Shig Sód. *h*Gúm	Sé	Syet	Séh	Sot	Séré	Sérú	Sát	Syá	Thód. Séd	Gnáp	Sat
Bring	*h*Khyóng. *s*Kyeh	Bá syó, give-come	Gyap	Bá syó, give-come	Bú dí, give-come	Phép-pé	Pá áng†	Bou	Ha ki	Pòg	Rákò	Pít
Take away	*h*Khúr. *b*Khyer	Bák sóng, give-go	Khúr syop, lift	Bak sóng, give-go	Bú nón, give-go	Téré	Khá tu	Pòr	Yénki	Bhòd	A'rhó	Lat
Lift up, raise / Bear, carry	*h*Degs. Slon. / *s*Nyob	Khúr	Khúr	Thú	Chún	Pôkhé. Po-ke	Khúyú	Púyò	Lhon Bu	Nòd	Bû	Pok
Hear	Nyan. *g*Son	Nyen	Nyen	Nyen	Nyen	Khépsé. Khep-se	Yénú	Gnan	Nyò	Théd	Thanásô	Nyen
Understand	Soms. Go.	Sóm	Syen	Som tang. Noh	Ching	Singnite. Sing-te	Sin tú	Gò	Síkí	Mhádid	Phéro	*caret*
Tell, relate	*b*Shód. *h*Chhod	Láp. Chwó	Lap	Lap	Dan	Chékhé. Che-ke	Kháng métú	Syát	Kòn	Bíd	Khángni	Den
Good	Bazang-po	Yappo	Lémú	Lémô	Aryúm. Ryúmbo[c]	Noh ba. Kenohba	Núhva	Jába	Bhing[i]	Saba	Gyópchó	Rimso
Bad	Nang-po	Dúkpo	Ma lémú	Ma lém	Azyen. Zyembo	Phem ba. Kephémba	Uva. Euva	Ajába	Ma bhing	A'saba	Mágyepche	Marin noso
Cold	Grang-po	Thammo	Thyángmo	Khyú mo	Ahyúm. Hyúmbo	Kesémba. Sémba	Kéng yáng‡	Símba	Khwá won	Simba	Ring cho	Chí so
Hot	Tshá-po, Dró-po[a]	Chábô	Tenmo	Teu mo	Arhúm. Rhumbo	Kegó ba. Góba	Kúyáng‡	Lépá	Khwá	Kro ba	Khan cho	Hoso
Raw	......	Zyembo	Zyenba	Má cho bo	Azeu. Zeubo	Kúleh'la. Lehla	U'chíva	Chinga	Kachi	A'míva	Mibil	Chêri pla
Ripe	Sminbo	Chémbo	Chobo	Chochopo	Amyen. Myenbo	Kúsôngva	Dau va	Min ba	Nhin gú	Míva	Mincho	Mí so
Sweet	......	Gnármo	Gnormo	Gná mo	Akliam. Kliambo	Kelímba[h]	Lemko	Kéké ba	Chakú	Gnába	Jyú cho	Jiji
Sour	......	*caret*	*caret*	Tek po	Kròp	Menlim mina[d]	Súvo	*caret*	Phakí	Soba	Thûp cho	Dû so
Bitter	......	Khá ko	Khaktí	Khakó	Akrím. Krimbo	Ke khik pa[h]	Khákka. Kháko	Kám ba	Khaiyú	Kámba	Khá cho	Kaso
Handsome	Desmo. *s*Túgpo	Jóbó	Lémo. Simbu	Lé mô	Aryúm. Ryumbo	Noh'va. Kenoh'va	Khangnúhva§	Brot khába	Bánlá	Saba	Shó chéja	Rimso
Ugly	Midsesma / Mistugpo	Men jébo	Ma lemu / Masimba	Málem	Maryúnne[d]	Phém bá Kephémba	Kháng úva	Brotá khába	Bámala	A'saba	Má sécho	Marimnoso‖
Straight	Dranpo	Tháng bo	Tángo	Tháng bo	Náng	Tóndô	U'dung twòng tong	Thácho	Tapyong	Kyún	Dhing cho	Shêjo
Crooked	*s*Gúrbô. Tudpo	Kákpo	Kók lók	Tyokkú	Manáng[e]	Kòk tú	U'dungú twòng tong	Kòkténg	Békò	Kúding	Gúm cho	Bángo
Black	Nagpo	Nákpo	Nakpo	Nákpo	Anók. Nokbo	Kúmaklá	Mákachakwa	Mlángai	Hákú	Mlòngyá	Chik chi dan-cho	Kêrá
White	*d*Karpo	Karpo	Karpo	Ká pô	A'dúm. Dumbo	Kúphòra	U'mpi yáng wa	Tára	Túyú	Tárkya	Bòcho	Bwí Syê
Red	*s*Mukpo	Márpo	Márpo	Má bô	A'heur. Heurbo	Kúhélla	Hálá láwá	Bála	Hyoun	Wòlkya	Gyá cho	Lala
Green	*h*Jangkhú	Jhangú	Númmo	Nhyam bo	Phûng phông	Léhla	Chak la	Pingai	Wá wòn	U'rkyá	Phiphi dancho	Gígí
Long	Ringpo	Rimbo	Rimbo	Rím bú	Arhen. Rhenbo	Kemba	Ménta	Reng ba	Tahá. tá-há	Rhimba	Lot cho	Joso
Short	Thúngpo	Thún dúng	Thinmo	Thúm bú	A'tán. Tánbo	Tángba	Dúng ta	Túmba	Chíhá. chí-há	Ruíba	Tûn cho	Tù pah
Tall man	......	Thombo	Thenbo	Thémbo. Tho	Athô. Thôbo	Kemba	Kòn tá	Nohba	Tadhí. ta-dhí	Cuhba	Ghiáng cho	Laiso
Short man	......	Mábo	Mámo	Mhámten. Mhour	Amáu. Máubo	Tángba	Sim ta	Meba	Chigdhi=chi-ki-dhi. Bágo	Cheunbo	Tem cho	Ho cho
Small	Chhúng. Phra	Chún chúng	Tippé	Chúng bo	Achím. Chímbo	Chúk pa	U'chuk yáng‡	Jájá	Chíga. chí-gu	Chûmba	Már cho	The baba
Great	Chhánpo *s*Bombo	Bombo	Girbú	Bombo	Atím. Tímbo[e]	Yom ba	U'tok yáng	Gná jáng	Taúga. tá-gu	The bá	Krán cho	Kol sotu
Round	*s*Lumpo	Riri	Gírmo	Gònto yeupo	Rér rérbo	Kúgak ma	Au bo	Ril to	Gogú. gò-gu	Phal dong	Dallo	Kûl kúl
Square	Grub zhi	Thúsi	Túpchi	Dúshí yeupo	Ton kyong phali	Kuyok túve lísh	Pheb dába / Léa kòna	Kúni pliba	Pékúng-lág, four corners having	Kona pli	Chou khá nya, H.	Chár pátya
Flat / Level	......	*caret*	Líblib	Le blep / Sab them	Alep. Lépbo / Asáp. Sápbo	Ku phélla	Phemdawa	*caret*	Pati	Phlébá	*caret*	*caret*
Fat	*r*Gyagspo	Thó thombo	Gyámo	Gyá mô	Asyúm. Syûmbo	Meh'rú	Tòk páng‡	Chòba	Lhóng	Choba	Dhêsho	Dúmso
Thin	Srobbo. Ridpo	Mábo	Nénma	Byé kô	Achím. Chímbo	Yòshú	Yòm. Yomyang	Jeutpa	Gónsi	Jhén ba	Rûcho	Gyé so
Weariness	......	Gyák	Yé ché	Tháng chhé	Pél	Nama	Hòtáng	Blap chi	Tyánu	Bhlá	Mhúncho	Dati
Thirst	*s*Kóm	Khákúm	Khákum	Khá kòm	U'ngnó	Wa mik ma	Wáit' má	Kwi phúi	Piás. H.	Kwi phi	Dí sona	Páng dati
Hunger	*l*Tógs	Tók	To kúng	Tò ki	Tidok. Kridok	Set lah ma	Sáá	Phoïdang	Pitya	Phókré	Tûkrésya	Amáija

[a] *Tshá, Chá, Tén, Teu, Rhú, Gó*, of the six first columns, exhibit the mutations of the verbal radical, it is hot, whence is universally derived the adjectival form of words by means of pre- and post-fixes, quite analogous throughout or very nearly so. The pre-fixes are often omitted, as *Ke-goba*, *Goba*, good, in Limbú.

[b] *Chung*, Tibetan, unde *chum, chí, já, chá, chí, chák*, through the columns. The rest of the differences belonging to the appendical letters or pre- and post-fixed.

[c] Both these forms are common. The latter shows plainly the Tibetan affinities of the Lepcha tongue.

[d] Double negatives, pre- and post-fixes, *Ma-ryun-ná*, *Men-limminá*, with the positive form euphonically altered. So also *Marim-noso* of Sunwar.

[e] Lepcha adjectives are like Burmese: *ryu* 'it is good,' and *kung*, ditto: *a-ryu-m* and *akung* 'good': *Nak* 'it is black': *Anak* 'black,' in both tongues.

[f] *é* final is the common imp. sign = the *a, o, u*, of other tongues. The preceding consonant denotes the class. It is the dental *t, d*, or it is the labial *b, p, v*, or guttural *k, g*, or nasal *n, ng*, or is the simple radical merely, or a liquid (*l, r*) consonant, or lastly, the sibilant forms the conjugational sign. If we turn to the Limbú and Kiránti columns, and note carefully the root, the resemblance of the words becomes often almost identity. —Thus *Singnite, Sintu*—root, *Sin*; *Phéppe, Pá-ang*—root, *Pha, Pa*; *Lockte, Lóyu*—root, *Lo*; *Bége, Biya*—root, *Be, Bi*; *Sere, Seru*—root *Se*. *Ang* penultimate is partitive. Thus *Piráng*, 'give

English.	Tibetan, written.	Tibetan, spoken.	Sérpa.	Bhútáni or Lhópa.	Lepcha.	Limbú.	Kiránti.	Múrmi.	Néwár.	Gúrung.	Magar.	Súnwár.
Plantain	caret	Gnálá	Lánga sí	Gnálá	Kardúng	Lá seh'	Gnák si	Moché	Mwai	Kala	Mocha	Múhji
River	g Tsang po	Chángpó	Hyúng	Chhu kyong	Ongkyong	Chua. Wo-hong	Hong-kú	Syóng	Khúsi	Khwóng	Kholá	Lí kû
Road	Lam	Lam	Lam	Lám	Laum	Lam	Lam	Ghyám	Lón	Kyán	Lam	Lá
Salt	Tshá	Chhá	Chhá	Chhá	Vom	Yúm	Yúm	Cháchá	Chhi	Cháchá	Ohá	Yû sí
Skin	Pagspa	Pág-pa	Kop-pa	Páko. Kompo	Athún. Kómbó	Horik. Saho	U'hók bá	Dí bhí	Syú. Chégú	Dhí	Chála	Kûsyûl
Sky	Nam khah	Nam	Nam	Nam	Ta liáng	Témsákpá	Nam cho	Mú	Sarag	Túndi. Mún	Sarang	Saréngi
Snake	s Brúl	Deu	Drúl	Beu	Beu	Osék	Pachám	Púkúri	Bí	Bhúgúri	Búl	Bû sa
Star	s Karma	Karma	Karma	Kám	Sahór	Khésévá	Sángyen	Karchin	Nagú	Pírá. Tárgya	Bhúga	Sorû
Stone	r Do	Dó	Doh	Doh	Lóng	Lúng	Lúng tá	Yúmbá	Lohong	Yúma	Lhúng	Phûnglû
Sun	Nyimá	Nyí ma	Nimo	Nyim	Sachak	Nam	Nam	Dini	Sújá, S.[g]	Dhini	Nám khán	Ná
Tiger	s Tag	Tak	Jik	Táh	Sathóng	Keh'va	Kíwá	Chyan	Dhuṇ	Chéṇ	Ránghú	Gúpsa
Tooth	So	So	So	Soh	Aphó	Hébo	Kang	Swá	Wá	Sak	Syak	Kryú
Tree	l Jonshing	Shingdóng	Dóngó	Shing	Kúng	Sing	Sangtáng	Dhong	Símá	Sindú	Sing	Rawa
Village	Yúl tsho	Thóng	Yúl	Kyong	Kyong	Bángphé	Téng	Namso	Dé. Gang, S.	Nása	Lángha	Gáún
Water	Chhú	Chhú	Chhú	Chhú	Ong	Chúá	Cháwa	Kwí	Lau. Lá	Kyú	Dí	Pánkhú
Yam	Dova	Thómá	Dhóá	Kyú	Búk	Khé	Sá kí	Témó	Hí	Taya	Námi	Rébé
I	Na	Gnyá	Gná	Gná	Go	Ingá	A'nka	Gná	Ji	Gná	Gná	Go
Thou	Khyod	Khé	Khyo	Chhú	Hau	Khéné	Khaná	Ai	Chha	Kéṇ	Náng	Gai
He, she, it	Kho	Khú	Khwo	Khó	Heu	Khúné	Moko	Thé	Wo	Thi	Hos	Hari
We	Na chag	Gnánjo	Níráng	Gná chá	Kayú. Ká	Anigé	Ankan	Gnáni	Ji ping	Gni mo	Kan kúrik	Gov ki
Ye	Khyod chag	Khenjo	Khyoráng	Khá chá	Háyú	Khénih'	Khananin	Aini	Chha ping	Kén mo	Náng kúrik	Gaiv ki
They	Khochag	Khónjo	Khworáng	Khóng	Hóyú	Khúnchi	Moko chi	Théni	Wo ping	Thi mo	Hos kúrik	Harév ki
Mine	Nahi. Nayi	Gná yi	Gná ti	Gné gi	Kaseusa	Ingá in	Ang ko	Gná lá	Ji[h] gú mha	Gná lá	Gnou	A'ké
Thine	Khyodkyi	Khé yi	Khyo ti	Chhé gi	Hadosa	Khene in	Am ko	Ai lá	Chhang gú	Kén lá	Núwo	I'ké
His, etc.	Khóyi. Khóhi	Khó yi	Khwoti	Kheu gi	Heusa	Khúne in	Mósó	Thé lá	Waya gú	Thí lá	Ho chú	Hareá ké. Méré ké
Our's	Nachaggi	Gnánjo yi	Nirátí	Gná chégi	Kayú pongsa	Anigen in	A'in ko	In ná	Jiping gú	Gni molo	Kan kúrikûm	Go ain ké
Your's	Khyod chaggi	Khenjo yi	Khyéráti	Kheu chégi	Hayú pongsa	Khenih' in	A'mno	An ná	Chhaping gú	Keme molo	Nang kúrikûm	Gai ain ké. In ké
Their's	Khochaggi	Khonjo yi	Khworáti	Khong gi	Hoyú pongsa	Khúnchi in	Myaucho. Moyo so	Thenná	Waping gú	Thaméla	A kúrikûm	Hari ain ké
One	g Chig	Chik	Chik	Chi	Kát[b]	Thit. Thí-t[b]	Ektai. kTai'	Ghrik. ghRi-k	Chhi' (One to Ten: All take hma gú, suffixed like other qualitives.)	Kri. kRí	Kát[b]	Ká[b]
Two	g Nyis	Nyi	Nyi	Nyi	Nyet	Nyetsh	Hasat	Gni. gNi	Ni'	Ni	Nis	Níshi
Three	g Súm	Súm	Súm	Súm	Sam	Syúm sh	Súmyá	Sóm	Són	Song	Song	Sáng
Four	b Zhi	Zhyi	Zhyi	Zhi	Pha lí[c]	Lí sh	Láyá	Blí[c]	Pi	Plí[c]	Buli. Bu-lí[c]	Lé
Five	Hna	Gná	Gná	Gná	Pha gnón[c]	Gná sh	Gnáyá	Gná	Gná	Gná	Bangá. Ba-gná	Gnó
Six	Drúk. dRu-k	Thú	Túk	Dhú	Tarok'	Túk sh	Túkyá	Dhú	Khú	Tú	caret	Rúk
Seven	b Dún	Dún	Dyún	Dún	Ka kyok'	Nú sh	Bhágyá	Nis	Nhé	Nis. Limbu, S.	caret	Cha ni
Eight	br Gyad	Gyé	Gyé	Gyé	Kakeú	Yet sh[e]	Réyá = rTibet	Pré = brTib.	Chyá	Pré. pRé	caret	Yoh
Nine	d Gú	Gúh	Gúh	Gú	Ka kyót	Pháng sh	Phángyá	Kúh	Gúṇ	Kúh	caret	Gúh
Ten	b Chú. Thámbá	Chúh	Chúh	Cha-tham	Ka tí	Bong. Thi bong. Gip[f]	Kip	Chíwai	Sá nhó	Chúh	caret	Sa shi
Twenty	Nyi shú	Nyi shú	Nyi shú	Nyi sho. Khéchik	Khakát	Ní bong	caret	Nhi shú	Sang sánho. Níé	Kúti	caret	Khalká
Thirty	Súmchú	Súm chú	Súm chú	Khé phédáni	Khakátsa katí	Súm bong, 3 tens	caret	Bokal ché shú	Ní sánho. Súyé	caret	caret	Sasi sán
Forty	b Zhibchú	Hip chú	Hip chú	Khé ní	Kha nyét	Lí gip, 4 tens	caret	Bokal nhi	Sú sánho. Pí-yé	caret	caret	Khák néshi
Fifty	Hna bchú[a]	Gnap chú	Gnap chú	Khè phedángsúm	Kha nyét sa ká tí	Gná gip, 5 tens	caret	Bokal ni shú chú	Pí sánho. Gniáye	caret	caret	Khák nishisasika
Hundred	br Gya thámbá	Gyá. Gya thámbá	Gyá	Khé gná	Khá pha gnón	Thi bong gip, one [10 + 10	caret	Bokal gná	Gún sánho. Sat chi	caret	caret	Swaiká
Of	Kyi. Gi. Hi. Yi.	Gi	Tí	Gi. Yé	Sa	Le. In	Wó	Lá	Yá	Yé. Lá. Bó	Yó. Wó	Kwé. Kyé
To	La. Tú[j] Dú. Ra. Sú. Lá	Lá	Lá	Ló	Ká. Rem	Mo. Nin	caret	Dá	Yáta	Dé	Kí	Kali
From	Nas. Las.	Né. Diné	Nébá	Nálá. Lé chálé	Nan. Liang[d]	Nú. Manú	Dánká	Yanché	Ni shyang. Ang. Yákéṇ	Wájé	In	Gná
By, instr.	Kyis. Gis. S: His. Yis.	I'	Elongation of terminal vowel	Ki. Dá	Nan. Sa	Illé. Nú	Yá	Syé	Ang[i]	Jí	Ye. I'	Mi

[a] *hNab Chu* = 5 tens, and so of 20, 30, 40, which also give the radicals of the decimal scale, and show how serviles are always dropt in compounds. See and compare all.

[b] Compare *Tá-ku*, Burmese *Ta*, with the neuter sign *ku* = Néwári *gu*, *G vel K*, final of Tibetan; Sérpa and Múrmi is the same. So also is initial *ka* of Lepcha 7 to 10. In Néwári the numeral adjuncts are invariable. *Hic et hæc*, *hma*; *hoc*, *gu*, suffixed as in the other qualitives. The varied position and optional use of these addenda create much false semblance of diversity.

[c] *Pha* pre-fixed here, like *sh* post-fixed in the Limbú columns, is not radical. *Pha* is equal to the silent *b* of written Tibetan. *Bu*, *Bá*, in the Magar column, are equivalent to the Lepcha *Pha*, that is, pre-fixed only, and so *B* in the Múrmi and *P* in the Gúrung columns. The mutation throughout is very instructive: compare *h na*, *b zhi*, of the written Tibetan.

[d] *Liang* added, forms the locative form. Its meaning is place, locus. *Nan* alone is instrumental.

[e] The terminal *sh vel si* is omissible, though usual; so also the Kiránti final *ya*.

[f] Limbú *gip* = Kiránti *kip*, got from 40, just as Gyarung *pli* for four got from *plisi* 40. *Bong* equals *Gip*.

[g] Corruption merely of Súryá.

[h] *Hma* mas. et fœm.; *Gu* neuter. *Jihma*, myself; *Jigu*, my goods. *Hma* and *gú* are affixed to every qualitive whatever. See note [i] *voce long*.

[i] *Anuswar* merely, and for instr. and abl. alike: also *yakén*, which likewise expresses with, or *sáth* of Urdu; the Latin *cum*.

[j] Chinese 到 *tau*, equals first, a verb to arrive at, second a prep. to (going to.)

Comparative Vocabulary of the languages of Hór Sôkyeul and Sifán, by B. H. HODGSON, *Esq.*

English.	Thóchú.	Sókpa.	Gyámi.	Gyárúng.	Hórpa.	Tákpa.	Mányak.
Air	mozyú	sálki	sphún	talí	púryú	rhót	mérdah
Ant	tú-khrá	khoró-khwé	•mai-thún	ko-rok	ʼkhró	rhok-pó	ba-rah
Arrow	jáh*	sé li mé	chen	ki-pi	ʼdá	mlá	má (rili, bow)
Bird	marwó	thá-kól	sphúï-chher	pyé-pyé	gyó	pyá	há
Blood	sáh	khoro-gwé	syé	tá-shí	syé, séh	khrá	sháh
Boat	phyá	sákersú	sí-thú,-thú	brú { tabrú, small shabru, great	grá	grú	gú
Bone	ripat	yá sô	kú-thó	syá-rhú	ré-rá	róspá	rúkhú
Buffaloe	caret	caret	swi-nyú	caret	caret	caret	dingmi
Cat	ló-chi	si-mí	mau, myau	ta-rhú	chúlah	syimbú	macheu
Cow	gwah (bull zyáh)	sá-lo	neu, nyeu	nyé-nyé	gnaumeh	báh	womi (gnázi, [bull)
Crow	nyágwo	khóré	láwa	ta-brok	kalé	ákpo	kali
Day	styákló	wúndúr	póth-yan	pish-né, nyé	nyé-lé	nyénti	nashchah
Dog	khwah	nhókhwé	kou	khí	katáh	khi	kehah
Ear	núkh	khikhé	airto	tir-né	nyó	nebláp	nápi
Earth	zip	wonnish	ti, thou	séh	kcha	sáh	malí, mlí
Egg	kiwóst	caret	chitun	kí-tan	sgangá	khálúm	rácha
Elephant	caret	lhábոché	syáng	láng-chhen	lámochhén	láng-chhén	„
Eye	kan	nútû	yen-chin	tai-myék } tam-myek }	mó	mélóng	mni
Father	ai	I'chiki	dhá-dá	ta-pé	ápá	ápá	ápá
Fire	méh	kwál	ak-khá	ti-mi	u-mah	meh	sa-méh
Fish	izháh	khólé	yúé	chú-ngyó	hyá	gná, nyá	yú
Flower	lámpáh	chichúk	khwá	tau-den	métô	ménto	ménto
Foot	jákó	khóil	chya a	tá-mi	kó	lémi	lipchhéh
Goat	tsáh	yá má	chúlyú	kús-só	chhé	rá	tsáh
Hair	hompá { kachû grong { of head	kéchigé	thou phwá	tár-ní	spú†	pú (krá of head)	múi (tsi of head)
Hand	jipah	kar	syú, syeu	ta-yak	lhá	lá	láp-chéh
Head	kapat	tholá-gwé	thau	ta-kó	ghó (hard)	gók-ti	wúlli
Hog	pi	khá-khai	dhú	kí	váh	phá	wáh
Horn	rak	yé-bour	tiko	ta-rú	k-rúm-bo	rú-ba	rí-bu
Horse	róh	má-ri	má	bo-róh	rhí, ryí	téh	bó-roh, bróh
House	kih	pá-syáng	shbangcha	chhém	hyó	khém	nyéh
Iron	sormo	thúmár	thé	shóm	chú	lékh	shi
Leaf	thrompi	náí	yé-cha	tai-mek	báláh	blap	nipchéh
Light	uik	caret	réyai	caret	sphó	wot	wúh
Man	náh	khún	rin	tir-mí	vzih	mih	chhoh
Monkey	wáissi	móchi	khouch	shé-prí, ti	zumdeh	prá	miyahâh
Moon	chháh	sárá	yoliáng	tsi-le, chileh	slikno	leh	lheh
Mother	ou	ákhi, yekhi	má	to-mó	ama	ama	amá
Mountain	spyáh	tává	sán, syan	ta-vet'	rihrap	ri	ᵐbi
Mouth	dsúkh	ámá	chwé	ti-khé	ya	khá	yebá
Moschito	beup	khó-khwé	wocha	caret	ʼvasa	pholi	bimo
Name	ʼmáh	nér	minn	tir-min'g	smen	myéng	ming
Night	áshá	sú	khélo	tó-di, tong-mor	spha	senti	kwakah
Oil	chingyú	má-chin-thóso	eué, yú	chin-swí, { mustard seed juice	marnak	kyamar	íchírá, ítira
Plantain	sarmi	caret	máchouker	„	„	lamrep	„
River	chabráh	wassú, úsú	shúi	ti-chí	hráh	chhi	dyáh
Road	grih	chám	lú	tri	chéh	lémdaug	ráh
Salt	cheh	távósó	yan	chhé	chháh	tsá	cheh
Skin	rápí	sárú	phi-cha	tidri	gla	phyekh (hard)	grah
Sky	mahto	théng-gré	khen	tú-món, teumeun	koh	namdúng	mah
Snake	brigi	thólé	shré	kha-bri	phri	mrúï	brú
Star	ghada	„	singh-syú	tsi-ni	sgré	karma	krah

English.	Thóchu.	Sókpa.	Gyámi.	Gyárúng.	Hórpa.	Tákpa.	Mányak.
Stone	gholopi { hard arabic gh	chhilo	hri-thou	rú-gú	ʼgámé	gôrr	wobi
Sun	mún	nára	ré-thou	ki-ni	gna	pláng	nyi-ma
Tiger	khoh (hard)	pár	khú	kóng	sták	tê ê	léphê
Tooth	swéh	syú-chi	yá	ti-swè	syô	wáh	phwih
Tree	gwozósi	mótó	hrú	shi	nah	shêng dong	sápoh
Village	wékhá	hótó	twáng-cha	wo-khyú, túk-	rhava	yú	hú
Water	chah	wassú, u'sú	shúi	tichí [hyú	hráh	chhi	dyáh
Yam	jyah	caret	yángsú	sé-ten	zó	khé	zgwáh
I	chi, ká	mi, bi, a'bú	gnô	gná, gná-yo	gná	gnê, nyê	á
Thou	kwá, kwé	chhá	ni	sán-ré	ni	í	nô
He, she, it	kwán-tá-cha	thá‡	thá	gná-pos, watú	ʼja, ʼjya	pê, bê	thi
We	chúklar chiki, pl. cheun, dual	mini‡	gnômé	yô	gnáni, gnáriggi or rigya	gna-rá	plural, ádúr dual, ajú
Ye	kwéniko, kwa-nik-lar	chhini‡	nimé	nyô	nini, ni-riggi	i-rá	nóndúr
They	táho, thák-lar	thání‡	thámé	ya pos [prefix‖	ʼjini, ʼji-riggi	pe-rá	thídúr
Mine	kák chi	caret	gnóti	gnà, conjunct	gná-á (elonga-	gnokú	aí
Thine	kwék chi	caret	niti	ní, the same	níí [tion)	í'kú	nöö
His, hers, its	thákchi, kwanák-chi	caret	tháti	wá, the same	ʼjyá a	pêkú	thíô
Our's	chikúk	caret	gnométi	caret	gnáárígya	gná rá kú	ádurí
Yours	kwánikúk	caret	nimeti	caret	nii rigya	í'rákú	nónduri
Their's	thakúk	caret	thámeti	caret	ʼjaa rigya	pêrákú	thidúri
One	a'ri	négé	í-ku, í	katí	rá	thi	tábí
Two	gnári	hóyúr	liáng-ku, ar'	ka-nés**	gnó	nai	nabi
Three	kshíri	kórbá	sang'-ku, san	ka-sám	sú	súm	síbi
Four	gzháré	tírbá	sí-ku, sí	ka-dí	hla	pli	rêbi
Five	wáré	thábá	wú-ku, wu	kung-gnó	gwé	liágnô	gnábi
Six	khatáré	chórka	leuku, leu	kú-tôk	chhô	kró	trúbi
Seven	stáré	tóló	chhí-ku, chhi	kúsh-nés	znó	nis	skwibí
Eight	khráré	némá	pá-ku, pá	or-yét	rhiéé	gyet	zibi
Nine	ʼgúre	yésó	chyú-ku,	kúng-gú	gô	dúgú	gúbi
Ten	hadúré§	árbá	ish'-sa [chyú	sih	sgá	ʼchi	chéchibi
Twenty	gninaso	hóré	air'-sa	kinis-si	naská	khali	náchábi
Thirty	kshyáso	kóchhen	san'-sa	ka-sam-si	súská	caret	sá chá bi
Forty	ghyiso	téché	syú-sa	kaplis-si	léska	caret	zyizabi
Fifty	wásso	shâché	wú-sa	kúngnôsi††	gwêská	caret	gná zábi
Hundred	akshi	chóvó	ípé	paryé	ʼhyá	caret	têjô
Of	k	na, né	ti	caret.§§ um?	i dang?	kú	í
To	shil	tú	khá	caret	gi? da	syá, lá	wô
From	k, to, gé	gásá	li	sbis. s	lháno, gha	í	tha, ni
By, instru.	í	rá	lá	gi	khá, wú	gi	lô
With, cum	ong	thángdi	kháng chhen	kri, khyás	áchô	núm láng	phâô
Without, sine	marúk	úg gwé	mómá, meyú	kameí	máchú	ma nôna	májú
In, on	kúkú, tik, ti	thú, tú	lá [má	s. pri	ná, no, chá	ná	khú, choh
Now	patino	ótô, wótô	chhá yé	púz-dúi	habdeu	dá	milê
Then	etáka, hatús	caret	lá khún	tís-dúí [tra	tabdeu	têné	thilê
When?	thisni	khech ché	ná khún	this-dúi, kwús-	sa deu	kashú	ninkhô
To-day	pashi	in dúr	chin thé	pish-nyi	pas-ni	tashi	tanyúr
To-morrow	sózyú	mágár	min thé	sôs-nyi	khasi	nogor	sôrúh
Yesterday	narr	nokhor	hou thé	púsyúr	naga, a'wesni	dáng	yáhá
Here	cho, kúzga, chaksi	in dé	thi mé	chidú	údú	wo cho	khopú, dait
There	háto, thúzga,	yá bú	lá mé	hadú	oúthú	wo tho	thúngá pu, kwa-
Where?	tano [thaksi.	thyerthar	lá li	katú	lórê	gá, gáhá	khadê [nait
Above	tíkh	téré	syáng thou	ûrkyé	chhá	gáng	chú

* h, underscored thus, *h*, marks the abrupt accent.

† Ghórmé, hair of head. Yâ-spú, hair of mouth or moustache. Pá-spú, hair of body.

‡ But for the analogy of the Hórpa plural in 'ni,' I should say these were genitives and possessives, and that the plurals were wanting.

§ Ri, re, final, is a servile. So ba and ka of Sókpa and ku of Gyámi.

‖ A disjunct and complete series of possessives formed by adding the suffix 'young' to the personals has been alleged to me, but it is so rarely used, I doubt its genuineness. Here it is Gnayong vel Gnong, Nayong vel Nong, Gnaposyong, Yoyong, Nyoyong, Yaposyong.

** In composition these names of the numerals are liable to variation, as tirmi-targé, one man; tirmi-tage, two men; but three men is tirmi-kasam, unchanged.

†† Ka, prefix, varying to ku, and taking a nasal or other euphonic appendage, ku-ng, ku-sh, is servile. It is the common and almost inseparable adjunct of nouns, verbs, etc., and it interchanges with ta. Sometimes both are used. Compare ta-pé, a father, with ka-pá, Kasia, and Ta-gappen, Tamil, etc.

§§ No declensional signs as the general rule; but úm has been obtained as an anomalous exception of very special and narrow use, as Láma-úm-boroh, the Láma's horse.

English.	Thóchú.	Sókpa.	Gyámi.	Gyárúng.	Hórpa.	Tákpa.	Mányak.
Below	kól	tóro	ti syá	wáki	wó	wá	zyé
Between	tigú	toung dú	túng jen	úle. tilé	kyúkú	‡ bút ká, képá	onglhé
Without, outside	khanyis	gáchá	wai thú	wónpo	pheu-so	phit ka	nwá
Within, inside	kúkú	tótar	lithú	úgú, wógú	náng	néngá	khú
Far	grikho	khóló	ywén	ka sri	chéchi	ringbú	rassá
Near	grin, grinista	nangni	jhin	kaching	tháne	thungbú	rini
Little	khwini	bágá	syóti (small)	kúh ché	ámché	chúti	tameh
Much	brobo	elvik	tá-ti (great)	kak-ti	kagaré	shibo	tabrá
How much?	nikal	in chhin yûbi	tó syó	this-ti	haisyi	gó	trimni
As, rel.	tek	caret	ah men-ti	caret	naya	dantang	mi
So, correl.	stáká	caret	lá men-ti	caret	nyú	dantarang	thûzyó [moh
Thus, poss.	cheu [chan	yénichhin	thi men-ti	caret	wodé	ustúm	thûsû, thûsû,
How?	nikanjú nika-	caret	thi má	thígúpso, this-pé	áchibi	katin gyá	hanus moh
Why?	niblin, nishi	tharichhin	syá chú	thús-pé	áchú gnó	sagyak	hámilé
Yes	gnówá, gno	bi	syó	do-mos	gnór*	in*	zyi
No	mángwá, mang	bi si	púsitiéyó	di-mek	nyér	men	má zyi
(Do) not	chí	puthi ké	púsyo	met	má, di†	má,** magyá	thá**
And, also	tah, dah	pichhé	orcha	caret	ré	"	"
Or	gnóá	"	thúng	kó, wóvé	ná	na, iná	lé
This	chá	ani. yéni	thikou	chidi	údé	wochû	thú
That	thá	théni	lákou	hadi	outhá, yé	wotho	quathú
Which, who, jon	"	"	hi mé	caret	caret	caret	"
Which, who, ton	"	"	lá mé	caret	caret	caret	"
Which, who, kon?	"	"	syá, himé	sú	sú, ló	sû	sú
What? kyá	sú	"	syácha, hima	thú	achin	si	hánó
Anything	ning [wan	"	hiong	tenzi, tizzé	áke	sirang	táká
Anybody	ningwan sóng-	"	ohki, hiong	sú	súyó	sirang	súyé
Good	náï	chháng béné	houkhou. houti	kasné	gáyé gnor	lihûni	deundah
Bad	gháï. ghé. mari	má béné	hou ti myú	ma-kasné	gáyé nyér	lihûmani	mánda
Cold	styú	khou thún	sidi	kavandró, kamishta	kúrkú	krang-mo	phemphé
Hot	si	há lon	ré-di	kassí. kavassi	ché ché	gromo	chéché
Ripe	an. min	bál chhén	phú-ti	ka-smán	núlúmsi	choso	demi
Raw	a-min	chhik thé	myúphû	ma-ka-smán	númálúmsi	machoso	demámi
Sweet	jam	am thé thé	syángdi	kam-gnar	thú-thú	nyok-pa	debi
Sour	chak	ammahálon	lá-ti	kúch-chúr	'gús-go	kyûr-pû	da-chá
Bitter	khák	" [(good)	khû-ti	kúch-chék	'nésné	khák-bó	dá-khá
Handsome	'kwi	chháng béné	houti (good)	kúm-chúr	kam-syúr	lihúmi. gnómánó	phyún phú
Ugly	márkwi	má béne (bad)	houti myú (bad)	ma-kumchhúr	mem-syúr	lihúmáni. gnómá	mám phyu
Straight	kasth [gwê	"	ting-di	ka-kas'to	kathông	tráng bó [-mano	chú chú
Crooked	jaggra. jablá	"	ting-di myû	ma-kas'to	gúngú	kyok po	kho kho
Black	nyik	"	khidi	ka-nak'	nyá nyá	nak po	dáná
White	phyokh	chhágán	pi-di	kaprom'	phrú phrú	khérû	dallú

English.	Thóchu.	Sókpa.	Gyámi.	Gyárúng.	Hórpa.	Tákpa.	Mányak.
Red	shidzi	ulán	khóng-di	kaver' ni	gingi	leu	dani
Green	zyángkú	khó khó	lig-di	karmyák.	jhángú	chángú	chúgindo
Long	drithu	úr thú	tháng-ti	kasri	kachi	ringbo	shásbá
Short	wóngchithá	"	thóng-ti	kachan'	kalgó	thongpo	dridra
Tall	bráthá	ún dúr	kou-ti	kasri'	gakhyé	zúgring	hra hra
Short	ᵏthátha	"	ti-ti	kachin'	gádé	zúg thung	dridrá
Small	bratsi tha	bágá	syou-ti	kachhai'	kamma	chúngbo. prû	yú
Great	pwí tha	íkhī	tá-ti	kahti'	kamthú	thénbo	kah kah
Round	ashyara	"	eang-di. yáng	kálárlar ‡‡	lóló	birhi	wáh wah
Square	ghzírú	"	pyáng-di [-di	zbirdo	súr zhi	túp-zhi	drazo
Fat	charwá	yokhwé thé	hou-ti (good)	kwipan	kalbo. galvo	gyák pa	dachúh
Thin	charghé	ókhú ná é	syou-ti (small)	kwichem	chú chú	kámrháng	kárí
Weariness	darvatch	yá tava	sphwá leu	disdúk	nerthá	"	ná bríde
Thirst	tirpitch	úléso	kháng-ti	taskom'	nasyá	"	depsyá
Hunger	ashpitch	wolúso	ó-ti. wó-ti	tomos	namjôngsi	"	vitengné
Eat	adz	éthé	thyé, khyé	ta-zô	nangi	zó	gnajeu
Drink	áthí	wúó	khwá	ta-mot	wathi	thong	gnachhóh
Sleep	ánan	wúm tha	swikyór	korman'	gúrgyún	nyet	khaiyah
Wake	toron (get up)	pós	khilé	tar-was	taryén	láng (get up)	dougwáh
Laugh	daran	enua	syó	ka-náré	khá khé	gyé	narir
Weep	arzan [kochin	wún na	shúhrin	da-ka-krú	nakabrá	gnu	dangwá
Be silent	'sgástan. dzúk	áh má hóp-	quápótho	nák-chún	yá-gúzi	thómá	thathadyú
Speak	kwor, kúrr	caret [chhi	caret	ta-chén	nap-shéh, tayin	syát	thadyu
Come	hai	iré	lé	ka-pún. papún‖	kwi-lhen	syô	lemo
Go, depart	dákan	yá bú? ††	chhi	yeyen, da-chin	ta-shin. wa-shin	gai	yú
Stand up	toron	posth	chhilé	tar-yúp ya-chin	zúryén	lang	khanjéh
Sit down	ajon	só	chó	ná-nen	únzun. wanzún	zúk	naijeu
Move, walk	dákan	áhyar yábo	chú. chhi	ye-yen. ya-chin	tashin	gai	yú
Run	dádran	thúr keng	théwo	danar-gyúk	tamgyo	pahet	tachimoyú
Give	dagsh (cuivis)§ kwúgsh (mihi)	wûg. euk	kí. yoho	davo (cuivis) da-vong (mihi)	tú-khyé (cuivis) tú-không (mihi)	bé. bin	wa-khi. ta-
Take	jádjh	caret	rákwô	da-ven	gwonkhé. túshthú	ya. §§ lóngá	dangó [khi
Strike	da-gatch	chhok ka	tá	ta-túp	nazbi	dúngé	dan-thá
Kill	ta-séh	"	sá	ná-sé	ta-shé	sòtá	na-sya
Bring	dzi-la	áhba-táhira	lá-le	ko-pet	wú-khyé	rotá	trúlhé
Take away	doukwa	áhba-chhi	lá-chhé	di-cháng	wúm-bé	khor	túyú
Lift up	ta-chi	wúra	máyú	ta-yok	rang-ké. rházi	longna	da-chi
Put down	kwaksh	caret	caret	na-tok	rale	nina	wúchi
Hear	kokshustan	súnú	thyén	kar-nyou	wul min	nyan	khabé ní
Understand	ákhchan	hériya	syá	ti-sen	sam tenchú	sém	najinjó
Tell, relate	kúrr	khala	shró	ta-chen	ta-yin. nap-shé	syat	thai-dyú

* These are the positive and negative forms of the substantive verb = the Persian hast, nést, exactly.

† Di, an in-fix, medial; ma, pre-fix.

‡ Horizontal and perpendicular betweenity. ** Initial and medial.

§ In all these tongues there is a special and general term, indicated by the Latin appendage.

†† Quare? Iré bú, come not, in Kalmak.

‡‡ Ka pre-fix of all the above words is the same as that added to the numerals. (See note at "ten.") To the verbs the analogous prefix ta vel da is usually added. But ka is also used with verbs, *e.g.*, dong = it is, he is, in Bodo (Du of Newari and Tibetan du-g) is ka-m-dong in Gyárúng.

‖ Ka prefix becomes pa, according to that alliterative principle which prevails so greatly, though irregularly.

§§ Bé, Yá, have a special sense. Give to *me:* take from me. Bin, Láng, a general sense. One solicits; the other commands.

English.	Tibetan, written.	Tibetan, spoken.	Sérpa.	Bhutáni or Lhópa.	Lepcha.	Limbú.	Kiránti.	Múrmi.	Néwár.	Gúrung.	Magar.	Súnwár.
With, cum, latin; Sath in Hindí and Urdú	Lhanchig	Lá, Dá	Táng	Chá. Dá. Cháro	Sá. Tyol	Nú. Téng	Dá	Ta	Yákên. Nápo	Dê. Dêyê	Lê tháng. Kháta	Núh
Without, sine	......	Thána	Mé tála	Té ló	Ti	Menné	Mádang	A'dá ná	Madaya kang	Ar esyá	Mámúlê	Mabáthú
In, on	La. Na	Lá	Lá	Ná	Ká. Plóng	Mo. khep mo. kúthung tho	Dá	Ri	E'. Tê[d]	Rí	Yáng. Ang	Mi
Now	Dengtsé. Da. Deng	Thándá	Tángdá	Dháto	Along	Alo	Handê	Dande	A'	Tasso	Chamlán	I'chí
Then	Dé tsé	Thi dwi	Támá	O'dé. O'di gáng	Othá	Khem pha lé	Khwomlo	Jámá	Wala	Chok lênê	Arnan	Méná
When?	Gang tsé. Nam.	Kha dwi	Tanam	Nam	Sathá	A'phá lé	Dêmkhê	Ká ï ma	Gola	Khai mo	Syen	Géná
To-day	Déring	Thíring	Táring	Dharing	Sarong	Ain	Ai	Tíní	Thawon	Tíní	Chini	Mún láti
To-morrow	Sang. Thoré	Sáng	Thoráng	Nábah	Lúk	Tándik	Mang koleng	Nangar	Ka nhai	Nhá gá	Pyúngúra	Dís
Yesterday	mDang	Dáng	Dáng	Kháchá	Tasso	Meh ma	Akhománg	Tílá	Mhigo	Tela	Tisyêngmi	Sináti
Here	Hadina	Dicho	Dirú	Dí té	Alim. Aba	Kót ná	Wadá	Jyásê	Thúkê	Chúri	Ilak	Watha
There	Héna	Hácho	Chúrú	Phá té	Pil. Woba	Ná	Míyánú	Kersyê	Ukê	Kyúri	Alak	Yéré
Where?	Gangna	Khácho	Káni	Káná té	Saba. Sabi	Atáng	Khádánú	Khán	Gúkhê	Khanírí	Kúlak	Dótha. Gétha
Above	sTengna	Téng. Chá. Yégi	Tyáng	Tén khá. Téng	Atún. Tal. Aplóng	Tháng	Madhani	Toyáng	Cho	Túnri	Dhênam	Ri
Below	Hog na	Wó. Syú. Magi	Wag	Wóh	Achúm, Cheul. Sadam	Mó	Móyúni	Moyang	Ko	Munri	Mháka	Yú
Between	Bar, du	Bhar	Par	Páná	Abik. Achúk	Kúlám	Lúndi	Gúngari	Dathú	Khíonri	Mi khiáng	Daté
Without, outside	Phyi, rohna	Chi	Yáng	Phí	Póng	Báhar	Udúng yá	Mang gyêr	Pine	Hú jéri	Báhar	Báhir
Within	Nang, na	Náng	Náng	Náng	Sagong	Kúsi gang. Hong	Ukúng yá	Túng gyêr	Dúnó	Nhóri	Bhitar	A'gá
Far	Né. Nyé	Tháring	Rimbo	Thá ring	Maram	Mánkhó	Mángsá	Tháring	Tápá	Rhêmó	Lós	Gnani
Near	Ring	Tháni	Thak nimbo	Thá ni	Athol	Neng dáng	Nêktá	Jyat na	Satti	Kêndo	Khwep	Néthá
Little	Nyúng	Nigúva	Cháyak chik	Nyúng bo	Amán	Mí sa	Chichi	U'dit	Bhati	Chigi dê	Chêk já	Iská
Much	Mang. Tumo.	Má gúá	A'lá	Máng bo	Ag yáp	Yorik	Badho	Lháná	Apá	Lhêyó	Dher	Itch ká
How much?	Tsam. Tsoma.	Khá chwé	Kajó	Kájéu	Satet	A'khen	Dê móyê	Gádê	Guli	Ká tê	Kúdit	Gísí
As, rel.	Hadétsúg	Khándá	Kándé	Katé	Salom	Aphá dong bá	Khaïn súko	Khájú	Gathing	Khaga líyon	Kúdáng cha	Dódiv
So, corr.	Détsúg	Théndá	Phándé	O'té	Olom	Khem phá dong ba	Khóïn súko	Waspa	Athing	Húchúga líyon	Adáng cha	Módiv
Thus, poz.	Jitsúg	Dindá	Dindé	O'dé. Dé	A'lom	Kon phá dong ba	Woïn súko	Chúspá	Tha thing	Chúga líyon	I'dáng cha	Akko
How?	Tsúg. Chitsúg.	Khāché. Khánda	Kánde. Kándá	Káte bé	Salom	A'pha	A'insúko	Khátpá	Gathé. Gé.	Khaga líyon	Kúdáng cha	Dódiv
Why?	......	Khá in	Káng	Kám bé	Shú mat	Thé áng, Thé jokma, Thé yambókle	Kháinsê	Tik	Chhá	Ta	Kútta	Maru
Yes	......	I'n	In	Túp. In	Ak. Euk	O'k	Angá	Minná. Yá	Khau. Da Ang.	Wói	Hó. Le Au	Mai
No	Ma. Mi	Men	Men	Mé túp. Men	Má né	Men. Ná	Máng	A'nin	Ma Khú. Mai Ahang	A'ní	Mále	Ma mai
(Do) not	......	Má	Má	Má	Má-nan[a]	Ma-ne[b]	Man	Thá	Matê	A'. Wachê	Má	Mo
Also, and	Yang	Yáng	Dáng. Ang	Dá	Lá	Ang	Ning	Yen. Den.	Ang. Nang	Yê	Ra	Nú
Or	......	Mo	Nam. Inam	Yáng. Mo	Yáng. Eu	Bi	Hê	Wá	Lá	Bani. Gi	Kí	Dó
This	Hadé	Dí	Diráng	Di. Didi	Aré	Kon	Wó	Chún	Thô	Chún yo	Ise ná	Yékwé
That	Dé	Phi-dí	Phi diráng	Phé. Phedi	Oré	Khen	Mô	Hô chan	Wô	Hô chun yo	O'sé ná	Makwé
Which, rel.	......	Thinda?[*]	Swín?	Kádi?	Sare?	A'ti?	Sá?	Khá chúá?	Gú. Sú?	Sú?	Kós?	Tékwé
Which, corr.	......	Thé	Thi dáng	U'di	Waré	Khen	Khó	Ho chúá	Wo	Thí	Hôs	Me kwé
Which?	Gang	Khangi	Swín	Ká di	Saré	Thé	Khá. Ko	Khá chúá?	Gú. Sú	Sú	Kós	Té kwe
What?	Ohi	Kháng	Kháng	Kang chi. Kan	Shú	Thé	Dí. Dê	Tigi	Chhú	Tá	Hí	Maro
Who?	Sú. Kha.	Khangi. Sú	Sú	Ká	To	Hát	Dí	Khá	Sú	Sú	Súra. Hira	Súká
Any thing	Chizhig	Khá in	Khai náng	Kándochi	Shúri. Tham	Thé ré	Dimin	Tigi	Ohhúng	Tayáng	Ihihi ko	Márká
Any body	Súzhig. Khachig	Sú in	Súi náng	Káyé. Ka imchi	Tólá	Hát ló	Aktai	Khá lái	Súng	Súyáng	Súr	Súká
Eat!	Zó	Só	Sé. Só	Sah	Zó. Thá	Ohé	Chô	Chou	Na	Chad	Jeú	Jau
Drink!	hThúng	Thúng	Thúng	Thóng	Thóng	Thúng nê	Dúng	Thúng	Tôn	Thú nú	Gau	Túng
Sleep	Nyah	Nyé	Nyól	Nyé	Dá	Ipsê[c]	Imsê[c]	Gnúng	Dyón	Rod	Mís	I'p
Wake	......	*caret*	*caret*	Lhóng	Sí	Phókê	*caret*. Póka.	*caret*	Don	Rêd	Swón	Bók
Laugh	bGad	Gá	Gwet	Gá	Then	Yêrê	I'yá. Isa	Nyet	Nhyú	Nyêd	Ret	Rís[c]

[a] *Ma* is a pre-fix and *nan* a conjunct post-fix, thus *ma mat nan*, 'do not.'

[b] In Lepcha and Limbú the double negative is used in composition precisely in the same way. Thus, from the Lepcha verbal root *ryu*, 'it is good,' is formed *ma-ryún-ná*, 'bad,' and from the Limbú root *noh*, 'it is good,' *men-noh-mina*, 'bad,' an extra-harmonic consonant being added in both tongues to the root.

[c] Of the verbs, the final silibant is the neuter sign: the final nasal, almost neutral, referring rather to root than grammar: the final dentals (*t*, *d*) and gutturals (*k*, *g*) and labials (*p*, *b*, *v*, *m*), with or without a consonant, transitive signs; but all more or less passing into oblivion, as well as all sense of that suffixed pronoun, which, in the more complex tongues, helps to difference the transitive and reflex forms, as in Hayu, Gyarung, etc. See on, to weep and give.

[d] Elongation of terminal vowel merely often expresses in *eu* in declension. Also the abl. and instr. sign *kên*, *yáken*.

[*] Though the list exhibits relative and correlative terms, pronominal, and others, it may be doubted if there be any such, or any conjunction equivalent to English "and." The correlative pronoun and the conjunction "also" may indeed be had.

PART II.

ON THE PHYSICAL GEOGRAPHY OF THE HIMALAYA.*

A CLEAR outline, illustrated by a sketch map of the principal natural divisions of the Himálaya§, is, and long has been, a great desideratum; for physical geography, which derives so many aids from the other physical sciences, is expected in return to render back to them, without unnecessary delay, a distinct demarcation of its own provinces, since by that alone researchers in so many departments are enabled to refer the respective phænomena they are conversant with to their appropriate local habitations, in a manner that shall be readily intelligible causally significant, and wholly independent of the shifting and unmeaning arrondissements of politics.

It is true, that our knowledge of the large portion of these mountains, lying beyond the limits of British dominion, is far from complete. But is our knowledge any thing like complete of our own hill-possessions? and, if we are to wait until Népál, Sikim, and Bhútán become thoroughly accessible to science, must we not indefinitely postpone a work, the most material part of which may (I think) be performed with such information as we now possess?

The details of geography, ordinarily so called, are wearisomely insignificant; but the grand features of physical geography have a pregnant value, as being alike suggestive of new knowledge, and facilitative of the orderly distribution and ready retention of old. I purpose to adhere to those grand features, and to exhibit them in that causal connexion which gives them their high interest with men of cultivated minds.

I had been for several years a traveller in the Himálaya, before I could get rid of that tyranny of the senses, which so strongly impresses almost all beholders of this stupendous scenery with the conviction that the mighty maze is quite without a plan. My first step towards freedom from this overpowering obtru-

*Extracted from the *Selections from the Records of the Government of Bengal*, No. xxvii, Calcutta 1857. (Reprinted from the Jour. As. Soc. Bengal for 1849.)

§ Hima 'snow,' Alaya 'place of.' The compouud is Himálaya, not Himaláya as usually pronounced. The synonymes Himáchala and Himódaya (whence the Classic *Æmodus*) mean, respectively, 'snowy mountain' and 'place of appearance of snow (udaya).'

AA

siveness of impressions of sense was obtained by steady attention to the fact, that the vast volume of the Himálayan waters flows more or less at right angles to the general direction of the Himálaya, but so that the numberless streams of the mountains are directed into a few grand rivers of the plains, either at or near the confines of the two regions. My next step was due to the singular significance of the topographic nomenclature of the Népálese, whose "Sapt Gandaki" and "Sapt Kausika"‡ rivetted my attention upon the peculiar aqueous system of the Himálayas, urging me thenceforward to discover, if possible, what cause operated this marked convergence of innumerable transverse parallel streams, so as to bring them into a limited series of distinct main rivers. My third and last step was achieved when I discovered that the transcendant elevation and forward position, at right angles to the line of ghát́s, of the great snowy peaks, presented that causal agency I was in search of; the remotest radiating points of the feeders of each great river being coincident with the successive loftiest masses* belonging to the entire extent of the Himálaya. It was in Népál that this solution of these problems occurred to me, and so uniformly did the numerous routes I possessed represent the points of extreme divergence of the great rivers by their feeders as syntopical with the highest peaks, that I should probably long ago have satisfied myself upon the subject, if my then correspondent, Captain Herbert, had not so decidedly insisted on the very opposite doctrine—to wit, that the great peaks intersect instead of bounding the principal alpine river basins.

Captain Herbert's extensive personal conversancy with the Western Himálaya, added to his high professional attainments, made me for a long time diffident of my own views. But the progress of events, and increasing knowledge of other parts of the chain, seeming to confirm the accuracy of those views, it occurred to me more carefully to investigate whether the facts and the reason of the case were not, upon the whole, demonstrative of the inaccuracy of that able and lamented officer's dogma. Doubtless the Western Himálaya§ presents appearances calculated to sustain Captain Herbert's opinion, whilst such persons only as are unaccustomed to deal with the classifications of science, will expect them to correspond point by point with those natural phænomena, which it is at least one chief merit of such arrangements, merely to enable us readily to grasp and retain. But that the entire body of facts now within our ken is upon the whole opposed to Captain H.'s doctrine,† and that that doctrine suits ill with the recognized axioms of Geology and Geography, is, I think, certain; and I shall with diffidence now proceed to attempt the proof of it.

‡ See Journal Asiat. Soc. of Bengal, No. 198, for December 1848, p. 646 &c.

* This expression is used advisedly, for every pre-eminent elevation of the Himálaya is not so much a peak as a cluster of peaks springing from a huge sustaining and connected base. But observe, *some* of the peaks are not advanced before the ghát-line, but thrown back behind it, as Chumalári and Devadhunga or Nyánám. These do not influence the aqueous system of the Indian slope of Himálaya; see on, to remark on Chumalári. This is a new inference from new facts in part.

§ The Western Himálaya, as it approaches the Belúr, is in many respects anomalous, owing, as I conceive, to the crossing of that meridional chain. The true and normal Himálaya is parallelic or runs west and east.

† Journal No. 126, extra pp. 20 and 22.

A tyro in geology, I shall not dwell further on the theoretical side of the question than may be requisite to facilitate and complete the apprehension of my readers; but the facts, *quoad* Népál at least, I trust, that my sketch map, rude as it is, and the following observations, may render sufficiently indisputable; it being always remembered that I deal with generals, not particulars, aiming to establish the general accuracy of my main proposition, *viz.*, that the great peaks, bound instead of intersecting the alpine river basins, and that, in truth, the peaks by so bounding *create* the basins, whereas their *intersection would destroy them.*

The whole Himálaya extends from 78 deg. to 94 deg. of longitude, comprising the following peaks and basins:—peak of Jamnoutri (a), peak of Nanda-dévi (A), peak of Dhoula-giri (B), peak of Gosain-thán (C), peak of Kangchan‡ (D), peak of Chumalhári (E), peak of the Gemini§ (e)—which peaks include and constitute the following alpine river basins, *viz.*, that of the Ganges, that of the Karnáli, that of the Gandak, that of the Cósí, that of the Tishta, that of the Mónas, and that of the Subhansri (pars). The subjoined table exhibits the elevation and the position of these dominant peaks, with the authority for both.

a	Jamnoutri	25,669	30° 55	78° 12‖
A	Nanda-dévi ..	25,598	30° 22	79° 50
B	Dhoula-giri ..	27,600	29° 10	83°
C	Gosain-thán ..	24,700	28° 20	86°
D	Kang-chan ..	28,176	27° 42	88° 10
E	Chumalhári ..	23,929	27° 52	89° 18
e	Gemini	21,600 21,476	27° 50	92° 50

The Himálaya proper is traced along the line of the gháts or passes into Tibet; and the principal passes of Népál and Sikim into Tibet, or Taklakhár, Mustáng, Kérúng, Kúti, Hatia, Wallúng, Láchén.

Along the last low range of hills are the Máris or Dhúns within the range, and the position of the Bháver and Tarai* without it.

Sallyán-mári, Gongtali-mári, Chitwan-mári, Makwáni-mári, and Bijaypúr-mári are so many Népálese samples of those singular *quasi* valleys, termed Dhúns to the westward.§

In the plateau of Tibet I have indicated the limits of the northern and southern divisions, and in the latter those of the three great Trans-Himálayan provinces,

‡ Kang 'snow'; chan 'abounding in,' 'having,' like the English suffix full in fearful, etc., Chumalhári, holy mountain of Chuma.

§ I have so named the two proximate peaks of nearly equal height, which are inserted without name in Pemberton's large map, in long. 92 deg. 50 min, lat. 27 deg. 50 min.

‖ Cf. J.A.S. Nos. 126 and 197; Asiatic Researches, vol. xii; also Pemberton's Report and Map.

* Tarai, tarei, or tareiáni, equal to 'lowlands,' 'swampy tract at the base of the hills,' seems to be a genuinely Turanian word, and were the map of India carefully examined, many more such pre-Arian terms would probably be discovered, to prove the universal spread over the Continent of that earlier race, which is now chiefly confined to the Deccan. Tar in Tamil, Tal in Canarese, means 'to be low,' and the affixes *ei* of Tar-ei, and *ni* of Tareia-ni, are, the former, Tamilian, and the latter, very general, *in* or *ni* being the genitival and inflexional sign of several Southern and Northern tongues of the Turanian group of languages. The 'Thal' of Cutch is a term precisely equivalent to our Tarei, and is the merely aspirate form of Canarese Tal above cited. (Another etymology was proposed by Lassen's Ind. Alt., i. 69.)

§ See J.A.S. No. 126, p. 33, *et seq.*, and p. 134.

or Gnári, extending (from the Belúr) easterly to the Gángrí boundary range of Lake Mapham; Utsáng, thence stretching to the Gakbo River beyond Lhasa; and Khám, which reaches from the Gakbo River to the Yúnling, or limitary range of China and Tibet.|| Thus reverting to the regions south of the line of gháts leading into Tibet, we have, clearly defined, the several natural provinces or divisions of the Himálaya, with their casual distribution, as follows, commencing from the westward—1st, the alpine basin of the Ganges, extended from the peak of Jamnoutri to the peak of Nanda-dévi (Juwar or Juwahir), or, in other words, from east long. 78° 12' to 79° 50'; 2nd, the alpine basin of the Karnáli, reaching from the peak of Nanda-dévi to that of Dhoula-giri, or from 79° 50' to 83°; 3rd, the alpine basin of the Gandak, stretching from the peak of Dhoula-giri to that of Gosain-thán, or from 83° to 86°; 4th, the alpine basin of the Cósí, extending from the peak of Gosain-thán to that of Kangchan, or from 86° to 88° 10'; 5th, the alpine basin of the Tishta, reaching from the peak of Kangchan to that of Chumalhári. or from 88° 10' to 89° 18'; 6th, the alpine basin of the Mónas, stretching from the peak of Chumalhári to that of Gemini, or from 89° 18' to 92° 50'; and, lastly, the alpine basin of the Subhansri, of which the western limit is the Gemini, but the eastern peak is unascertained. It should be sought somewhere about 94° 50', between which point and the extreme eastern limits of the Himálaya must be the basin of the Dihong. That the above distribution of the Himálaya into natural districts is, upon the whole, as consistent with the facts as it is eminently commodious and highly suggestive, I have no hesitation of asserting. Lest, however, I should extend my present essay to undue limits, or trench upon the province of Colonel Waugh and the other able professional men who are now engaged upon the western hills, I shall say nothing further of the alpine valley of the Ganges and those west of it, nor upon those lying east of Sikim.*

If my main assumption be valid, it will be easily worked out by abler hands and better furnished ones than mine: wherefore the following more detailed expositions will be chiefly confined to the three great central basins of the Karnáli, the Gandak, and the Cósí. In the first of these basins we have (successively from west to east) the Sárju, the Góri, the Káli, the Swéti-ganga, the Karnáli proper, the Bhéri, and the Jhingrak or Rápti.† And it is certain that, whereas

|| See Routes from Káthmándú to Peking in sequel and paper on Horsok and Sifán. Sifán is the eastern boundary of Khám, which commences, on the line of route from Népál at Sangwa, the 51st stage, and extends to Tachindo, the 104th and political boundary of Tibet and China. The Yúnling chain seems to run along the western verge of Sífán.

* In the sequel I shall give the river basins of the Western Himálaya upon the authority of Dr. Thomson, in order to complete the enumeration of Himálayan districts, but simply as results, and without discussion. Dr. T.'s river distribution proceeds on the same principle as mine, which was published three years prior to his. I think he has needlessly increased the number of basins and thereby almost marred the effect of the causal connection of them with the geological structure of the mountains.

† This identification is probably erroneous, though adopted by Buchanan. The Jhingrak with a higher source is turned into the Karnáli by the Dhoula-giri ridge; the proximate Ráputi is not so influenced, owing to its lower source, and hence has an independent course through the plains to the Ganges, like the Gumti, etc., as enumerated in the sequel.

these streams drain the whole alpine valley of the Karnáli, so their most westerly source and course is confined on the west by the Nanda-dévi peak, as their most easterly is limited on the east by that of Dhoula-giri. These rivers do not wholly unite within the hills, though their tendency to union is so decided, that they are known by one name, even in the plains, where their collective appellation is Sárju or Káli or Ghógra. In the hills the whole of them are universally denominated by the collective name of Karnáli (corrupted by Rennell and his followers into Kenár). Karnáli is the proper name of this noble river, the Karnáli branch being by far the largest, the central, and most remote of origin. It rises in Tibet, not far from one of the sources of the Sutlej, and has a considerable Trans-Himálayan course to the westward of the Taklakhár pass, where it quits Tibet. No natural district can be more distinct than the alpine basin of the Karnáli, as above defined. It includes the political divisions of Káli-Kúmáun, belonging to Britain, and of the Báisi, or twenty-two Rájes of Népál, with Yúmila or Júmla, Dóti, and Sallyán. In the second basin, or that of the Gandak, we have, successively from the west, as before, the Barigár, the Náráyani, the Swéti-gandaki, the Marsyángdi, the Daramdi, the Gandi, and the Trisúl. These are the "Sapt Gandaki" or seven Gandaks of the Népálese, and they unite on the plainward verge of the mountains at Tirbéni above Sáran. They drain the whole hills between Dhoula-giri and Gosain-thán, the Berigár, and one head of the Náráyani, rising from the former barrier, and the Trisúl, with every drop of water supplied by its affluents from the latter. Nor does a single streamlet of the Trisúl arise east of the peak of Gosain-thán, nor one driblet of the Berigár deduce itself from the westward of Dhoula-giri. We have thus in the alpine basin of the Gandak another admirably defined natural division comprised within two great proximate Himálayan peaks. This division is named, vernacularly, the Sapt Gandaki. It includes the old Choubisi or twenty-four Rájes, and belongs to the modern kingdom of Népál.

Our third sample of a Himálayan natural province, conterminous with the utmost spread of the feeders of a large river, and bounded on either hand by a prime snowy peak, is the basin of the Cósí, which, like the Gandak ,has seven principal feeders These are as follows:—the Milamchi, the Bhótia Cósí, the Támba Cósí, the Líkhu, the Dúd Cósí, the Arun, and the Tamór.* Of these, the Milamchi, rising from Gosain-thán, is the most westerly, and the Tamór, rising from Kangchan, is the most easterly feeder.† And those two great peaks, with the pre-eminent ridges they send forth southwards, include every drop of water that reaches the great Cósí of the plains through its seven alpine branches. All these branches, as in the case of the Gandak, unite at (Varáha Kshétra above Náthpúr) within the hills, so that the unity of this alpine basin also is as clear, as are its limitary peaks and its extent.

* Tamór, Hindi equivalent to Tamvar, Sanskrit. So Dhoula-giri for Dhawala-giri, and Jamnoutri for Jamnavatari. I have throughout adopted the vernacular forms of words as being more familiar and quite as correct.

† See J. A. S. No. 189. Route from Káthmándú to Darjeeling.

The alpine basin of the Cósí is denominated by the Népálese the Sapt Kausika, or country of the seven Cósís. It comprises the old Rájes of the Kirántis,* Limbús, and Kála Makwánis, and is included, like the two prior basins, in the modern kingdom of Népál.

The country drained by the above three rivers (Karnáli, Gandak, and Cósí) includes the whole of Népál and the proximate part of Kúmáun, or, in other words, 800 miles of the central and most characteristic portion of the Himálaya. Wherefore it is legitimately presumable that, whatever is true of its natural divisions, is true of those of the residue, *quoad* ruling principle and geological causation.

Now if the above facts relative to these three rivers be justly represented (and that they are so, in the main, I confidently assert), we are led irresistibly to inquire *why* the numerous large feeders of the rivers, instead of urging their impetuous way from the snows to the plains by independent courses, are brought together upon or near the verge of the plains? *how* unity is effected among them, despite the interminable maze of ridges they traverse, and despite the straight-downward impulse given them at their sources?—I answer, it is because of the superior elevation of the lateral barriers of these river basins, between which there are synclinal slopes of such decided preponderance, that they over-rule the effect of all other inequalities of surface, how vast soever the latter may sometimes be.

These lateral barriers of the river basins are crowned by the pre-eminent Himálayan peaks, that the peaks themselves have a forward position in respect to the ghát-line or great longitudinal watershed between Tibet and India, and that from these stupendous peaks, ridges are sent forth southwards proportionably immense. Thus from the peak of Kangchan is sent forth the ridge of Singilélá, which towers as loftily over all the other sub-Himálayan ridges of Eastern Népál and Western Sikim, as does Kangchan itself over all the other Himálayan peaks.

This Singilélán prolongation (so to speak) of Kangchan entirely separates the waters of the Cósí and of the Tishta. A similar ridge, that of Dayabhang,† stretching south from the great peak of Gosain-thán, as entirely divorces the waters of the Cósí and of the Gandak. Another like ridge rising from Dhoula-giri as effectually sunders the waters of the Gandak and of the Karnáli. Another starting from Nanda-dévi in like manner wholly separates the proximate feeders of the Karnáli and of the Ganges; whilst yet another originating with Jamnoutri wholly separates the Ganges from the Jumna.

Equally effective with the divergent power of each of these supremely peaked ridges, which run parallel to each other and at right angles to the ghát-line of the snowy range, upon two river-basins, as just noticed, is of course the convergent

* The classical *Cirrhatæ*, and a once dominant and powerful race, though they have long since succumbed to the political supremacy of other races—first the Makwánis and then the Gorkhális.

† Hence the name Dhaibúng, erroneously applied by Colonel Crawfurd to the peak Dayabhang, 'the destroyer of pity,' from the severity of the ascent.

power of two ridges upon the single contained river-basin. The synclinal lines from the inner faces of the two adjacent ridges draw the waters together; and, because these ridged peaks are the *loftiest masses of the entire mountains*, the effect of all their other masses, even that of the spine of Himáchal or the ghát-line of the snows, is over-ruled or modified, so that in the most rugged region on earth a very limited series of distinct main rivers appears in the plains from innumerable independent alpine feeders, in the manner which all behold, but few indeed think of referring to its cause.*

It is inconsistent with all we know of the action of those hypogene forces which raise mountains, to suppose that the points of greatest intensity in the pristine action of such forces, as marked by the loftiest peaks, should not be surrounded by a proportionate circumjacent intumescence of the general mass; and, if there *be* such an intumescence of the general surface around each pre-eminent Himálayan peak, it will follow, as clearly in logical sequence as in plain fact it is apparent, that these grand peak-crowned ridges will determine the essential character of the aqueous distribution of the very extended mountainous chain (1,800 miles) along which they occur at certain palpable and tolerably regular intervals. Now, that the infinite volume of the Himálayan waters is, in fact, pretty regularly distributed into a small number of large rivers, we all see; and, whereas the fact is thoroughly explicable upon my assumption, that the great peaks bound, instead of intersecting, the river-basins, it is wholly inexplicable upon Captain Herbert's assumption that the said peaks intersect the basins.

The above are normal samples of Himálayan water-distribution, and it is very observable that, whereas all those principal streams which exhibit the unitizing principle so decidedly, take their origine in the alpine region, at or near the snows, so the inferior streams, which rise from the middle region only, show no such tendency to union, but pursue their solitary routes to the Ganges; as for example, the Máhánada, the Konki, the Bágmatti, the Gumti, the Ráputi, the Cósilla, and the Rámganga. Here is both positive and negative evidence in favour of the doctrine I advocate, as furnishing the key to the aqueous system and natural divisions of the Himálaya; for the upper rivers do, and the lower rivers do not, stand exposed to the influence of the great peaks.

The petty streams of the lower region, or that next the plains, which water the Dhúns or Máris, traverse those valleys lengthwise; and as the valleys themselves run usually parallel to the ghát-line of the snows, such is also the direction of these petty streams. In the central, as in the western,* hills, they usually disembogue into the rivers of the first class.

* Since this was written a new peak of transcendant height has been determined, which yet does not influence the river basins of the Indian slope. The reason is that this peak is thrown back behind the ghát-line like Chumalhári, as to which see on. Such facts need not affect the justice of what is written above, but must be regarded as exceptional, at least for the present.

* J. A. S. No. 126, p. xxxiii.

I have observed that the three great river basins of the Karnáli, Gandak, and Cósí extend throughout Népál, and truly so; for a river basin includes the widest space drained by its feeders. But it results necessarily from the manner in which the deltic basins of the Himálayan rivers are formed, that there should be intervals between the plainward *apices* of these deltic basins. Of these intervals the most conspicuous in Népál is that which intervenes between the Cósí and Gandak. This tract, watered by the Bágmatti, deserves separate mention on many accounts, and it may be conveniently styled the valley region, since it contains not only the great valley of Népál proper, but also the subordinate vales of Chitlong, Banépá, and Panouti.

It has been already remarked, that the classifications of physical geography, as of the other sciences, do not constitute a perfect "open sesamè" to the mysteries of nature, but only a material help to their study. This observation I will illustrate by a few comments on the basin of the Tishta, lest the somewhat anomalous instance of that basin should be captiously quoted to impugn the doctrine I contend for; but contend for, not as exhibiting in every instance an absolute conformity with natural arrangements, but as doing all that can be reasonably expected in that way, and as furnishing, upon the whole, a generally truthful, causally significant, and practically useful, indication of those arrangements.

I have stated above, that the basin of the Tishta extends from the peak of Kangchan to that of Chumalhári. Between these two peaks there occurs what miners call "a fault" in the ghát-line of the snows, which line, after proceeding N. Easterly from the Láchén pass to Powhanry,‡ dips suddenly to the south for nearly forty miles, and then returns to Chumalhári. A triangular space called Chúmbi is thus detached from the Himálaya and attached to Tibet; and the basin of the Tishta is thus narrowed on the east by this salient angle of the snows, which cuts off the Chúmbi district from the Tishtan basin, instead of allowing that basin to stretch easterly to the base of Chumalhári. Chúmbi is drained by the Máchú of Campbell, which is doubtfully referred to the Torsha of the plains, but which may possibly be identical with the Háchú of Turner and Griffiths,§ and consequently with the Gaddáda of the plains. But besides that these points are still unsettled, one of the transnivean feeders of the Tishta rounds Pow-

‡ *Vide* Waugh's outline of the snowy range of Sikim, J. A. S. *loc. cit.*

§ Embassy to Tibet and J. A. S. Nos. 87 and 88, with sketch maps annexed. Also Pemberton's large map of the Eastern frontier. Rennell is not easily reconcilable with them. I had identified the lakes of Cholámú, which give rise to the Tishta, with Turner's lakes. But I now learn from Hooker, that the latter lie a good deal east of the former, and I am satisfied that Campbell's Máchú is distinct from Turner's Háchú. We need, and shall thus find, space in the hills, correspondent to that in the plains watered by Rennell's Torsha and Saradingoh and Gaddáda and Súncósí. The Máchú, (Maha tchieu of Turner) rises from the west flank of Chumalhári. The Háchú of Turner is a feeder joining his Tehin chú from the west; the Chaan chu of Turner is the Suncósí (the Eastern Suncósí, for there are two there, besides that of Népál,) of Rangpur, his Tehin chu is the Gaddáda, and his Maha chu the Torsha. The Arun has its rise in the broken country of Tibet lying north-east and west of the sources of the Tishta and south of the Kambalá, or great range forming the southern boundary of the valley of the Yáru; this broken country Dr. Hooker estimates at from sixteen to eighteen thousand feet above the sea. It is a good deal terraced near Himáchal.

hanry and rises from a lake (Cholamú) approximating to Chumalhári; so that, one way or another, the Tishta may be said, without much violence, to spread its basin from Kangchan to Chumalhári.

Chúmbi and all the adjacent parts of the plateau of Tibet constitute a region as singular as is the access to it from Sikim by the Láchén pass. That pass surmounted, you at once find yourself, without descent, upon an open undulated swardy tract, through which the eastern transnivean feeders of the Tishta and of the Arún sluggishly and tortuously creep, as though loath to pass the Himá laya, towards which indeed it is not easy to perceive how they are impelled; the plateau of Tibet generally sloping on their right to Digarchi, and seeming to invite the streams that way. This is however of course a water-shed, though by no means a palpable one; and we know by the signal instances of the vast rivers of South America and those of North-eastern Europe, how inconspicuous sometimes are the most important water-sheds of the globe. The sources and courses of the feeders of the Tishta will shortly be fully illustrated by Dr. Hooker, my enterprizing and accomplished guest, to whom I am indebted for the above information relative to the Láchén pass and its vicinity, and whose promised map of Sikim, which state is the political equivalent for the basin of the Tishta, will leave nothing to be desired further on that head.*

But the Himálaya must necessarily be contemplated in its breadth as well as its length; and we have therefore still to consider what regional divisions belong to these mountains in relation to their breadth, or the distance between the ghát-line of the snows and the plains of India.

The Himálayan mountains extend from the great bend of the Indus to the great bend of the Bráhmapútra, or from Gilgit to Bráhmakúnd, between which their length is 1,800 miles. Their mean breadth is (reckoning from the gháts and purposely omitting the questions of axis and counterslope) about ninety miles; the maximum, about 110, and the minimum, seventy miles. The mean breadth of ninety miles may be most conveniently divided into three equal portions, each of which will therefore have thirty miles of extent. These transverse climatic divisions must be, of course, more or less arbitrary, and a microscopic vision would be disposed to increase them considerably beyond three, with reference to geological, to botanical, or to zoological, phænomena. But upon comparing Captain Herbert's distribution of geological phænomena with my own of zoological, and Dr. Hooker's of botanical, I am satisfied that three are enough. These regions I have already‡ denominated the lower, the middle, and the upper. They extend from the external margin of the Tarai to the ghát-line of the snows. The lower region may be conveniently divided into — I. the sand-stone range with its contained Dhúns or Máris—II. the Bháver or Saul forest—III. the Tarai. The other two regions require no sub-divisions. The following appear to be those demarcations by height which most fitly indicate the three regions:—

* The reader will observe that this paper was written in 1846.
‡ J. A. S. for December 1847 and June 1848.

Name.	*Elevational limits.*
Lower region	Level of the plains to 4,000 feet above the sea.
Central region	4,000 to 10,000 feet above the sea.
Upper region	10,000 to 16,000‡ feet above the sea: highest peak measured is 29,002 feet.

It is needless to remind those who are conversant with physical geography, that in passing in a tropical country, by a long and gradual ascent, from near the sea level to several (4-6) miles above it, one must necessarily meet with regions equivalent, *quoad* organic phænomena, to the three great zones of the earth, or the tropical, the temperate, and the arctic; and, in fact, our three regions above indicated correspond in the main with those zones, and might be named after them, but that it is desirable to avoid terms involving theory, when those designating mere facts will suffice. But to resume. It is thus made apparent that the Himálaya, or, to be more precise, the Indian slope of the Himálaya, admits of a double series of natural and convenient divisions, those of length being coincident with the basins of the main rivers, and those of breadth with a triple division on the scale of elevations, from that of the plains to that of the perpetual snow, which latter tallies pretty nearly with the mean height of the passes into Tibet, or sixteen to seventeen thousand feet. But, as the plains are customarily divided into the upper, central, and lower provinces, so the Himálaya, in reference to its length, may be conveniently divided, when larger divisions than those of the river basins need to be spoken of, into the western, embracing the basins of the Jhilum, Chinab, Bias, Ravi, Satluj, Jamna, Ganges, Ghagra, within the British territories; the central, including the basins of the Karnáli, the Gandak, and the Cósí, within those of Népál; and the eastern, embracing the basins of the Tishta, Mónas, Subhansri, and Dihong, which include Sikim, (now half British), Bhútan, and the territories of the disunited lawless tribes lying east of Bhútan. And it is very observable that, in respect of climate, the above suggested analogous divisions of the plains and mountains correspond, for the more you go westward in plains or mountains, the greater becomes the dryness of the air and the extremes of heat and cold.

But the grand determiner of climate, as dependent on heat, in all parts of the Himálaya, is elevation, which acts so powerfully and uniformly, that for every thousand feet of height gained, you have a diminution of temperature equal to 3° or 3½° of Fahrenheit: consequently the transverse regions, notwithstanding their proximity, show, upon the whole, a much more palpable variety of climate than is incident to the lengthwise divisions of the chain, how remote soever they may be. But in reference to moisture, the next element of climate, the case is somewhat altered, for every movement towards the west (N.W.) along the lengthwise development of the Himálaya, carries you further and further out of

‡ This is about the average height of the gháts and of the perpetual snow. It is also nearly the limit of possible investigation, and of the existence of organic phænomena. But the upward limit need not be rigorously assigned—4,000 is the limit of snow-fall to the south, well tested in thirty years—4,000 is also that point which best indicates the distinction of healthful and malarious sites.

the line of the rainy monsoon, which is the grand source of supply of moisture. The third determining and very active cause of climate operates throughout the chain, determining chiefly the specific differences. It consists in the number, height, and direction of the ridges interposed between any given position and the direction of the S. W. or rainy monsoon; for, each of these ridges, crossing more or less directly the course of the vapour from the ocean, has a most marked effect in diminishing the quantity of rain and moisture behind such covering ridge, so that, inasmuch as by receding from the plains towards the snows, you interpose more and more of these ridges, you find not only temperature falling with elevation gained (as a general rule,) but also greater dryness of air, less moisture, more sunshine, (and so far more heat); and, as a general consequence, a gradual diminution of that excessive natural vegetation, arboreal and other, which is the universal characteristic of these mountains; yet still with greater power in the climate of these remoter districts of ripening grains and fruits of artificial growth, owing to the diminished rain and increased sunshine of summer, and in spite of the general decrease of the temperature of the air. That combination of tropical heat and moisture to which we owe the generally "gorgeous garniture" of mountains so stupendous has, at low elevations, the bad effect of generating a malaria fatal to all but the peculiar tribes, whom ages untold have been inured to it, and whose power of dwelling with impunity in such an atmosphere is a physiological fact of very great interest. The tribes adverted to are called Awalias, from *áwal*, the name of malaria.

The whole of what I have denominated the "lower-region," as well as all the deep beds of the larger rivers of the "central region," lying much below what I have given as the elevational demarcation of the two regions, or four thousand feet, are subject to the *áwal*.

After what has been stated, it will be seen at once, that tables of temperature, rain-fall, and moisture, could, if given, only hold true of the exact spots where they were registered.

The latitude in a small degree, but in a far greater, the longitude, or position with reference to the course of the rainy monsoon—the number of interposed ridges crossing that course—and the elevation, are the circumstances determining the heat and moisture, that is, the climate, of any given spot of the Eastern, Central, or Western Himálaya. There are amazing differences of climate in very proximate places of equal elevation, caused by their relative position to covering ridges, and also, as has been proved experimentally, by the effects of clearance of the forest and undergrowth, and letting in the sun upon the soil.

The general course of the seasons is the tropical, with cold and dry weather from October to March, and wet and hot weather from April to September, correspondent to the duration of the N.E. and S.W. monsoons. The springs and autumns, however, are more clearly marked than the latitude would promise, and from the middle of March to the middle of May, and again, from the middle of September to the middle of December, the weather is delightful. From the middle of December to the end of February is the least agreeable portion of the year, being cloudy and rainy or snowy, with cold enough to make the wet tell disagreeably, which it does not do in the genial season of the rains. The general character of the climate is derived from its combined and great equability and temperateness. For months the thermometer hardly ranges 5° day and night, and that about "temperate" of Fahrenheit, or the perfection of temperature; and altogether, the climate is one of the safest (I here speak of the central and normal region) and most enjoyable in the world. The wind is generally moderate, except in March, when the "Phagwa" of the N.W. plains reach us, but shorn of its fervour. The quantity of electricity is, on the whole, small, and storms are nearly confined to the setting in and close of the rainy season. Epidemics are very rare; endemics almost unknown; so that it would be difficult to cite a Himálayan disease, unless such must be called dyspepsia. Goitre is more or less prevalent, but not often accompanied by cretanism. The general character of the surface in all parts of the Himálaya is a perpetual succession of vast ridges, highly sloped, and having very narrow interposed glens. Valleys properly so called are most rare. There are, in fact, only two throughout the great extent from Gilgit to Bráhmakúnd, or those of Cashmere and Népál, the latter only sixteen miles in either diameter.

Lakes also are small and very infrequent. Three or four in Kúmáún, and two or three in Western Népál (Pókra), in both cases juxta-posed, constitute the whole nearly. But it seems certain that lakes were more frequent in some prior geological era, and that the present valleys of Cashmere and Népál once existed in a lacastrine state.

The Himálayan ridges are remarkable for the absence of chasm and rupture, and their interminable uniform lines, with the similarity of tone in the verdure of the ceaseless forests, (owing to the rarity of deciduous trees), detract somewhat from those impressions of grandeur and beauty, which mountains so stupendous and so magnificently clothed are calculated to convey. The transverse or climatic division of the Himálaya, though of course most noticeable and important in reference to organic phænomena, is also worth attention, in regard to inorganic ones. I shall however say little of the geology or of the botany of the Himálaya, abler pens than mine having now treated the subject. A little more space may be given to the ethnology and zoology, both as matters I myself am more conversant with, and which still have a deal of novelty in reference to geographical distributions particularly.

Every part of the chain abounds in minerals, particularly iron and copper; lead, zinc, sulphur, plumbago, in less degree. Mineral springs, both hot and cold, sapid and insipid, are generally diffused, and I am aware of other instances of lambent flame issuing in the fashion of the well-known Jwálamúkhi of the Punjáb, which superstition has consecrated. There is no lime-formation, and the mineral is very rare as a deposit: salt is unknown, though it abounds across the snows. So also the precious metals. Minerals and mineral springs are most frequent in the central region, so likewise the iron and copper veins: organic fossil remains and the small traces of coal, almost or quite peculiar to the lower region, and far more abundant to the N.W. than to the S.E. In geology the upper region may be called the locale of granites and gneisses; the middle region that of gneisses and schists; the lower region that of the sandstone formation and of diluvial débris. It may be added that granite is much more extensively developed in the upper region than had been supposed, and that igneous rocks are by no means so entirely wanting: indeed, igneous action is displayed to a stupendous extent in the hypogene rocks, both stratified and unstratified, of the upper and central regions. There are no volcanos, active or extinct. Slight earthquakes are very frequent: severe ones, rare; very severe ones, unknown.

In botany the upper region is that of Junipers, Cypresses, Cedars, Larches, Yews, Poplars, Boxes, Dwarf Rhododendrons, Hollies, Willows, Walnuts, Birches, and, in general, of the superior Conifers, particularly to the S.E., for to the N.W. they descend into the middle region, even the stately Cedar, which however is unknown east of Kúmáun. In the second or central region* Birches, Hollies, and Willows recur. It is the region of Oaks, Chesnuts, Horse Chesnuts, Magnolias, Laurels, Alders, Tree Rhododendrons, Cherry and Pear Trees (large and wild), Oleas (forest trees), Maples or Sycamores, Thorns, Ashes, Elms, Horn-beams, Elders, Paper and Wax Trees, Tea Allies, (Eurya and Thea also,† as an importation which has succeeded to perfection, but chiefly below 4,000, Tree Ferns, some few and peculiar Palms (Chamœrops, etc.), and the inferior sorts of Pines.

The third or lower region is that of Sauls (Shorea) Sissus (Dalbergia), Acacias and Mimosas, Tunds (Cedrela), Cotton Trees (Bombax), Tree Figs (Elasticus, Indicus, Religiosos, etc.), Buteas, Dillenias, Duabangas, Erythrinas, Premnas, some common Palms (Phœnix), etc., but rare and poor, with recurring Tree Ferns, but more rarely than above perhaps, though the Tree and common Ferns, like the great and small bamboos, may be said to be borderers, denoting by their point of contact the transition from the lower to the central region. Pinus longifolia recurs in the lower region, descending to the plains nearly in Népál, but most of the other Conebearers in Népál, and still more, east of it, eschew even the central region, abundant as they are therein in the Western Himálaya. So likewise the Tree Rhododendrons in the Eastern Himálaya are apt to retire to the northern region, though in the Central Himálaya they abound in the central region.

In zoology, again, to begin with man, the northern region is the exclusive habitat of the Bhótias (Cis-Himálayan, called Palusén, Róngbo, Sérpa, Siena, Káthbhótia, etc.,) who with their allies the Thakoras and Pákías extend along the whole line of the ghâts, and who, with the name, have retained unchanged the lingual and physical characteristics, and even the manners, customs, and dress, of their transnivean brethren. To the central region are similarly confined, but each in their own province from east to west, the Mishmis and Mirris, the Bors and Abors, the Kapachors, the Akas, the Daphlas, (east of Bhútan), the

* *N.B.*—Central in length is called, central only, or central Himálaya; central of breadth, central region.

† Both tea and coffee plantations are now well advanced in the Eastern Himálaya, with the surest prospect of success. In the Western Himálaya that success is now a fact accomplished.

Lhópas (in Bhútan), the Lepshas or Deunjongmaro (in Sikim), the Límbus or Yakthúmbas, the Yakhas, the Khómbos or Kirántis, the Múrmis or Tamars, the Pahi or Padhi, the Néwárs, the Súnwárs, the Chépángs, the Kúsúndas, the Vayus or Kayus, the Gúrungs, the Magars, the Khas or Khasias (in Népál), the Kohlis, the Doms, the Rajhis, the Haris, the Garhwalis, the Kanets, the Dogras,* the Kakkas, the Bambas, the Gakars, the Dardus, the Dúnghars (west from Népál). To the lower region again, and to similarly malarious sites of the middle region, are exclusively confined, the Kocches, the Bodos, the Dhimáls, (Sikim and east of it), the Kíchaks, the Pallas, the Thárus, the Dénwárs, the Kúmhas, the Bhrámus, the Dahis or Daris, the Kuswárs, the Thámis, the Bótias (not Bhótia) (in Népál), the Bóksas (in Kúmáun), the Khátirs, the Awans, the Janjohs, the Chibs, and the Báhóas (west of Kúmáun to the Indus).

The Himálayan population is intensely tribe-ish, and is susceptible of a threefold division of pregnant significance, and quite analagous to what holds true of the aboriginal Indian and Indo-Chinese populations, *viz.*, *first*, into the dominant or unbroken tribes, such as the Khas, Magar, Gúrung, Néwár, Múrmi, Lepsha, Bodpa, etc.; *second*, into the broken tribes, such as nearly all those termed Awalias,† as well as the Chépáng, Kusúnda, and Háyu; *third*, into the tribes of helot craftsmen :—

Of the mountains of Nepal.	Of the valley of Nepal.
Chunára, carpenters.	Pó, executioners and workers in bamboo.
Sárki, curriers.	Kulu, curriers.
Kámi, blacksmiths.	Náy, butchers.
Sunar, gold and silver smiths.	Chamakhala, scavengers.
Gáïn, musicians.	Dong, Jugi, musicians.
Bhánr, ditto, but prostitute their women.	Kou, blacksmiths.
Damai, tailors.	Dhusi, metallurgists.
Agri, miners.	Awa, architects.
Kumhal or Kinari, } potters.	Báli, agriculturists.
	Nou, barbers.
	Kuma, potters.
	Sangat, washermen.
	Tatti, makers of shrouds.
	Gatha, gardiners.
	Sáwo, bleeders & suppliers of leeches.
	Chhipi, dyers.
	Sikami, carpenters.
	Dakami, house builders.
	Lóhóngkami, stone cutters.

* The late Captain Cunningham (in epist.) refers the Dardurs (Darada) and the Donghers to the upper region, as also the Kanets, who extend northward, beyond the Himálaya, where they even form "the mass" on either side the Satluj. They are of mixed origin, like the Khas of Népál, the Dogras of Punjáb, and the Gadhi of Chamba.

† A list of Awalias;—1 Kocch, 2 Bodo, 3 Dhimal, 4 Garo, 5 Dólkhali, 6 Batar or Bór, 7 Kudi, 8 Hájong, 9 Dhanuk, 10 Maraha, 11 Amát, 12 Kébrat, 13 Kíchak, 14 Palla, 15 Tháruh (not own name in Sallyan), 16 Bóksa (Kumaon), 17 Dahi or Darhi (allied to Bhramu), 18 Thámi, 19 Pahi or Pahri (allied to Newar and Murmi), 20 Kumha (not own name), 21 Botia (allied to Kuswár), 22 Kuswár, 23 Denwar (allied to two last), 24 Bhrámu (allied to Dahi), 25 Váyu (not Awalias but broken tribe), 26 Chepang, and 27 Kusunda (ditto).

The position and affinities of the last are still to me an enigma, as they were when I adverted to them in my work on the Koech, Bodo, and Dhimál. As blacksmiths,* carpenters, curriers, etc., their services are, and ever have been, invaluable; yet they are degraded to the extent of being outcasts. Their manners have little, and their tongues nothing, and their physical attributes not much, to denote their race and lineage. Of the other two masses of the population, the unbroken tribes are clearly the more recent immigrants from the north, and in general they are distinguished by languages of the simpler Turanian type, whereas the languages of the other or broken tribes are of the complex or pronomenalized type, tending, like their physical attributes, towards assimulation with the Dravirian or the Ho, Sontal or Munda, sub-families of the sons of Túr. These broken tribes are demonstrated by their relative position to be of far older date in the Himálaya as in Indo-China, and perhaps also in India, than the unbroken; and altogether, the phænomena of ethnology in the Himálaya warrant the conclusions, that the Himálayas were peopled by successive swarms from the great Turanian hive, and that its tribes are still traceably akin alike to the Altaic branch of the north and to the Dravirian and Munda of the south.‡ The Khas, Kanéts and Dógras, and several others of the Western Himálaya, are clearly of mixed breed; aboriginal Tartars by the mother's side, but Arians (Bráhman and Kshétriya) by the father's, as I have shown in my memoir on the military tribes of Népál. (J.A.S.B. May 1833.)

In reference to those European speculations touching the peopling of the Indian continent which have been lately raised, chiefly on the basis of my vocabularies, I may remark generally, that very remotely sundered periods of immigration, from the north by no means involve totally different *routes* of immigration, and still less races so trenchantly demarked from all the priorly recognized ones as have been lately assumed and denominated Gangetic, Lohitic, Taic, &c. Every day multiplies the proofs of affinity between the Himálayans and the recognized sub-families of Altaia, Indo-China, and Draviria; whilst, abating the single fact of the Brahoi tribe having lingual affinities with the Turanians, I see no safe ground for assuming that the sons of Túr entered India generally or exclusively by the well-known route of the immigrant Arians, or by any yet more southerly route. The hundred gates of the Himálaya and of its off-shoots have stood open in all ages; beyond them, in all ages, have dwelt the diversely tongued and featured tribes of the vastest, and most erratic, and most anciently widespread, but still single branch of the human race; and, as I find similar diversities of tongue and feature, characterising that branch alike in the Cis and Trans-Himálayan countries, so I believe that the former have been peopled from the latter by successive incursions along the whole Himálayan ghát line, of races and tribes which there is yet no sufficient ground for contra-distinguishing from all the here-

* Of all the unbroken tribes, the Magar alone have their own miners and smiths. See and compare what is told of the old mines and miners of the Altai. See also a note in my work on the Kocch, Bodo and Dhimál.

‡ See paper on Nilgirians, J. A. S. B., and also two essays on the Vayu and Bahing tribes, in the same Journal (1857).

tofore recognized ones of the north.* African immigration at any time, and by any route, appears to me a sheer assumption. But it may well be, that some of the sons of Túr entered by the Arian route, and that these were among the earliest immigrants, whose more westerly abode and point of entrance into India is still indicated by the higher structured tongues of their presumed descendants. But we must not forget that there are complex tongues at the eastern as well as at the western extremity of the Altaic region (in its wide sense); that many of these tongues are most imperfectly known; that Sífán and Central Himálaya and Indo-China are now known to be tenanted by races speaking tongues of the complex type, some even more complex than the Dravirian, and more allied to the Gónd, Hó and Sontal type; § and, above all, that the essential character (including differences and resemblances) of the above adverted to several sub-types of language, embracing the true affiliation of the races using them, is yet to be determined. So that we can only now safely say that the general relationship of all the sons of Túr in and beyond India is as certain as their more special and close affinities are uncertain.†

But to proceed with our zoological enumerations. To the upper region exclusively belong, among the ruminants, the bisons (poephagus) and musks, the wild goats (ibex, hemitragus) and wild sheep (pseudois, caprovis); among the rodents, the marmots and pikas (lagomys); among plantigrades, the bears proper (ursus). In the middle region, true bovines (bos) take the place of the bisons of the upper region; bovine and caprine antelopes (budorcas, capricornis, nemorhedus) replace its musks and wild goats and sheep; common rats and mice, and hares and porcupines and hedgehogs its marmots and pikas; and sun bears (helarctos) its true bears; whilst the deer family, unknown to the upper region, is here represented only‡ by the anomalous stilt-horns (stylocerus). In the lower region the ox-family is represented by bibos and bubalus (splendid wild types); the deer family, here abundant, by rusas, rucervi, axises, and stilt-horns to boot; the

* I allude more particularly to the writings of Prof. Max Müller and Dr. Logan No one can more freely than myself admit the scholastic attainments and skill in the science of grammar of the former, or the immense and skilful industry of the latter. But I demur to their inductions, nor can I see the advantage of multiplying nominal, that is to say, undefined or crudely defined ethnological groups. We must have first a just definition of the family, and thereafter, by and bye, definitions of the several sub-families already recognized, when the definition of the rest may follow.

§ See the essays on the Vayu and Bahing now published in the Jour. As. Soc. Bengal, [A.D. 1857].

† In my papers on the Nilgirians and in those on the Vayu and Bahing, above alluded to, I have classed the Himálayans under the two great divisions, of such as use pronomenalized and complex and such as use non-pronomenalized and simple tongues. In the memoirs on the Vayu and Bahing, I have analysed their languages as exemplars of the complex type of speech in Himálaya. The double pronomenalization of those two tongues, indicates their close affinity to the Ho-Sontal group of languages of the plains.

‡ I am fully aware that Rusas (sámber) are found in the western hills, but a careful consideration of the facts in that part of the Himálaya, with due advertence to the known habits of the group, satisfies me that these Deer have been driven into the western hills by the clearance of the Tarai and Bháver. For some remarks on this subject, see J.A.S. of Bengal No. 211, for January 1850, page 37.

antelopes by tetracerus, or the four-horned kind; the rodents by the bambú rats (rizomys) and spiny hares (caprolagus); and the bear family by the honey-bears (melursus); add to all which that to this region are exclusively confined all the large pachydermes, such as the elephant and rhinoceros; and the monkeys also (semnopithecus et macacus), though not so exclusively in their case. The carnivora, again, are represented in the upper region by ounces, by foxes of a large sort (montanus), by the weasels proper, and by the ailuri or catlories; in the middle region, by the wild dogs (cyon), the marten-weasels, leopards, thick-tailed leopards (macroceloides), wild cats (murmensis, pardochrous, ogilbii), chauces or Lybian lynxes (Lybicus), zibets, screwtails (paradoxurus), and priono-dons; and in the lower region by tigers, leopards, hyenas, wolves, jackals,* insectivorous foxes (kokri), bear-badgers (ursitaxus), sand-bears (arctonyx), urvas, mangooses, helictes or Oriental gluttons, small civets (viverrula), hirsute screw-tails, and sharp-faced cats (celidogaster). Zibets and chauses recur in this region frequently, and one small species of mangoose (auropunctata) is found in special spots of the central region. The otters in the upper region are represented by the small golden and brown species (aurobrunnea); in the central, by monticola and indigitata; in the lower, by the large Chinese species (Sinensis). Among the squirrels, the great thick-tailed and large purple species (macruroides et purpureus) belong solely to the lower region; the small lokries (locria et locro-ides) to the central; and the Siberian, to the upper; whilst flying squirrels, a numerous group, (magnificus, senex, chrysothrix, alboniger), are confined to the central region, so far as appears. In the bat group, the frugivorous species, or pteropines, all are limited to the lower region, whilst the horse shoes (rhinolophinæ) specially affect the central region; and the bats proper (vespertilioninæ) seem to be the chief representatives of the family in the northern region. From the class of birds, we may select, as characteristic of the three regions, the following:—

The true pheasants [phasianus], the tetrougalli, the sanguine pheasants [ithaginis], the horned and crested pheasants [ceriornis, lophophorus] of the upper region, are replaced by fowl-pheasants [galophasis]† in the mid-region, and by

* Jackals have made their way (like crows and sparrows) to the most populous spots of the central region, but they are not proper to the region, nor Indian foxes, though some of the latter turned out by me in 1827 in the great valley of Népál have multiplied and settled their race there. *Ab his disce alia.* Tigers, for example, are sometimes found in the central and even northern region. But ample experience justifies my asserting that they are wandering and casual intruders there, whereas leopards are as decidedly fixed and permanent dwellers. As a sportsman during twenty years, I have, whilst shooting pheasants and cocks, fallen in with innumerable leopards, whose fixed abode in numberless *locales* was pressed on my attention involuntarily. But I never fell in with a single tiger, and I know them to be wanderers and intruders.

† The influence of longitude on geographic distribution might be singularly illustrated, did space permit, from numerous Himálayan groups, Galline and other: thus, for example, a black-breasted Ceriornis is never seen east of the Káli, nor a red-breasted one west of it. So of the black and white-crested Gallophasis; whilst a black-backed one is never seen west of the Arún, nor a white-back east of it. With reference to the more dominant influence of latitude, or what is the same thing. elevation, I may add that the Rasores of the three transverse regions exhibit an exquisite sample of gradation from a Boreal or Alpine to a tropical type; Phasianus, Gallophasis and Gallus

fowls proper (gallus) in the lower. In like manner, among the partridges (perdicinæ), the grouse and snow-partridges (lerva and sacfa) belong exclusively to the upper region ;§ the chakórs (caccabis) and the tree partridges (arboricola) to the central ; and the francolines (francolinus) to the lower, though the black species of this last form are also found in the mid-region. In the pigeon group the blanched pigeons (leuconota) belong solely to the upper region ; the vinous pigeons (Hodgsoni) to the central ; and the green, the golden, and the banded (treron, chalcophaps, macropygia) almost as entirely to the lower ; the trerons alone partially entering the central tract from the lower.

The splendid Edolian shrikes (chibia, chaptia, edolius) belong exclusively to the lower region. They are replaced in the central tract by plain dicrurines, and in the upper by plainer lanians. The cotton-birds (campephaga) of the south are replaced by gaudy ampelines (cochoa) and leiothricinians (leiothrix, pteruthius, cutia) in the middle region; but both groups seem excluded from the north. Among the fly-catchers the gaudy or remarkable species and forms belong wholly or chiefly to the lower region, as tchitrea, rhipidura, cryptolopha, myiagra, hemichelidon, chelidorhynx ; whilst those which approach the warblers (niltava, siphia, digenea) belong to the mid-region ; and the plainer and more European types are alone found in the northern.

Among the fissirostres, goat-suckers and swallows are pretty generally distributed; but rollers, bee-eaters, eurylaimi, trogons, and all such gaudy types belong to the south, with only occasional alpine representatives, as bucia is of merops. The tenuirostral birds belong distinctly to the lower region, yet they have representatives or summer visitants in all three, even among the sun-birds. Upon the whole, however, it may be safely said that the sun-birds (nectarinia) belong to the south ; the honey-suckers (meliphagidæ) to the centre and south ; and the creepers, honey-guides, nut-hatches, and wrens‡ to the north and centre. The sylvians or warblers are too ubiquitarian, or too migratory for our present purpose even Boreal types being common in the lower region in the cold weather. Hornbills, barbets, parroquets (palæornis, psittacula) belong to the lower region, though they have a few representatives in the central; none in the upper. Woodpeckers abound in the lower and central regions, but are rare in the upper. True cuckoos (cuculus) are as common and numerous (species and individuals) in the central region as walking cuckoos (phænicophaus, centropus, &c.) are in the southern, where also the golden (chrysococcyx) and dicrurine cuckoos (pseudornis) have their sole abode ; whilst what few of the group belong to the upper region are all allied to the European type. Of the conirostral group, the ravens, pies, choughs, nut-crackers, and conostomes of the upper region are replaced in the central region by tree pies (cissa, dendrocitta), jays, rocket-birds (psilorhinus),

being thoroughly normal forms of their respective regions, and Gallophasis being as intermediate in structure and habit as in *locale*.

§ Sacfa and Crosoptilon are more properly Tibetan.

‡ I have in this paper followed, without entirely approving Mr. Gray Junior's classification of my collections in the printed catalogue. The geographic distribution is now attempted for the first time. But I will recur to the subject in a separate paper devoted to it.

pie-thrushes (garrulax), timalias, and hoopoe thrushes (pomatorhinus); and in the lower region by the common Indian crows (culminatus et splendens), grackles, pastors,* stares, vagabond-pies and dirt-birds (malococercus). Thrushes proper, with rock-thrushes, ousels, myophones, zootheres, tesias, and hypsipetes are as abundant in the central and upper region as bulbuls, orioles, pittas are in the central and lower.

In the finch family the haw-finches, bull-finches, gold-finches and cross-bills (loxia) are as strictly confined to the upper region as are the corvine-conostomes, nut-crackers, choughs and ravens. The former are replaced in the central region by the buntings, wood-finches (montifringilla) and siskins; and in the lower region, by the weavers and múnias. The raptorial-birds are in general to cosmopolitan to subserve the purposes of geographic distribution. Still it may be remarked that the archibuteos and true eagles belong, *quoad* breeding at least, to the upper region; the crested eagles (circæetus,) the neopuses and hawk eagles (spizætus) to the central; and the pernes *(haliætus et pandion)* and haliasturs to the lower. Among the vultures the distinction is more marked; for the eagle vultures *(gypaetus)* belong exclusively to the upper region; the large European vultures *(fulvus et cinereus)* to the central; and the neophrons and the small Indian vultures *(Bengalensis et tenuirostris)* to the lower. The Himálaya abounds in falconidæ, all the occidental types and species being found there, and many more peculiar and oriental ones; and it deserves special remark that whereas the former *(imperialis, chrysætos, lanarius, peregrinus, palumbarius, nisus, etc.)* affect the upper and central regions, the oriental types *(hypotriorchis, haliastar, ierax, hyptiopus vel baza, elanus, poliornis)* are quite confined to the lower region.

Those perfect cosmopolitans, the waders and swimmers, migrate regularly in April and October, between the plains of India and Tibet, and, in general, may be said to be wanting in the mountains, though most abundant in the Tarai. The great herons *(nobilis et cinereus;)* the great storks *(nigra et purpurea,)* and great cranes (the *cyrus*, *culung*, and *damoiselle*) of the Tarai are never seen in the mountains, where the egrets alone and the little green and the maroon-backed represent the first group. But the soft-billed smaller waders (*scolopacidæ*) are sufficiently common in the mountains, in which the woodcock‡ abounds, breeding in the upper region and frequenting the central, and rarely the lower region, from October till April. Geese, ducks, and teals swarm in the Tarai, where every occidental type, so to speak, for they are ubiquitous, may be seen from October till April; and many oriental non-migratory types; whereas in the mountains the mergansers (*orientalis*) and the corvorants (*Sinensis et pygmæus*) only are found, and that very scantily; with a few rails, ibisbills, porphyrios, hiaticulas, gallinules,

* When Darjeeling was established, there was not a crow or pastor or sparrow to be seen. Now there are a few crows and sparrows, but no pastors. Enormously abundant as all are in the lower region, this sufficiently proves they are not native to the central tract, though common in the great valley of Népál. Sparrows first seen in 1855. Crows soon made their appearance.

‡ H. Schlagintweit procured a woodcock with its nest and young in June at an elevation of about 12,000 to 13,000 feet. They are frequently got, and snipes also, in the scrub rhododendron thickets near the snows.

and sandpipers, out of the vast host of the waders.‡ In the way of general remark I may observe that the zoology of the Himálaya is much richer in the multitude of its divers forms (genera and species) than in individuals of the same form, and that it is remarkably allied to the zoology of the Malayan islands, as may be seen at once by a reference to the excellent work of Horsfield. As you pass northwards, towards and across the snows, the forms and species tend much to approximation with those of our European home; but the species are not often absolutely identical.

But I must hasten from these zoological details to make some remarks on the sub-divisions of the lower region, a subject which, though in many ways interesting and important, is so little understood, that the celebrated Mrs. Somerville, in her excellent treatise of Physical Geography, has represented the Tarai as being *within*, not only the Bháver, but the Sandstone range.§

All observant persons who have proceeded from any part of the plains of India into the Himálaya are sensible of having passed through an intermediate region distinguished by many peculiarities; and, if their route have lain to the north-west, they can hardly have failed to notice successively the verdant Tarai, so unlike the arid plains of Upper India; the vast primæval Saul forest, so every way unique; and the Dhúns or valleys, separated from the last tract by a low range of hills. The natives of the plains have in all ages recognized these several distinct parts of the lower Himálayan region, which they have ever been, and are still wont to frequent periodically, as strangers and foreigners, in order to graze innumerable herds of cows and buffaloes in the Tarai, or to procure the indispensable timber and elephants peculiar to the Bháver, or to obtain the much-prized drugs and dyes, horns and hides, (deer and rhinoceros,) ráls and dhúnas (resin of Saul and of Cheer), and timber of the Dhúns. Nor is there a single tribe of Highlanders between the Cósí and the Sutlej which does not discriminate between the Tarai or Tari, the Jhári or Bháver, and the Dhúns or Máris. Captain Herbert has admirably described* the geological peculiarities and external aspect of each of these well-known tracts. His details are, indeed, confined to the space between the Káli and the Sutlej; but the general characteristics of these tracts he affirms to be equally applicable to all the country between the Méchi and the Sutlej; and Captain Parish, whilst confirming Herbert's statements, makes them so likewise as far westward as the Beas.† What Captain Herbert states as holding good from his own personal researches in regard to the Western Himálaya (Sutlej to Káli), I can confirm from mine in regard to the Népálese portion (Káli to Méchi), but with this reservation that no more in the Western than in the Né-

‡ For an ample enumeration of the mammals and birds of the Himálaya, (150 sp. of the former, and 650 of the latter,) see separate catalogue printed by order of the Trustees of the British Museum in 1845. The distribution is not there given. For additions to the catalogue since 1845 see A and M of *Natural History and Zoology Journal* of London, and Bengal Asiatic Society's Journal, and second catalogue of British Museum, published in 1863.

§ Physical Geography, vol. i., p. 66.

* J. A. S. B., number 126, extra pp. 33 and 133, *et seq.*

† J. A. S. B., numbers 190 and 202, for April 1848-49.

pálese Himálaya does the Sandstone range, with its contained Dhúns, prevail throughout or continuously, but only interruptedly or with intervals; and thus the Sallyán-mári, the Gongtali-mári, the Chitwan-mári, the Makwánpur-mári, and the Bijaypur-mári of Népál (which are mostly separate), represent with perfect general accuracy the Deyra, Kyarda, Pinjor, Pátali, and other Dhúns to the westward. The accompanying sectional outline will give a more distinct idea than any words could do of the relations of the several parts of the lower Himá-

Disposition of parts in the lower region of the Himálaya.

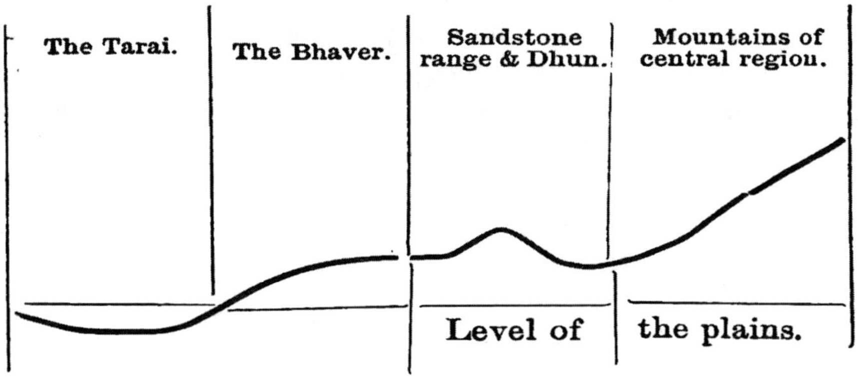

layan region to the plains on the one hand, and to the mountains on the other, according to Captain Herbert's views. The continuous basal line represents the level of the plains; the dip on the left, the Tarai; the ascending slope in the centre, the Saul forest; the dip on the right, the Dhúns or Máris. It is thus seen that the Tarai sinks below the level of the plains; that the forest forms a gradual even ascent above that level; that the Dhúns continue the ascent to the base of the true mountains, but troughwise, or with a concave dip; and, lastly, that the Dhúns are contained between the low Sandstone range and the base of the true mountains. *The Tarai* is an open waste, incumbered rather than clothed with grasses. It is notorious for a direful malaria, generated, it is said, by its excessive moisture and swamps—attributes derived, first, from its low site; second, from its clayey bottom; third, from innumerable rills percolating through the gravel and sand of the Bháver, and finding issue on the upper verge of the Tarai (where the gravelly or sandy débris from the mountains thins out), without power to form onward channels for their waters into the plains. *The forest* is equally malarious with the Tarai, though it be as dry as the Tarai is wet. The dryness of the forest is caused by the very porous nature of that vast mass of diluvian detritus on which it rests, and which is overlaid only by a thin but rich stratum of vegetable mould, everywhere sustaining a splendid crop of the invaluable timber tree *(shorea robusta)*, whence this tract derives its name. The *Sandstove range* is of very inconsiderable height, though rich in fossils. It does not rise more than three to six hundred feet above its immediate base, and is in some places half buried (so to speak) in the vast mass of débris through which it penetrates.* *The Dhúns* are as

* The low range which separates the Dhún and Bháver, on the high-road to Káthmándú, consists almost wholly of *diluvium*,[a] rounded pebbles loosely set in ocherous

malarious and as dry as the Bháver. They are from five to ten (often less, in one instance more) miles wide, and twenty to forty long, sloping from either side towards their centre, and traversed lengthwise by a small stream which discharges itself commonly into one of the great Alpine rivers—thus the Ráputi of Chitwan-mári falls into the Gandak, and that of Bijaypúr-mári into the Cósí. The direction of the Máris or Dhúns is parallel to the ghát line of the snows, and their substratum is a very deep bed of débris, similar to that of the Bháver, but deeper, and similarly covered by a rich but superficial coating of vegetable mould which, if not cultivated, naturally produces a forest of Saul equal to that outside the Sandstone range, and then in like manner harbouring elephants, rhinoceroses, wild bulls (*bibos*), wild buffaloes, rusas, and other large deer (*rucervi*), with creeping things (*pythons*) as gigantic as the quadrupeds. The height of the Sandstone range Captain Herbert estimates at 3,000 feet above the sea, or 2,000 above the plains adjacent; and that of the Dhúns (at least the great one), at 2,500 above the sea, and 1,500 above the plains. These measurements indicate sufficiently the heights of the lower region, and it is observable that no elevation short of 3,000 to 4,000 feet above the sea suffices to rid the atmosphere of the lower Himálaya from malaria. Thus, the Tarai, the Bháver and the Dhúns are alike and universally cursed by that poisonous atmosphere. And this (by the way) is one among several reasons* why I have assigned 4,000 feet of elevation as the southern limit of the healthful and temperate mid-region; that above it being the arctic or boreal, and that below it, the tropical region, though it must never be forgotten that much of the tropical characters, especially in the course of the seasons, pervades the whole breadth (and length likewise) of the Himálaya, whatever be the decrement of heat, and also that, from the uncommon depth of the glens in which the great rivers run, and which, in the central and even upper region often reduces the height of those glens above the sea below the limit just assigned for salubrity, such glens are in both these regions not unfrequently as malarious as is the whole lower region.†

clay, such as forms the great substratum of the Dhún and Bháver. The sandstone formation only shews itself where the rain torrents have worn deep gullies, and it there appears as white weeping sand, imperfectly indurated into rock. Crude coal, shale, loam, are found in this quarter, but no organic fossils, such as abound to the westward.

[a] By "diluvium" I merely mean what Lyell expresses by "old alluvium." I advert not to *the* deluge, but simply imply aqueous action other than recent, ordinary and extant.

* That 4,000 feet of elevation form a good demarcation of the tropical and temperate regions of the Himálaya, is well denoted by the fact, that this is the point where snow ceases to fall, as I have ascertained in the Central and Eastern Himálaya by the observations of thirty years. What I mean is, that snow just reaches that limit and never falls beyond it or below it. It may be otherwise in the Western Himálaya, where snow is more abundant at equal elevations. The small or hill species of bamboo, which prevail from 4,000 to 10,000 of elevation, mark with wonderful precision the limits of the central healthful and normal region of the Himálaya. These most useful species (there are several) would doubtless flourish in Europe.

† Thus the valleys of the Great Rangit and of the Tishta, near and above their junction, are not more than 1,000 feet above the sea, at a distance nearly intermediate between the plains and the snows, and in the midst of the central region; and

But the above characteristics of the sub-divisions of the lower Himálayan region, how noticeable soever to the west of the Méchi, are by no means so to the east of that river, where a skilled eye alone can painfully detect the traces§ of the sandstone formation (without which there can be, of course, no Dhúns,) and where the Tarai, considered as a trough running parallel to the mountains, form no marked feature of the country, if indeed in that sense it can be said to exist at all. And as, even to the westward, the Sandstone range, with its contained Dhúns, is by no means constant, it may be desirable to attempt to characterise the lower region considered as a whole, without reference to local peculiarities or too rigidly defined sub-divisions. Now I conceive that the lower region owes its distinctive character, as a whole, to the vast mass of diluvial detritus, which was shot from the mountains upon the plains, like gravel from a cart, at some great geological epoch, and which has been, since its deposit, variously and often abraded both in degree and direction, by oceanic, and, in a far less degree, by ordinary floods. Where there was, at the epoch in question, no sandstone range to intercept the downward spread of the débris, this débris would necessarily be carried further south, and be of less thickness; where there *was* such a barrier, it would be carried less far southward and be accumulated in greater thickness, especially within the barrier; and, in like manner, where no sandstone range existed, but only spurs, sent forth, like bent arms, upon the plains from the mountains, the embayed detritus would still be deeply piled and lofty within such spurs,* and thinly and unequally spread without them, by reason of the action of the spurs on the currents. Again, where, as from Gowhatty to Saddia, there was not room upon the plains for the free spread and deposit of the descending Himálayan detritus, owing to large rapid rivers and to other chains, both parallel and proximate to the Himálaya, the phænomena created elsewhere by the more or less unrestricted spread of the Himálayan detritus over the plains would necessarily be faintly, if at all, traceable. Lastly, if at the time of the descent of the débris, there existed a great dip in the Gangetic plains

those valleys are consequently as malarious as the Tarai. So also the valleys of the Sunkósi at Damja and of the Trisúl below Nayakot, and many others well known to me.

§ In my recent expedition in the Tarai east of the Méchi, with Dr. Hooker, that accomplished traveller first detected traces of the sandstone formation, with imperfect coal, shale, etc., in a gully below the Pankabari Bungalow, as well as at Lohagarh. The sandstone rock barely peeped out at the bottom of the gully lying in close proximity with the mountains, so that nothing could be more inconspicuous than it was as a feature in the physiognomy of the country.

* There is a signal example of this on the road to Darjeeling *viâ* Pankabari, where the débris, embayed by a curving spur, is accumulated to several hundred feet, and where, moreover, there is outside the spur a conspicuous succession of terraces, all due to oceanic forest, and clearly shewing that the subsidence of the sea[b] was by intervals, and not at once. Constant observation has caused the people of the Tarai to distinguish three principal tiers of terraces, from the prevalent growth of trees upon each. The highest is the Saul level, the middle the Khair level, and the lowest the Sissu level; Shorea, Acacia and Dalbergia being abundantly developed on the three levels as above enumerated.

[b] I do not imply by this phrase any reference to the theory that the sea has sunk and not the land risen. I think the latter much the preferable hypothesis, but desire merely to infer a change in the relative level of the two, and to link my facts upon the string of an intelligible system.

from north-west to south-east, the lithologic character, as well as the distribution, of the débris, would be materially affected thereby; for the subsiding oceanic current would have a set from the former to the latter quarter, and would continue to lash the gravel into sand, and here to deposit both in a series of terraces, there perhaps utterly to displace both, in the latter quarter long after the former had emerged from the waves. Now, that the Himálaya really was, at one time, in great part submerged; that the vast mass of detritus from the Himálaya at present spread over the plains in its vicinity was so spread by the ocean when the founts of the deep were broken up; that this huge bed of detritus, every where forthcoming, is now found in unequal proportion and distribution and state of comminution; as for example, deeper piled within than without the Sandstone range and the embaying spurs, and also, more gravelly and abundant to the north-west, more sandy and scant to the south-east;* and, lastly, that the Gangetic plain really now has a great oblique dip† from the Sutlej at Ruper to the Bráhmapútra at Gwálpárá, whereby all the Himálayan feeders of the Ganges are in the plains so much bent over to the eastward—these are presumptions relative to the past, as legitimate as the extant facts suggesting them are incontrovertible; and we have but to observe how, at the grand epoch adverted to, the action of general causes was necessarily modified by the peculiar features of the scene, as above indicated, in order to come at a just conception of the aspect and character of the lower Himálayan region, all along the line of the mountains. Thus the longitudinal trough parallel to the mountains, and exclusively denominated the Tarai by Captain Herbert, may to the north-west have been caused by the set of the subsiding oceanic current from north-west to south-east; but however caused, it exists as a palpable definite creature, only beneath the Thakorain and Kúmáun, is faintly traceable beneath Népál, and is wholly lost beneath Sikim and Bhútán. But the great bed of débris is *everywhere* present, and with no other distinctions than those pointed out, whether it be divided into Bháver and Dhún, by the Sandstone range, as is usually the case west of the Méchi, or be not so divided owing to the absence of that range, as is always the fact east of the Méchi. Again, *every*

* Captain Herbert has given statements of its depth to the westward, where there is a Sandstone range. To the eastward, where there is none, I found it on the right bank of the Tishta, under the mountains, 120 feet; at fifteen miles lower down, 60 to 70 feet; at fifteen miles still further off the mountains, 40 to 50 feet. There was here no interruption to the free spread of the detritus, and I followed one continuous slope and level—the main high one. The country exhibited, near the rivers especially, two or three other and subordinate levels or terraces, some marking the effect at unusual floods of extant fluviatile action, but others unmistakeably that of pristine and oceanic forces. I measured heights from the river. I could not test the sub-surface depth of the bed. There was everywhere much more sand than gravel, and boulders were rare.

† Saharunpúr is 1,000 feet above the sea; Múradábád 600; Gorakpúr 400; Dumdanga 312; Rangpúr 200; Gwálpárá 112. My authorities are As. Res. vol. xii., J.A.S.B. No. 126, Royle's Him. Bot., Griffith's Journals, and J. Prinsep in epist. The oblique dip to the plains towards the east seems to be increasing, for all the Himálayan rivers descending into the plains, as they quit their old channels, do so towards the east only. I would propose, as an interesting subject of research, the formal investigation of this fact, grounding on Rennell's maps and noting the deviations which have occurred since he wrote. The Tishta which fell into the Ganges now falls into the Bráhmapútra.

where there is, at that point where this vast bed of gravel and sand thins out, a constantly moist tract, caused by the percolation of hill-waters through the said bed, and their issue beyond it; and that constantly moist tract is the Tarai, whether it runs regularly parallel to the line of mountains and be distinctly troughed, as to the Westward is the case, or whether there be no such regularity, parallelism, or of troughing, as to the Eastward is the case.

Why that vast mass of porous débris, which every where constitutes the appropriated domain of the Saul forest, and that imporous trough outside of it, which every where constitutes its drain, should as far Eastward as the Méchi, be both of them developed parallelly to each other and to the line of the mountains, whilst beyond the Méchi Eastward to Assam (exclusive) they should exhibit little or no such parallellism, but should rather show themselves plainwards, like an irregular series of high salient and low resalient angles resting on the mountains, or like small insulated plateau,* or high undulated plains,† surrounded in both the latter cases by low swampy land analogous to the Tarai, it would require a volume to illustrate in detail. I have given a few conspicuous instances in the foot-notes. For the rest, it must suffice to observe that such are the general appearances of the Bháver and Tarai to the Westward and to the Eastward; and that the general causes of the differences have been pretty plainly indicated above, where the necessary effects of the sandstone range, of the mountain spurs, and of the Eastern dip of the plains upon those oceanic forces, to which all phænomena of the region owe their origin, have been suggested.

Throughout Assam, from Gwálpárá to Saddia, Major Jenkins assures me there is neither Bháver nor Tarai; and if we look to the narrowness of that valley between the Himálaya and the mighty and impetuous Bráhmapútra, and consider moreover the turmoil and violence of the oceanic current from the N.W., when its progress was staid by the locked-up valley of Assam, we shall be at no loss to conceive how all distinctive marks of Bháver and Tarai should here cease to be traceable.‡

It will be observed that, in the foregone descriptions of our Himálayan rivers, I have not adverted (save casually in one instance, in order to correct an error

* Parbat Jowár, on the confines of Assam and Rangpúr, is one of the most remarkabel of these small plateau. It is considerably elevated, quite insulated, remote from the mountains, and covered with saul, which the low level around exhibits no trace of. Parbat Jowár is a fragmentary relic of the high level, or Bháver, to which the saul tree adheres with undeviating uniformity.

† Conspicuous instances occur round Dinajpúr and north-west and north-east of Siligori in Rangpúr, where are found highly undulated downs, here and there varied by flat-topped detached hillocks, keeping the level of the loftiest part of the undulated surface. Looking into the clear bed of the Tishta, it struck Dr. Hooker and myself at the same moment, how perfectly the bed of the river represented in miniature the conformation of these tracts, demonstrating to the eye their mode of origination under the sea.

‡ The climate of that portion of the Eastern Himálaya, which is screened from the south-west monsoon by the mountains Sonth of Assam, is less humid than the rest, precisely as are the inner than the outer parts of the whole chain. The fact, that much less snow falls at equal heights in the humid Eastern than in the dry Western Himálaya, depends on other causes. Darjeeling has uot half as much snow as Simla.

as to the true name of the Káli) to their partial Trans-Himálayan sources. And I confess it seems to me, that perspicuity is by no means served by undue insistency on that feature of our rivers. Captain Herbert was thus led to travel beyond his proper limits with a result by no means favourable; for, it appears to me, that he has confounded rather than cleared our conceptions of Central Asia as the Bám-i-dúnya (dome of the world) by attempting to detach therefrom that most characteristic part of it, the plateau of Tibet, *because* certain Indian rivers have (in part) Tibetan sources! My theory of water-sheds does not incline me to venture so far into regions too little known, to allow of the satisfactory settlement of the question, and the less so, inasmuch as the rivers I have to speak of would not afford so plausible an excuse for so doing, as if I had to treat of the Indus, Sutlej,* and Bráhmapútra *alias* Sánpú.† The Arun and the Karnáli, though they draw much water from Tibet, draw far more from the "*pente meridionale*" of the Himálaya, or the ghát line and all South of it; and this is yet more true of the Ganges, the Mónas and Tishta, though they also have partial Trans-Himálayan sources. To those sources of the several Himálayan (so I must call them) rivers above treated of, I will now summarily advert:—

The Mónas.—It is by much the largest river of Bhútán, which state is almost wholly drained by it. It has (it is said) two Tibetan sources, one from Lake Yámdotsó vel Palté vel Yarbroyum, which is a real lake, and not an island surrounded by a ring of water as commonly alleged—the other, from considerably to the West of Palté. These feeders I take to be identical with Klaproth's Mon-tchú and Nai-tchú vel Lábnak-tchú, strangely though he has dislocated them.

The Tishta is also a fine river, draining the whole of Sikim, save the tracts verging on the plains. The Tishta has one Tibetan source, also, from a lake, *viz.*, that of Chólamú. To speak more precisely, there are several lakelets so named, and they lie close under the N. W. shoulder of Powhánry, some thirty miles W. and forty S. of Turner's lakes.

The Arún is the largest of all the Himálayan rivers, with abundant Cis-Himàlayan and three Trans-Himálayan feeders. One, the Western, rises from

* Recte Salúj vel Satrudra.

† Dr. Gutzlaff, once read a paper before the Geographical Society of London, and reverted to Klaproth's notion, that the Sánpu is not the Bráhmapûtra. But Mr. Gutzlaff overlooked J. Prinsep's important, and I think decisive argument on the other side, *viz.*, that the Bráhmapûtra discharges three times more water than the Ganges, which it could not do if it arose on the north-east confines of Assam, notwithstanding the large quantity of water contributed by the Mónas. Yárû or Yerû (Erû) is the proper name of the river we call Sánpû, which latter appellation is a corruption of the word Tsang-po,[a] referring either to the principal province (Tsang) watered by the Yáru, or to the junction therewith, at Digarchi, of another river called the Tsang, which flows into the Yáru from the Nyenchhen chain or Northern boundary of Southern Tibet. Erû vel Arú is the proper spelling. But words beginning with the vowels á and é, take initial y in speech. I take this occasion to observe, in reference to the Yámdo lake above mentioned, that it is not, as commonly described and delineated in our maps, of a round shape, but greatly elongated and very narrow. It is stated to me on good autho rity to be eighteen days' journey long (say 180 miles), and so narrow in parts as to be bridged. It is deeply frozen in winter, so as to be safely crossed on the ice, whereas the Erú river is not so, owing to the great force of its current — a circumstance proving the rapid declivity of the country watered by this great river.

[a] [Tsang po means simply 'river,' and should not be called Sanpu but Tsang po.—J.S.]

the "*pente septentrionale*" of the Himálaya,in the district of Tingri or Pékhu; another, the Northern, from a place called Dúrré; and a third, the Eastern, from the undulated terraced and broken tract lying N. and a little W. of Chólamú and S. of Kambala, or the great range which bounds the valley of the Yárú* on the S. from W. of Digarchi to E. of Lhása.

The Karnáli is much larger than the Alpine Ganges, and nearly equal to the Arún, perhaps quite so. It drains by its feeders the whole Himálaya between the Nanda-dévi and Dhoula-giri peaks, and has itself one considerable Tibetan source deduced either from the north face of Himáchal near Momonangli, or from the east face of that crescented sweep, whereby Gángrí nears Himáchal, and whence the Karnáli flows eastward to the Taklakhár pass.

The Ganges also has of late been discovered to have one Tibetan feeder, *viz.*, the Jáhnavi, which after traversing a good deal of broken country in Gnári, between the Sutlej and the Himálaya, passes that chain at the Nilang Ghát to join the Bhágarathi.‡

I will conclude this paper with the following amended comparative table of Andean and Himálayan peaks, Baron Humboldt having apprised me that Pentland's measurements, as formerly given by me, have been proved to be quite erroneous, and Colonel Waugh having recently fixed Kangchan and Chumalhári with unrivalled precision and accuracy :—

Chief peaks of Andes.	Feet.	Chief peaks of Himalaya.	Feet.
Aconcagua	23,000	Jamnoutri	25,669
Chimbarazo	21,424	Nanda-dévi	25,598
Sorato	21,286	Dhoula-giri	27,600
Illimani	21,149	Gosain-thán Dévadhúnga	24,700 29,002
Descabasado	21,100	Kangchan	28,176
Desya-cassada	19,570§	Chumalhári	23,929

N.B.—Dévadhúnga vel Bhaíravthán vel Nyanam, half way between Gosain-thán and Kangchan, is 29,002, ft. determined in 1856. Kang-chan, 'abounding in snow.' Chúmálhári, 'holy mountain of Chúmá.' These are Tibetan words; the other names are Sanskritic, but set down in the Prakritic mode, *e.g.*, Jamnavatari equal to Jamnoutri, etc.

——o——

Postscript.—That sensible and agreeable writer, Major Madden, in a letter (May 1846) to Dr. Hooker, notices "the disgraceful state of our maps of the

* The valley of the Yárú is about sixty linear miles from the Sikim Himálaya (Láchén and Donkia passes); but the intermediate country, called Damsen, is so rugged, that it is ten stages for loaded yáks from the one terminus to the other. Damsen is stated to be one of the most rugged and barren tracts in the whole of Utsáng or Central Tibet, a howling wilderness.—*Hooker*.

‡ Moorcroft's Travels, J. A. S. B. No. 126, and I.J.S. Nos. 17-18.

§ Humboldt, in his *Aspects of Nature*, has given some further corrections of those heights. There are three peaks superior to Chimbarazo, but inferior to Aconcagua.

Himálaya, which insert ridges where none exist, and omit them where they do exist; and moreover, in regard to all names, show an utter ignorance of the meaning of Indian words." It is the express object of the above essay to contribute towards the removal of the weightier of those blemishes of our maps, without neglecting the lesser, by exhibiting, in their true and causal connexion, the great elevations and the river basins of the Himálaya. Major Madden supposes that the term Hyúndés, which he applies to Tibet, points to that region as the pristine abode of the Huns. But this is a mistake. Hyún-dés is a term unknown to the language of Tibet. It is the equivalent in the Khas or Parbatia language* for the Sanskrit Himyádés, or land of snow. Its co-relative term in the Parbatia tongue is Khas-dés, or land of the Khas. The Khas race were till lately (1816) dominant from the Satlúj to the Tishta: they are so still from the Káli to the Méchi. Hence the general prevalence of geographic terms derived from their language. By Hyún-dés the Parbatias mean all the tracts covered ordinarily with snow on both sides of the crest or spine of Hemáchal, or the ghát line; and by Khas-dés, all the unsnowed regions south of the former, as far as the Sandstone range.

The Bráhmans and those who use Sanskrit call the Hyún-dés, Bhútánt or appendage of Bhót, and hence our maps exhibit a Bhútánt in what Traill denominates (A. R. vol. 16) the Bhote perganahs of Kúmáun. But Bhútánt is not restricted by the Bráhmans to such perganahs in Kúmáun merely, far less to any one spot within them. It includes all the districts similarly situated along the entire line of the Himálaya. We might create confusion however by recurring to his extended meaning of the word, since it has long been restricted by us to the Déb Rájah's territory, or Bhútán (recte Bhútánt). Moorcroft's Giannak in Western Tibet is the *ne plus ultra* of abuse of words. Far to the east, some Bhótia must have told him, lie the Gyannak or Chinese, and thereupon he incontinently gives this term as a name of place.

The Tibetans call their neighbours by the generic name Gya, to which they add distinctive affixes, as Gya-nak, black Gyas, *alias Chinese;* Gya-ver, yellow Gyas, *alias* Russians; and Gya-gar, white‡ Gyas, *alias* Hindús. With reference to the Huns, if I were in search of *them* in Tibet, I should look for them among the Hór of that country, as I would for the Scythians among the Sóg vel Sók. Sogdiana or Sógland was, I conceive, the original Zakeia, the first known historic seat of the Indian Sákás and Tibetan Sóg vel Sók. Hórsók, as one term, means Nomade, in Tibetan such being still the condition of those two tribes in Tibet.

* For a sample of this tongue, which has a primitive base, but overlaid by Pracrit, see J. A. S. B. No. 191, June 1848.

‡ Observe that these epithets do not refer to the colour of men, but only to that of their dress; the Chinese are fond of black clothes and the Indians universally almost wear white ones. The like is probably equally true of similar designations of Turanian tribes in various other parts of the vast Tartaric area (*e.g.* Red Karens), though Ethnic theories have been spun out of the other interpretation of these distinctive terms.

2. ON THE ABORIGINES OF THE HIMALAYA.

The following paper was written in 1847. It was then presented to the Asiatic Society of Bengal, as a summary view of the affinities of these tribes as deduced from a tolerably copious comparison of their languages or dialects.

Accordingly, I submitted a comparative vocabulary of twelve of the dialects found in the Central* sub-Himálayas, inclusive; for comparison's sake, of the written as well as spoken language of Tibet, it being of much importance to give this language in both forms; first, because it is employed in the former state with many unuttered letters, and second, because all the dialects or tongues with which it is to be compared exist only (with two exceptions§) in the latter or unwritten and primitive state.

With regard to the English vocables selected, I have adopted those of Mr. Brown, in order to facilitate comparisons with the Indo-Chinese tongues, as exemplified by him; but, to his nouns substantive, I have added some pronouns, numerals, verbs, adverbs, prepositions, conjunctions, and adjectives, under the impression that nothing short of such a sample of each of the parts of speech could at all suffice for the attainment of the end in view. Geographically or topically, I have confined myself to the East of the river Káli or Ghagra, as well because the dialects prevailing to the Westward of that river are for the most part extremely mixed, and indeed almost merged in the ordinary tongues of the plains of Hindústhan, as also because I have no immediate access to the people of the West. The case is very different in the Eastern sub-Himálayas, where I was domiciled, and where, as will be seen, the Indian Prakrits have hardly been able to make a single cognizable impression upon any of the numerous vernaculars of the people, with the sole exception of the Khás or Parbattia Bhásha, which, as being a mongrel tongue I have omitted. I have likewise, for the present, omitted some interesting tongues of a genuinely aboriginal character, which are spoken East of the Káli, either by certain forest tribes existing in scanty numbers, nearly in a state of nature, such as the Chépáng and Kúsúnda, or by certain other peculiar and broken tribes, such as the Háyu, the Kúswár, the Botia, the Dénwár, Dúrré or Dahri, Bhrámu, Tháru, and Bóksa, who cultivate those low valleys from which malaria drives the ordinary

* I formerly spoke of the Himálaya, as divided lengthwise (north-west to south-east) into Western and Eastern. I now regard it as divided into Western (Indus to Káli), Central (Káli to Tishta), and Eastern (Tishta to Bráhmakúnd) portions. The present paper treats of the Central Himálaya. Breadthwise the chain is regarded as divided into the Northern, Middle, and Southern regions, the word region being always added to contra-distinguish the latter demarcation. Himálaya properly speaking is the perpetually snowed part of the chain. I used to contra-distinguish the lower part or southern slope by the term sub-Himálayas. But objections having been raised, I now acquiesce in the term Himálaya as applied to the whole.

§ The exceptions are the Néwári and Lepsha, which form the topic of my second essay.

population, or, lastly, by several races of helotic craftsmen* whose habitat is general. That ordinary population, exclusive of the now dominant Khas or Parbattias Proper,† above alluded to, consists, between the Káli and the Dhansri, in Népál, Sikim, and Bhútan of; 1st. Cis-Himálayan Bhótias vel Tibetans, called Rongbo, Siéna or Káth Bhótia, Palu Sén,‡ Sérpa or Sharpa etc.; 2nd. Súnwar; 3rd. Gúrung, 4th. Magar; 5th. Múrmi; 6th. Néwár; 7th. Kiránti; 8th. Limbu vel Yak thúmba; 9th. Lepcha or Deúnjong-maro; 10th. Bhútanese or Lhópa vel Dúkpa.||

I have enumerated the races as they occur, in tolerably regular series, from west to east, in given and definite locations of old standing: but the first named are found pretty generally diffused throughout the whole extent, west and east, of my limits, though confined therein to the juxto-nivean tracts or Cachár region; whilst the participation of the Gúrungs and Magars, or military tribes, in the recent political successes of the now dominant Khás, has spread them also, as peaceful settlers, in no scanty numbers, easterly and westerly, from the Káli to the Méchi. The rest of the tribes have a more restricted fatherland or *janam bhúmi*, and indeed the *locale* of the Magars and Gúrungs, not a century back, or before the conquests of the House of Górkha, was similarly circumscribed; for the proper habitat of these two tribes is to the west of the great valley, which tract again, (the valley), and its whole vicinity, is the region of the Múrmis and Néwárs; whilst the districts east of the great valley, as far as Sikim, are the abode of the Kirántis and Limbús, as Sikim is that of the Lepshas, and Déva Dharma or Bhútan that of the Lhópas or Dúkpas, usually styled Bhútanese by us. These constitute, together with the Súnwárs, who again are mostly found west of the great valley and north of the Magars and Gúrungs, near and among the cis-nivean Bhótias,§ the principal alpine tribes of the sub-Himálayas between that western point (the Káli) where the aboriginal tongues are merged in the Prákrits and that eastern limit (the Dhansri) where they begin to pass into so-called monosyllabic-tongued races of presumed Indo-Chinese origin.‡[a] The sub-Himálayan races I have just enumerated inhabit all the central and temperate parts of these mountains, the juxta-nivean or northernmost tracts being left to the Rongbo vel Sérpá vel Pálu

* See p. 14, part ii. of this volume, *supra*, and note.

† Parbattia means 'Highlander,' but this general sense of the word is restricted by invariable usage to the Khás.

‡ The Néwárs of Népál Proper call the cis-nivean Bhótias, Pálú Sén, and the trans-nivean, Thá Sén. The Chinese call the Mongolian Tartars, Thá Thá.

|| Lhópa is a territorial designation, Dúkpa a religious, that is, the country is called Lhó, and the sect of Lamaism prevailing in it, Dúk. Klaproth's Lokabadja, and Ritter's Lokba, are both equivalent to Bhótan vel Lhó. The postfix *ba* means 'of, or belonging to,' so that Lokba, recte Lhópa, is 'a Bhútanese man or native of Lhó.'

§ Bhótia is the Sanskrit, and Tibetan the Persian, name, for the people who call themselves Bodpa, or Bod, a corruption possibly of the Sanskrit word Bhót.

‡[a] More recent researches induce me to demur entirely to a trenchantly demarked monosyllabic class of tongues, and to adopt the opinion that India (Dravirian) and the countries around it on the north aud north-east were peopled by successive incursions of affiliated tribes of Northmen, among whom I see no sufficient reason to segregate from the rest, as is commonly done, the Bodpas of Tibet, the Eastern Himálayans, nor even the proximate Indo-Chinese or people of Western Indo-China.

Sén, and the southernmost parts, as well as the low valleys of the interior and central region, being abandoned to the Kúswárs, Dénwárs, Dúrrés, and other malaria-defying tribes, which, for the present, I do not purpose to notice. The people under review therefore may be said to occupy a highly healthful climate, but one of exact temperatures as various as the several elevations (four to ten thousand feet) of the ever-varied surface; and which, though nowhere troubled with excessive heat,§ is so by excessive moisture, and by the rank vegetation that moisture generates, with the aid of a deep fat soil, save in the Cachár or juxta-nivean region, where the lower temperature and poorer scantier soil serve somewhat to break the prodigious transition from the thrice luxuriant sub-Himálayas to the thrice arid plains of Tibet.

That the sub-Himálayan races are all closely affiliated, and are all of northern origin, are facts long ago indicated by me,* and which seem to result with sufficient evidence from the comparative vocabularies now furnished. But to it lingual evidence in a more ample form will however in due time be added, as well as the evidence deducible from the physical attributes, and from the creeds, customs and legends of these races. It must suffice at present to observe that the legends of the dominant races indicate a transit of the Himálaya from thirty-five to forty-five generations back—say 1,000 to 1,300 years, and that I prefer the remoter period because the transit was certainly made before the Tibetans had adopted from India the religion and literature of Buddhism, in the seventh and eighth centuries of our era. This fact is as clearly impressed upon the crude dialects and cruder religious tenets‡ of the sub-Himálayans as their northern origin is upon their peculiar forms and features, provided these points be investigated with the requisite care; for superficial attention is apt to rest solely upon the Lamaism recently as imperfectly imported among them, and upon the merely exceptional traits of their mixed and varying physiognomy. That physiognomy exhibits, no doubt, generally and normally, the Scythic or Mongolian type (Blumenbach) of human kind, but the type is often much softened and modified, and even frequently passes into a near approach to the full Caucasian dignity of head and face, in the same perplexing manner that has been noticed in regard to the other branches of the Allophylian tree,§ though among the Cis- or Trans-Himálayans there is never seen any greater advance towards the Teutonic blond complexion than such as consists in occasional ruddy moustaches and grey eyes among the men, and a good

§ In the great valley of Népál, which has a very central position aud a mean elevation of 4,500 ft., the maximum of Fahr. in the shade is 80°.

* Illustrations of the Languages &c. of Népál and Tibet, and Res. A.S.B, Vol. XVI. 1827.

‡ Of these religious tenets, the full description given in my work on the Kócch, Bodo, and Dhimál, may be accepted as generally applicable. The Bonpa faith of Tibet (the old creed of that country) and the Shamanism of Siberia are both more or less cultivated types of the primitive creed, subsequently largely adopted into Bráhmanism and Buddhism. The exorcist of the Mûrmi or Tamar tribe is still called Bonpa, aud every tribe's chief priestly agent is an exorcist, variously named.

§ See Prichard, Vol. IV. pp. 323, 344, 356, and Humboldt's *Asie Centrale* 2.62 and 133. Who could suppose the following description refered to a Scythic race?—'*Gens albo colore est atque pulchritudine et forma insigne.*"

deal of occasional bloom upon the cheeks of the children and women. A pure white skin is unknown, and the tint is not much less decided than in the high caste Hindus; but *all* are of this pale brown or isabelline hue in Tibet and the sub-Himálayas, whilst the many in the plains of India are much darker. The broken or depressed tribes above alluded to passed the Himálaya at various periods, but all long antecedent to the immigration of the dominant tribes, and prior to the least whisper of tradition; and the lingual and physical traits of these broken tribes, as might be expected, constitute several links of connexion between the Altaic tribes on the north and the Dravirian on the south. The general description of the Himálayans, both of earlier and later immigration is as follows:—head and face very broad, usually widest between the cheek-bones, sometimes as wide between the angles of the jaws; forehead broad, but often narrowing upwards; chin defective; mouth large and salient, but the teeth vertical and the lips not tumid; gums, especially the upper, thickened remarkably; eyes wide apart, flush with the cheek, and more or less obliquely set in the head; nose pyramidal, sufficiently long and elevated, save at the base, where it is depressed so as often to let the eyes run together, coarsely formed and thick, especially towards the end, and furnished with large round nostrils; hair of head copious and straight; of the face and body deficient; stature rather low, but muscular and strong. Character phlegmatic, and slow in intellect and feeling, but good-humoured, cheerful and tractable, though somewhat impatient of continuous toil. Polyandry yet exists partially, but is falling out of use. Female chastity is little heeded before marriage, and drunkenness and dirtiness are much more frequent than in the plains. Crime is much rarer, however, and truth more regarded, and the character on the whole amiable. The customs and manners have nothing very remarkable, and the creed may be best described by negatives. Indifferency is the only, but heretofore effective obstacle to indoctrination by Bráhmanical, Buddhist, or Christian teachers, so that the Scottish phrase "we cannot be fashed" seems best to describe the prevalent feeling of the Himálayans on this, as on many other matters. The whole population is intensely tribual, some races still bound together by a common appellation, as the Kirántis for example, being nevertheless divided into several septs, distinguised from each other by strongly marked dialects, non-intermarriage, and differences of customs, whilst the tribes which bear distinct names are still more palpably separated in those respects. But the barrier of caste, in the true sense, is unknown, and on the other hand there exists not in any tribe, race or nation, any notion of a common human progenitor, or eponymous deity.* The general status of all the tribes and races is that of nomadic cultivators. "*Arvā in annos mutant et superest ager*" is as true now of the Himálayans as it was of our ancestors when they burst the barriers of the Roman Empire. A few tribes, such as the Néwár, have long become stationary cultivators; and the Gúrungs are still, for the most part, pastoral. There are no craftsmen, generally speaking,

* The instance of the Gorkhalis, who undoubtedly derive their appellation from the demi-God Górakh (Goraksha) Náth, is only a seeming exception, recent and borrowed.

proper to these tribes, stranger and helot races, located among them for ages untold, being their smiths, carpenters, curriers, potters, &c., and the women of each tribe being its domestic weavers. The Néwárs alone have a literature, and that wholly exotic; and they alone have made any attempts at the fine arts, in which they have followed chiefly Chinese, but also Indian, models.

Before concluding this notice of the Alpine Indian aborigines, it may be as well to define summarily the limits and physical characters of their original and adopted abodes, or Tibet and the Sub-Himálayas. Tibet is a truncated triangular plateau, stretching obliquely from south-east to north-west, between 28° and 36° of north latitude and 72° and 102° of east longitude. It is cold and dry in the extreme, owing to its enormous elevation, averaging 12,000 feet above the sea, to the still vaster height of those snowy barriers which surround it on every side, and which on the south reach 29,000 feet, to an uncommon absence of rain and cloud, to the extreme rarification of its atmosphere, to its saline and sandy soil, and as a consequence of all these and a reciprocating cause too, to the excessive scantiness of its vegetation. It is bounded on the south by the Hemáchal, on the north by the Kuenlun, on the west by the Belúr, and on the east by the Yúnling—all for the most part perpetually snowclad, and of which the very passes on the south average 16,000 to 17,000 feet of elevation. Tibet is, for the most part, a plain and a single plain, but one extremely cut up by ravines, varied much by low bare hills, and partially divided in its length by several parallel ranges approaching the elevation of its barriers, and between the third and fourth of which ranges stand its capitals of Lhása and Digarchi.* These capitals are both in the central province of the Utsáng, all west of which, to the Belúr, composes the province of Nári, and all east of it, to Sífán, the province of Khám, provinces extending respectively to Bukharia and to China. Tibet, however arid, is nowhere a desert,§ and however secluded, is on every side accessible; and hence it has formed in all ages the great overland route of trade, and may even be called the grand ethnic, as well as commercial, highway of mankind; its central position between China, India, and Great Bukharia having really rendered it such for ages, before and since the historic æra, despite its snowy girdle and its bleak aridity. Hence we learn the supreme importance of Tibet in every ethnological regard. Its maximum length is about 1,800 miles, and maximum breadth about 480 miles; the long sides of the triangle are towards India and Little Bukharia; the short one, towards China; the truncated apex towards Great Bukharia, where the Belúr, within the limits of Tibet, has an extent of only one degree, or from 35° to 36° N. lat.; whereas the base towards China, along the line of the Yúnling, reaches through 8° or from 28° to 36° N. lat. Just beyond the latter point, in the north-east corner of Khám, is Siling or Tangut, the converging point of all the overland routes, and which I should prefer to include ethnologically within

* De Körös from native written authority *apud* J. A. S. B.

§ In the next plateau of High Asia, or that of Little Bukharia, the vast desert of Cobi or Gobi, which occupies the whole eastern half of that plateau, has ever formed, and still does, a most formidable obstruction to transit aud traffic.

Tibet, but for the high authority of Klaproth, who insists that we have here a distinct* language and race, though certainly no such separating line in physical geography,|| Siling or Tangut being open to the plateau of Tibet as well as to those of Little Bukharia and Songaria though demarked from China both on the north and east by the K'i-lian and Peling respectively.

South of the whole of Tibet, as above defined, lie the Sub-Himálayas, stretching from Gilgit to Bráhmakúnd, with an average breadth of ninety miles, divided climatically into three pretty equal transversal regions, or the northern, the central, and the southern, the first of which commences at the ghat line of Hemáchal, and the last ends at the plains of Hindostan; the third lying between them, with the great valley of Népál in its centre. That valley is of a lozenge shape, about sixteen miles in extreme length and breadth, cultivated highly throughout, and from 4,200 to 4,700 feet above the sea. The only other valley in the whole eastern half of the Sub-Himálayas is that of Júmlá, which is smaller and higher, yielding barley (Hordeum celeste,) as the great valley, rice. To the west is the large but single vale of Cashmere and the Dúns, both too well known to require further remark. The sub-Himálayas form a confused congeries of enormous mountains, the ranges of which cross each other in every direction, but still have a tendency to diverge like ribs from the spine of the snows, or a south-east and north-west diagonal, between 28° and 35°. These mountains are exceedingly precipitous and have only narrow glens dividing their ridges, which are remarkable for continuity or the absence of chasm and rupture and, also for the deep bed of earth everywhere covering the rock and sustaining a matchless luxuriance of tree and herb vegetation, which is elicited in such profusion by innumerable springs, rills, and rivers, and by the prevalence throughout all three regions of the tropical rains in all their steadiness and intensity. There are three or four small lakes in Kúmáun situated near each other, and three or four more in Pókrá similarly juxtaposed. But in general the absence of lakes (as of level dry tracts) is a remarkable feature of the Sub-Himálayas at present, for anciently the great valleys of Cashmere and Népál, with several others of inferior size, were in a lacustrine state. The great rivers descend from the snows in numerous feeders, which approach gradually and unite near the verge of the plains, thus forming a succession of deltic basins, divided by the great snowy peaks as water-sheds, thus:—

Basins.	*Peaks.*
1. Alpine Gangetic basin.*	Nanda-dévi.
2. ,, Kárnálic basin.	Dhavalagiri.
3. ,, Gandacean basin.	Gosain-thán.
4. ,, Cósian basin.	Kángchánjúnga.
5. ,, Tishtan basin.	Chumalhári.
6. ,, basin of the Mónas.	The Gemini, two unnamed peaks.

* Siling or Tangut is in Sók-yul or the country of the Mongol tribe.

|| It must be admitted, however, that the Payam Khar of Klaproth seems to divide Khám from Tangut. Klaproth cites Chinese geographers.

* See the article on "Geography of the Himálaya."

In the two first of these five regions, all of which are plainly indicated by the distribution of the waters, the people are mongrel and mixed, save in the north-west parts, where the Palu Sén or cis-nivean Bhótiás, the Garhwális, and the inhabitants of Kanáver and Hangrang are of Tibetan stock. The third, or Gandacean basin (Sapt Gandaki in native topography, from the seven chief feeders,) is the seat of the Sunwárs, Gúrungs, and the Magars. The fourth, or Cósian basin, (Sapt Cousika in native topography, after the seven chief feeders,) is the abode of the Kirántis and Limbús. The fifth or Tishtan basin, again, is the fatherland of the Deunjongmaro, and the sixth that of the Pru or Lhópá, that is,§ Lepshas and Bhútanese, respectively. And, lastly, the high and level space—(a system of valleys around the great one, which is nearly 5,000 ft. above the sea)—between the basins of the Gandak and Cósí is the seat of the Néwárs and Múrmís. But observe that the terms level space and system of valleys, applied to this last tract, are merely relative, though as such significant, nor meant to be contradictory of what has been above remarked, more generally, as to the whole Sub-Himálayas. And here I should add that the best representation of the Himálayas and Sub-Himálayas is by a comparison with the skeleton of the human frame,*a in which the former are analogous to the spine and the latter to the ribs. The Sub-Himálayas therefore are transverse rather than parallel ridges, as above stated, or, at all events, their main ridges diverge more or less rectangularly from the ghát line, so as to unitise the several great streams, but still with an irregularity which close observance of the aqueous system can alone reveal. The ruggedness of the surface, by preventing all inter-communication of a free kind, has multiplied dialects: the rank pasture, by its ill effect on herds and flocks, has turned the people's attention more exclusively than in Tibet to agriculture, though even in Tibet the people are mostly non-nomadic,* heat and moisture, such as Tibet is utterly void of, have relaxed the tone of the muscles and deepened the hue of the skin, making the people grain-eaters and growers rather than carnivorous tenders of flocks. Thus the Cis-Himálayans are smaller, less muscular, and less fair than the Trans-Himálayans; but the differences are by no means so marked as might have been expected; and though there are noticeable shades of distinction in this respect between the several tribes of the Cis-Himálayans according to their special affinities, as well as between most of them and the North-men, according to their earlier or later immigration, yet if they all be (as surely they

§Pru is the Lepsha name of the Bhútanese, whom the Hindu Shastras designate Plava, and themselves, Lhopá.

*a Professor Müller (*apud* Bunsen's Philosophy of Language), grounding on my Essay on the Physical Geography of the Himálaya, has likened the whole to the human hand with the fingers pointing towards India. The ghát line with its great peaks is assimilated to the knuckles, the dips between being the passes; and the three transverse Sub-Himálayan regions, extending from the gháts to the plains, are likened to the three joints of the fingers.

* Within the limits of Tibet are found abundance of nomades of Mongol and Turkish race, called respectively Sókpo and Hór by the Tibetans, who themselves seem much mixed with the latter race, which has long exercised a paramount influence in North Tibet: witness the facts that all its hill ranges are *taghs*, and all its lakes *núrs*, both Túrki words.

are) of the same Turanian origin, it must be allowed that very striking differences of climate and of habits, operating through very many generations, can produce no obliterative effects upon the essential and distinctive signs of race. But this is, in part, speculation, and I will terminate it by remarking that, for the reasons above given, my investigations have been limited to that portion of the Sub-Himálayas which lies between the Káli and the Dhansri, or say $80\frac{1}{2}°$ to $92\frac{1}{2}°$ of East longitude and $26\frac{1}{2}°$ to $30\frac{1}{2}°$ of North latitude.

3. ORIGIN AND CLASSIFICATION OF THE MILITARY TRIBES OF NEPAL.

(Read before the Bengal Asiatic Society, 9th January, 1833.)

The great aboriginal stock of the inhabitants of these mountains, east of the river *Káli*, or in *Népál*, is Turanian. The fact is inscribed, in characters so plain, upon their faces, forms, and languages, that we may well dispense with the superfluous and vain attempt to trace it historically in the meagre chronicles of barbarians.

But from the twelfth century downwards, the tide of *Mussulmán* conquest and bigotry continued to sweep multitudes of the *Bráhmans* of the plains from *Hindústán* into the proximate hills, which now compose the western territories of the kingdom of *Népál*. There the *Brahmans* soon located themselves. They found the natives illiterate, and without faith, but fierce and proud.

Their object was to make them converts to *Hindúism*, and so to confirm the fleeting influence derived from their learning and politeness. They saw that the barbarians had vacant minds, ready to receive their doctrines, but spirits not apt to stoop to degradation, and they acted accordingly. To the earliest and most distinguished of their converts they communicated, in defiance of the creed they taught, the lofty rank and honors of the *Kshatriya* order. But the *Brahmans* had sensual passions to gratify, as well as ambition. They found the native females—even the most distinguished—nothing loath, but still of a temper, like that of the males, prompt to repel indignities. These females would indeed welcome the polished *Brahmans* to their embraces, but their offspring must not be stigmatised as the infamous progeny of a *Bráhman* and a *Mléchha*—must, on the contrary, be raised to eminence in the new order of things proposed to be introduced by their fathers. To this progeny also, then, the *Brahmans*, in still greater defiance of their creed, communicated the rank of the second order of *Hindúism;* and from these two roots, mainly, sprung the now numerous, predominant, and extensively ramified, tribe of the *Khas*—originally the name of a small clan of creedless barbarians, now the proud title of the *Kshatriya*, or military order of the kingdom of *Népál*. The offspring of original *Khas* females and of *Brahmans*, with the honors and rank of the second order of *Hindúism*, got the patronymic titles of the first order; and hence the key to the anomalous nomenclature of so many stirpes of the military tribes of *Népál* is to be sought in the nomenclature of the sacred order. It may be added, as remarkably illustrative of the lofty spirit of the *Parbattias*, that, in spite of the yearly increasing sway of *Hindúism* in *Népál*, and of the various attempts of the *Brahmans* in high office to procure the abolition of a custom so radically opposed to the creed both parties now profess, the *Khas* still insist that the fruit of commerce (marriage is out of the question) between their females and males of the sacred order shall be ranked as *Kshatriyas*, wear the thread, and assume the patronymic title.

The original *Khas*, thus favoured by it, became soon and entirely devoted to the *Brahmanical* system.* The progress of *Islám* below daily poured fresh refugees among them.

They availed themselves of the superior knowledge of the strangers to subdue the neighbouring tribes of aborigines, were successful beyond their hopes, and, in such a career continued for ages, gradually merged the greater part of their own habits, ideas, and language (but not physiognomy) in those of the *Hindús*.

The *Khas* language became a corrupt dialect of *Hindí*, retaining not many palpable traces (except to curious eyes) of primitive barbarism.

An authentic anecdote told me at Káthmándú confirms the origin above assigned to the modern Khas tribe of Népál. In the reign of Ram Sáh of Górkhá, an ancestor of the present dynasty of Népál, an ambassador was sent from the Durbar of Górkhá to that of Méwár, to exhibit the Górkháli Rajah's pedigree and to claim recognition of alleged kindred. The head of the renowned Sesodians, somewhat staggered with the pedigree, seemed inclined to admit the relationship, when it was suggested to him to question the ambassador about his own caste as a sort of test for the orthodoxy or otherwise of the notions of caste entertained in the far distant, and, as had always at Chitor or Udaypur been supposed, barbarous Himálaya. The ambassador, a Khas, who had announced himself as belonging to the martial tribe, or Kshatriya, thus pressed, was now obliged to admit that he was nevertheless a Pándé, which being the indubitable cognomen of a tribe of the sacred order of Hindúism, his mission was courteously dismissed without further enquiry.

The *Ekthériahs* are the descendants more or less pure of *Rájpúts* and other *Kshatriyas* of the plains, who sought refuge in these mountains from the *Moslem*, or merely military service as adventurers. With fewer aims of policy, and readier means in their bright swords of requiting the protection afforded them, than had the *Brahmans*, they had less motive to mix their proud blood with that of the vile aborigines than the Brahmans felt the impulse of, and they did mix it less. Hence, to this hour, they claim a vague superiority over the *Khas*, notwithstanding that the pressure of the great tide of events around them has, long since, confounded the two races in all essentials. Those among the *Kshatriyas* of the plains, who were more lax, and allied themselves with the *Khas* females in concubinage, were permitted to give to their children, so begotten, the patronymic title only, not the rank. But their children, again, if they married for two generations into the *Khas*, became pure *Khas*, or real *Kshatriyas* in point of privilege

* That is, they agreed to put away their old gods, and to take the new; to have *Brahmans* for *Gurus;* and not to kill the cow: for the rest they made, and still make, sufficiently light of the ceremonial law in whatever respects food and sexual gratification. Their active habits and vigorous character could not brook the restraints of the ritual law, and they had the example of licentious *Brahmans* to warrant their neglect of it. The few prejudices of the *Khas* are useful, rather than otherwise, inasmuch as they favour sobriety and cleanliness.

and rank, though no longer so in name! They were *Khas*, not *Kshatriyas*, and yet they bore the proud cognomina of the martial order of the *Hindús*, and were, in the land of their nativity, entitled to every prerogative which *Kshatriya* birth confers in *Hindústán!*

Such is the third and less fruitful root of the *Khas* race.

The *Ekthàriahs* speak the *Khas* language, and they speak no other.

The *Thákuris* differ from the *Ekthàriahs* only by the accidental circumstance of their lineage being royal. At some former period, and in some little state or other, their progenitors were princes.

The *Sáhi* or *Sáh* are the present royal family.

The remaining military tribes of the *Parbattias* are the *Magar* and *Gúrung*, who now supply the greater number of the soldiers of this state.

From lending themselves less early and heartily to *Brahmanical* influence than the *Khas*, they have retained, in vivid freshness, their original languages, physiognomy, and, in a less degree, habits.

To their own untaught ears their languages differ entirely the one from the other, and no doubt they differ materially, though both belonging to the unpronominalized type of the Turanian tongues. Their physiognomies, too, have peculiarities proper to each, but with the general caste and character fully developed in both. The *Gúrungs* are less generally and more recently redeemed from *Lamáism* and primitive impurity than the *Magars*.

But though both the *Gúrungs* and *Magars* still maintain their own vernacular tongues, Tartar faces, and careless manners, yet, what with military service for several generations under the predominant *Khas*, and what with the commerce of *Khas* males with their females,* they have acquired the *Khas* language, though not to the oblivion of their own, and the *Khas* habits and sentiments, but with sundry reservations in favor with pristine liberty. As they have, however, with such grace as they could muster, submitted themselves to the ceremonial law of purity and to *Brahman* supremacy, they have been adopted as *Hindús*. But partly owing to the licenses above glanced at, and partly by reason of the necessity of distinctions of caste to *Hindúism*, they have been denied the thread, and constituted a doubtful order below it, and yet not *Vaisya* nor *Sudra*, but a something superior to both the latter—what I fancy it might puzzle the *Shastrís* to explain on *Hindú* principles.

The *Brahmans* of *Népál* are much less generally addicted to arms than those of the plains; and they do not therefore properly belong to our present subject. The enumeration of the *Brahmans* is nevertheless necessary, as serving to elucidate the lineage and connexions of the military tribes, and especially of the Khas.

The martial classes of *Népál* are, then, the Khas, *Magar*, and *Gúrung*, each com-

* Here, as in the cases of the *Brahman* and *Khas*, and *Kshatriya* and *Khas*, there can be no marriage. The offspring of a *Khas* with a *Magarni* or *Gúrungni* is a titular *Khas* and real *Magar* or *Gurung*. The descendants fall into the rank of their mothers and retain only the patronymic.

prising a very numerous race, variously ramified and sub-divided in the manner exhibited in the following tabular statement.

The original seat of the Khas is ordinarily said to be *Górkhá*,* because it was thence immediately that they issued, some years ago, under the guidance of Prithvi Narayan, to acquire the fame and dominion achieved by him and his sucessors of the *Górkháli* dynasty.

But the Khas were long previously to the age of Prithvi Narayan extensively spread over the whole of the *Chaubísya*, and they are now found in every part of the existing kingdom of *Népál*, as well as in Kúmáun, which was part of *Népál* until 1816. The Khas are rather more devoted to the house of *Górkhá*, as well as more liable to *Brahmanical* prejudices than the *Magars* or *Gúrungs;* and, on both accounts, are perhaps somewhat less desirable as soldiers for our service than the latter tribes. I say somewhat, because it is a mere question of degree; the Khas having, certainly, no religious prejudices, nor probably any national partialities, which would prevent their making excellent and faithfnl servants in arms; and they possess pre-eminently that masculine energy of character and love of enterprize which distinguish so advantageously all the military races of *Nepál*. The original seat of the *Magars* is the *Bára Mangránth*, or *Satahung*, *Páyung*, *Bhírkót*, *Dhor*, *Garahúng*, *Rísing*, *Ghiring*, *Gálmi*, *Argha*, *Kháchi*, *Músikót*, and *Isma ;* in other words, most of the central and lower parts of the mountains, between the *Bhéri* and *Marsyándi*|| Rivers. The attachment of the *Magars* to the house of *Górkhá* is but recent, and of no extraordinary or intimate nature. Still less so is that of the *Gúrungs*, whose native seats occupy a line of country parallel to that of the *Magars*, to the north of it, and extending to the snows in that direction. Modern events have spread the *Magars* and *Gúrungs* over most part of the present kingdom of *Népál*. The *Gúrungs* and *Magars* are, in the main, *Hindús*, *only because it is the fashion ;* and the *Hindúism* of the Khas, in all practical and soldierly respects, is free of disqualifying punctillios.

These highland soldiers, who despatch their meal in half an hour, and satisfy the ceremonial law by merely washing their hands and face, and taking off their turbans before cooking, laugh at the pharisaical rigour of our *Sipáhis*, who must bathe from head to foot and make *pújá*, ere they begin to dress their dinner, must eat nearly naked in the coldest weather, and cannot be in marching trim again in less than three hours.

In war, the former readily carry several days' provisions on their backs: the latter would deem such an act intolerably degrading. The former see in foreign service nothing but the prospect of glory and spoil: the latter can discover in it nothing but pollution and peril from unclean men and terrible wizards, goblins, and evil spirits. In masses, the former have all that indomitable confidence, each in all, which grows out of national integrity and success: the latter can have no

* Górkhá, the town, lies about sixty miles W.N.W. of Káthmándú. Górkhá, the name, is derived from that of the eponymous deity of the royal family, *viz.* Górakshanáth or Górkhánáth, who likewise has given his name to our district of Górakpur.

|| The *Marichangdi* of our maps.

idea of this sentiment, which yet maintains the union and resolution of multitudes in peril, better than all other human bonds whatever; and, once thoroughly acquired, is by no means inseparable from service under the national standard.

I calculate that there are at this time in *Népál* no less than 30,000 *Dákhréahs*, of soldiers off the roll by rotation, belonging to the above three tribes. I am not sure that there exists any insuperable obstacle to our obtaining, in one form or other, the services of a large body of these men; and such are their energy of character, love of enterprize, freedom from the shackles of caste, unadulterated military habits and perfect subjectibility to a discipline such as ours, that I am well assured their services, if obtained, would soon come to be most highly prized.*

In my humble opinion they are by far the best soldiers in Asia; and if they were made participators of our renown in arms, I conceive that their gallant spirit, emphatic contempt of Madhesias (people of the plains,) and unadulterated military habits, might be relied on for fidelity; and that our good and regular pay and noble pension establishment would serve perfectly to counterpoise the influence of nationality, so far as that could injuriously affect us.

The following table exhibits a classified view of the *Brahmanical* and military tribes, with their various sub-divisions.

Tabular View of the Tribes.

BRAHMANS.

Arjal.	Dohál.	Dhákal.	Bikrál.
Pondyál.	Lamsál.	Adhikári.	Ukniyál.
Khanál.	Rimál.	Doeja.	Bhattwál.
Regmi.	Dêvakotya.	Rukái.	Gajniyál.
Bhattrái.	Parbatya Vash.	Sywál.	Chavala Gái.
Niróla.	Parbatya Misr.	Rijál.	Vasta Gái.
Achárya.	Davári.	Dhúngyál.	Banjára.
Bhatt.	Koikyál	Loiyál.	Dágí.
Sápan kotya.	Nepálya.	Dotiyál.	Sôti.
Maharáshtra.	Barál.	Kandyál.	Osti.
Kôirála.	Pokaryál.	Katyál.	Utkúlli.
Pakonyál.	Rúpákheti.	Dangál.	Kandariah.
Sattyál.	Khativára.	Singyál.	Ghart mél.

* Since this paper was written the value and the availability to us of the Górkháli soldier tribes have been well tested; and it is infinitely to be regretted that the opinions of Sir H. Fane, of Sir C. Napier, and of Sir H. Lawrence, as to the high expediency of recruiting largely from this source, were not acted upon long ago. So long as my voice carried any weight, I often pressed the subject on the attention of those in authority. But the then prejudice in favour of Brahman and Kshatri Sipáhis neutralized all my efforts, though the danger of so homogeneous an army of foreign mercenaries was, among other arguments, earnestly dwelt upon. (1857.)

Ghartyál.	Timil Sina.	Panêru.	Dulál.
Nivapánya.	Káphalya.	Loityál.	Parajuli.
Temrákoti.	Gaithoula.	Sigdhyál.	Bajgái.
Uphaltopi.	Gairaha Pipli.	Barál.	Satôla.
Parijai Kavala.	Ghimirya.	Gotamya.	Ghúrchóli.
Homya Gái.	Simkhára.	Ghorasaini.	Kéláthoni.
Champa Gái.	Phúnwál.	Risyál.	Gilal.
Gûra Gái.	Chamka saini.	Chálîsya.	Lahóni.
Subêri.	Pûra saini.	Dhôngána.	Muthbari.
Pandit.	Dhurári.	Bharári.	
Têva pánya.	Bhúrtyál.	Bágalya.	

KHAS.

1st Sub-division of the Khás, called Thápa.

Bagyál.	Gágliyá.	Powár.	Khapotari.
Takuryál.	Súyál.	Ghimirya.	Parájuli.
Palámi.	Maharáji.	Khulál.	Deoja.
Gûdár.	Lámichanya.	Sunyál.	

2nd Sub-division of the Khás, called Bishnyát.

Khulál. Khaputari. Sripáli. Puwár.

3rd Sub-division, called Bhandári.

Raghubansi. Láma. Sijapati.

4th Sub-division, called Kárki.

Sutár. Láma. Múndala. Khûlál.

5th Sub-division, called Khánka.

Powár. Maharáji. Partyal. Lakánggi. Lámichanya.
Khulál. Kálikotya. Khaputari. Palpáli.

6th Sub-division, or Adhikari.

Thámi. Tharirái. Pokriál. Musiah.
Dhámi. Khadhsena. Thákúri.

7th Sub-division, or Bisht.

Kálikotya. Puwár. Dahál.

8th Sub-division, or Kunwár.

Bagálya. Khulál. Khanka. Arjál.

9th Sub-division, or Baniah.

Sijapati.

10th Sub-division, or Dáni.

Sijapati.	Powár.

*11th Sub-division, or Gharti.**

Kalikotya.	Sijapati.

12th Sub-division, or Khattri.

Pánde.	Khulál.	Lámichánya.	Tewári.	Suveri.
Dhakál.	Panth.	Poryál.	Phanyál.	Adhikári.
Sakhtyál.	Burál.	Arjál.	Sápkotya.	Silwal.

True Khás not yet classified.

Dhongyál.	Sijal.	Satouya.	Rûpakhetí.
Loyál.	Parsái.	Khatiwata.	Chouvala Gái.
Lamsál.	Am Gái.	Chalatáni.	Bhatt Rái.
Khukriyál.	Baj Gái.	Kilathoni.	Naopánya.
Dangál.	Dahál.	Satya Gái.	Muri Bhús.
Sikhmiyál.	Deakota.	Alpháltopi.	Sóti.
Bhiryál.	Garhtóla.	Osti.	Parijái Kawala.
Pouryál.	Sèóra.	Bhatt Ojha.	Bamankotya.
Bikrál.	Bálya.	Tewári.	Kadariah.
Kanhál.	Gilál.	Porseni.	Kála Khattri.
Batyál.	Chonial.	Homya Gái.	Dhúngána.
Ganjál.	Regmi	Túmrakot.	Pungyál.

EKTHARYA, *or insulated Tribes ranking with Khás.*

Búrathoki.	Chohan.	Bohara.	Kutál.	Ráya.	Boghati.	
Chiloti.	Dikshit.	Ravat.	Khatit.	Dángi.	Pandit.	
Katwál.	Bávan.	Raimanjhi.	Parsái.	Kháti.	Mahat.	
Bhukhandi.	Chokhál	Maghati.	Barwál.	Bhusál.	Chohara.	Durrah.

THAKURI, *or Royal Lineages, ranking with Khás.*

Sáhi.	Singh.	Chand.	Jiva.	Malla.	Maun.
Hamál.	Rakhsya.	Séna.	Chohan.	Ruchál.	

MAGARS.

I.—Sub-division of the Magars, called Râná.

Bhusál.	Gyángmi.	Byángnási.	Kyápcháki.	Aslámi.
Pulámi.	Phyûyáli.	Durra Lámi.	Yahayo.	Gácha.
Lámichanya.	Máski.	Sárû.	Pusál.	Gandharma.
Charmi.	Arghounlé.	Tháda.	Dûtt.	

* Manumitted slaves are called Párgharti, if of Khas lineage. They form a separate and rather numerous class, and so also do the Khawás or manumitted slaves of royalty.

II.—*Sub-division of Magars, called Thápa.*

Gránjá.	Chumi.	Keli.	Bareya.	Namjáli.
Lûngeli.	Jhángdi.	Máski.	Darrlámi.	Sunári.
Yángdi.	Phyûyali.	Marsyangdi.	Chitouriah.	Jhári.
Arghounli.	Gelung.	Sinjali.	Sárû.	Rijái.

III.—*Sub-division of Magars, called Alaya.*

Yángmi.	Sarángi.	Pûng.	Lamjál.	Súrya Vansi.
Gónda.	Sripáli.	Sûyál.	Kháli.	Dukhcháki.
Sijapati.	Panthi.	Thokcháki.	Méng.	Gharti.
Rakhál.	Sithûng.	Maski.	Lámichánya.	Palámi.
Lahakpá.	Arghounle.	Khaptari.	Phyûyáli.	Kyapcháki.
Dûrrá.	Khulál.	Chermi.	Pacháin.	

GURUNGS.

Gúrúng.	Lámichánya.	Khaptari.	Tangé.
Ghalle.	Siddh.	Ghûndáne.	Ghónyá.
Byápri.	Karámati.	Dhárén.	Paindi.
Vumjan.	Gósti.	Jimel.	Méngi.
Láma.	Bagálva.	Lopáte.	Dah Láma.
Chandú.	Tháthúng.	Lotháng.	Kurángi.
Góthi.	Chárki.	Bûlûng.	Khulál.
Gondúk.	Kháti.	Shakya Láma.	Surya Vansi Láma.
Gohori.	Guáburi.	Golángya.	Madán.
Baráhi.	Pengi.	Khangva.	Palámi.
	Ghárti	Dhakaren	

4. ON THE CHEPANG AND KUSUNDA TRIBES OF NEPAL.

(See Journal Asiatic Society Bengal, 1857.)

Amid the dense forests of the central region of Népál, to the westward of the great valley, dwell, in scanty numbers, and nearly in a state of nature, two broken tribes, having no apparent affinity with the civilized races of that country, and seeming like the fragments of an earlier population.

"They toil not, neither do they spin:" they pay no taxes, acknowledge no allegiance, but, living entirely upon wild fruits and the produce of the chase, are wont to say that the Rajah is Lord of the cultivated country, as they are of the unredeemed waste. They have bows and arrows, of which the iron arrow-heads are procured from their neighbours, but almost no other implement of civilization, and it is in the very skilful snaring of the beasts of the field and the fowls of the air that all their little intelligence is manifested.

Boughs torn from trees and laid dexterously together constitute their only houses, the sites of which they are perpetually shifting according to the exigencies or fancies of the hour. In short, they are altogether as near to what is usually called the state of nature as anything in human shape can well be, especially the Kusúndas, for the Chépángs are a few degrees above their *confréres*, and are beginning to hold some slight intercourse with civilized beings and to adopt the most simple of their arts and habits. It is due, however, to these rude foresters to say, that, though they stand wholly aloof from society, they are not actively offensive against it, and that neither the Government nor individuals tax them with any aggressions against the wealth they despise, or the comforts and conveniences they have no conception of the value of.

They are, in fact, not noxious but helpless, not vicious but aimless, both morally and intellectually, so that no one could, without distress, behold their careless unconscious inaptitude. It is interesting to have opportunity to observe a tribe so circumstanced and characterized as the Chépángs, and I am decidedly of opinion that their wretched condition, physical and moral, is the result, *not* of inherent defect, but of that savage ferocity of stronger races which broke to pieces and outlawed both the Chépáng and the Kusúnda tribes during the ferocious ethnic struggles of days long gone by, when tribe met tribe in internecine strife, contending for the possession of that soil they knew not how to fructify! Nor is there any lack of reasonable presumptions in favour of this idea, in reference to the Chépángs at least; for the still traceable affiliation of this people (as we shall soon see), not less than the extant state of their language, demonstrates their once having known a condition far superior to their present one, or to any that has been their's for ages.

That the primitive man was a savage has already appeared to me an unfounded assumption; whereas that broken tribes deteriorate lamentably, we have several well-founded instances in Africa.* Quitting, however, these speculations, I proceed with my narrative. During a long residence in Népál, I never could gain the least access to the Kusúndas, though aided by all the authority of the Durbar; but, so aided, I once, in the course of an ostensible shooting excursion, persuaded some Chépángs to let me see and converse with them for three or four days through the medium of some Gúrungs of their acquaintance. On that occasion I obtained the accompanying ample specimen of their language; and, whilst they were doling forth the words to my interpreters, I was enabled to study and to sketch the characteristic traits of their forms and faces. Compared with the mountaineers among whom they are found, the Chépángs are a slight but not actually deformed race, though their large bellies and thin legs indicate strongly the precarious amount and innutritious quality of their food. In height they are scarcely below the standard of the tribes around them||—who however are notoriously short of stature—but in colour they are decidedly darker. They have elongated (fore and aft) heads, protuberant large mouths, low narrow foreheads, large cheek-bones, flat faces, and small eyes. But the protuberance of the mouth does not amount to prognathous deformity,‡ nor has the small suspicious eye much, if anything, of the Mongolian obliqueness of direction or set in the head. Having frequently questioned the Durbar, whilst resident at Káthmándú, as to the relations and origin of the Chépángs and Kusúndas, I was invariably answered, that no one could give the least account of them, but that they were generally supposed to be *autochthones*, or primitive inhabitants of the country. For a long time such also was my own opinion, based chiefly upon their physical characteristics as above noted, and upon the absence of all traceable lingual or other affinity with the tribes around them; so that I took the Chépángs, the Kusúndas, and also the Haiyus§—a third tribe, remarkably resembling the two former in position and appearance—to be fragments of an original hill population prior to the present dominant races of inhabitants of these mountains, and to be of Tamulian extraction, from their great resemblance of form and colour to the aborigines of the plains, particularly the Kóls or Uraons, the Múndas, and the Malés. It did not for several years occur to me to look for lingual affinities beyond the proximate tribes, nor was I, save by dint of observation, made fully aware that the Turanian type of mankind belongs not only to the races of known Northern pedigree, such as the

* Prich. Phys. Hist. Vol. II. passim. Scott's exquisite novels throw much light on this subject.

|| Magar, Múrmi, Khas, Gúrung, Néwár.

‡ It tends that way, however: and the tendency is yet more strongly marked in some of the broken Turanian tribes of Central India; so that the general effect upon the Northmen of their descent into the least healthy and malarious jungles and swamps of the tropics, would seem to be to cause the Turanian type of human kind to assimilate with the African type, but with a long interval: degradation and hardship may in these broken tribes facilitate the effects of bad climate.

§ Haiyu, Háyu vel Váyu. See full treatise on this people in *Jour. As. Soc. Bengal*. Also vocabularies of the Chépáng and Kusúnda tongues.

mass of the sub-Himálayan population, but equally so to all the aborigines of the plains, at least to all those of Central India. Having, of late, however, become domiciled much to the eastward of Káthmándú, and having had more leisure for systematic and extended researches, those attributes of the general subject, which had previously perplexed me, were no longer hindrances to me in the investigation of any particular race or people. I now saw in the Turanian features of the Chépángs a mark equally reconcileable with Tamulian *or* Tibetan affinities; in their dark colour and slender frame, characteristics at first sight, indeed, rather Tamulian than Tibetan, but such as might, even in a Tibetan race, be accounted for by the extreme privations to which the Chépángs had for ages been subject; and in their physical attributes taken together, I perceived that I had to deal with a test of affinity too nice and dubious to afford a solution of the question of origin.* I therefore turned to the other or lingual test; and, pursuing this branch of the enquiry, I found that, with the Southern aborigines, there was not a vestige of connection, whilst to my surprise I confess, I discovered in the lusty§ Lhópas of Bhútan the unquestionable origin and stock of the far removed, and physically very differently characterized, Chépángs! This lingual demonstration of identity of origin, I have, for the reader's convenience, selected and set apart as an appendix to the vocabulary of the Chépáng language; and I apprehend that all persons conversant with ethnological enquiries will see in the not mere resemblance, but identity, of thirty words of prime use and necessity extracted from so limited a field of comparison as was available for me to glean from, a sufficient proof of the asserted connextion and derivation of the Chépángs, notwithstanding all objections deducible from distance, dissolution of intercourse, and physical non-conformity. But observe, the last item of difference is, as already intimated, not essential, but contingent, for both Lhópas and Chépángs are of the same, essentially Turanian stamp, whilst the deteriorations of vigour and of colour in the Chépángs, though striking, are no more than natural, nay inevitable, consequences of the miserable condition of dispersion and outlawry to which the Chépángs have been subject for ages anterior to all record or tradition. And, again, with regard to local disseveration, it should be well noted, in the first place, that

* See addendum on Bhútán.

§ I am now satisfied that the source of my perplexity lay in the common Turanian origin of all the tribes adverted to, which differ physically or lingually only in degree —physically, according to their earlier or later immigration and more or less healthful and temperate new abodes; lingually, also, according to their more special affinity with the less or with the more simple-tongued tribes or sub-families of the North. The oldest tribes of Himálaya, as sufficiently proved by their relative condition and location, are the broken tribes driven to the inclement summits or malarious glens of the Himálaya; and these in general have languages of the pronomenalized or complex sub-type, so that Müller is wholly wrong in assuming that Himálaya has no lingual traits of Draviria[a]—wrong also, I think, in the importance assigned to these contradistinctive marks of race. In proof see Poole on Egyptian language *Jour. Royal As. Society*, Vol. xx., part 34, p.p 313, *et seq:* the two dialects of the one tongue have a different arrangement of the pronom. adjunct of nouns and verbs. It must be, after this, almost needless to add that the relationship of the Chépángs to the Lhópas is general, not special.

[a] Neither Tamil nor Telugu nor Kannadi possesses in like perfection this diagnostic pronomenalization of noun and of verb (*viz.*, prefixed to noun, and suffixed to verb.)

by how much the Chépángs are, and have long been, removed from Bhútan, by so much exactly do conformities of language demonstrate identity of origin, because those conformities cannot be explained by that necessary contact with neighbours to which the Chépáng language owes, of course, such Hindi, Parbattia, and Néwár terms as the vocabulary exhibits; and, in the second place, we must recollect, that though it be true that 300 miles of very inaccessible country divide the seat of the Chépángs from Bhútan, and moreover, that no intercourse therewith has been held by the Chépángs for time out of mind, still in those days when tribes and nations were, so to speak, in their transitional state, it is well known that the tides of mankind flowed and ebbed with a force and intensity comparable to nothing in recent times, and capable of explaining far more extraordinary phænomena than the disruption of the Chépángs, and their being hurried away, like one of the erratic boulders of geologists, far from the seat of the bulk of their race and people. Indeed, the geological agents of dislocation in the days of pristine physical commotion may throw some light, in the way of analogy, upon the ethnological ones during the formative eras of society; and though we have no record or tradition of a Lhópa conquest or incursion extending westward, so far as, or even towards, the great valley of Népál, we may reasonably presume that some special clan or sept of the Bhútanese was ejected by an ethnic cataclysm from the bosom of that nation and driven westward under the ban of its own community alike, and of those with which it came in contact in its miserable migration,—for misfortune wins not fellowship.

The lapse of a few generations will probably see the total extinction of the Chépángs and Kusúndas, and therefore I apprehend that the traces now saved from oblivion of these singularly circumstanced and characterized tribes, now for the first time named to Europeans, will be deemed very precious by all real students of ethnology. Their origin, condition and character are, in truth, ethnic facts of high value, as proving how tribes may be dislocated and deteriorated during the great transitional eras of society.

ADDENDUM ON BHUTAN.

Lhó is the native name for Bhútán, and Lhópa and Dukpa (written Brúkpa) are native names for an inhabitant of Bhútan—whereof the former is the territorial, the latter, the religious, designation. In other words, a Lhópa is one belonging to the country of Bhútan, and a Dúkpá (*rectè* Brúkpa), a follower of that form of Lamáism which prevails in Bhútán, and which has become equally distinctive with the local designation for an inhabitant of the country, since the people of Bhót or Tibet were converted to the new or Gélúkpa form of that faith. Bhútan is a Sanskrit word, and is correctly Bhútánt, or 'the end of Bhót' (inclusively), the Brahmans, like the natives, deeming the cisnivean region an integral part of Tibet, which it is ethnographically, though by no means geographically. Had Klaproth and Ritter been aware that Lhó is Bhútan, and Lhópa an inhabitant of Bhútan, we should not have had their maps disfigured by a variety of imaginary regions placed east of Bhútan and termed Lokabadja, etc., a sheer variorum series of lingual error, resting on the single local name Lhó and its derivatives of a personal kind, as correctly and incorrectly gathered by them. Originally, some Bengáli rendered Lhó by the—to him—familiar word Lôk *(regio);* and then, being unaware that the Tibetan affix *bá vel pá* means 'belonging to,' 'inhabitant of,' he subjoined to the *bá* his own equivalent of *já* (born of), and thus was deduced Klaproth's furthest error (I omit others short of this one) of Lokabadja. To trace an error to its source is the best way to prevent its repetition, an aphorism I add, lest any person should suppose me wanting in respect for the eminent persons whose mistakes I have pointed out. Klaproth was possibly misled by Hastings' letters to and from Téshúlúngba.* But he and Ritter are fairly chargeable with constant creation of new regions out of mere synonyma! I could give a dozen of instances from their splendid maps.

——o——

VOCABULARY OF THE LANGUAGE OF THE CHEPANG.

English.	*Chepáng.*	*English.*	*Chepáng.*
The world	. . . .	A bridge	Tá
God	‡Nyam Ding	Husband	Palam
Man	Pûrsi	Wife	Malam
Woman	Mírû	Father	Pá
Quadruped	Svá	Mother	Má
Bird	Mó-wá	Brother	Hou
Insect	Pling	Sister	Hou dhiáng
Fish	Gna T	Grand-father	Tő
Fire	Mí T	Grand-mother	Aie
Air	Máró	Uncle	Páng
Earth	Sá T	Aunt	Mûm
Clay, plastic	Sá lena	Child	Chő
Water	Ti	Boy	Chő
Light, (lux)	Angha	Girl	Chő riáng
The sun	Nyam T	Kinsfolk	Laikwo
The moon	Lámê T	Strangefolk	Sáing
The stars	Kar T	Day	Nyi Gni T
A mountain	Rías T	Night	Yá
A plain	Dání	Dawn	Wágo
A river	Ghoro	Noon	Syáwa
A ferry	Titachaparna ? (ford)	Evening	Nyam rama

* See Turner's Embassy and native account of Bhútan, in the Transactions of the A.S.B. The affix *lhung* means 'valley,' and Lhása also, being 'in a valley,' it is often called Lhása-lhungpa or lhumba, that is, Lhása of the Valley.

‡ Nyam is the Sun, which is no doubt worshipped, and hence the identity of terms. Nyĭ in Chinese.

English.	*Chepáng.*
To-day	Ten
Yesterday	Yon
To-morrow	Syang
A week	....
A fortnight	Bákha yatlá
A month	Yatlá
A year*	Yatang
Summer	Lhapa
A quail	Umbá-wá
A kite or hawk	Mó-wá†
A fly	Yang
Winter	Namjûng
The rains	Nyamwá
Grain	Yam
Rice, unhusked	Yáng
Rice, husked	Chûí
Wheat	Kan
Barley‡	....
Plantain	Maisê
Pear	Pá-sai
Tobacco	Mingo
Pepper	Marich H
Red pepper	Raksai
Garlick	Bin
Oil	Sátê
A tree	Sing-tak T
A leaf	Lò T
A flower	Rò
A fruit	Chai
Wood	Sying T
Fuel	Jháro sying
Grass	....
Straw	Won
Bran	Rock
A horse	Serang
An ox	Shya
A bull	You shyá
A cow	Mó shyá
A buffalo	Misha T
A dog	Kúï T
A cat	....

English.	*Chepáng.*
A monkey	Yúkh
A jackal	Karja
A tiger	Já
A leopard	Mayo já
A bear	Yóm
A goat	Micha
A sheep	....
A hare	....
A hog, pig	Piak T
An elephant	Kísí N
A deer	Kasya
A rat	Yú
A mouse	Mayo yú
A manis	Chángjúng
A fowl (gallus)	Wá
Its egg	Wá-kún
A pigeon	Bak-wá
A crow	Káwá
A sparrow	Yúrkúnwá
A lark	Bajú wá
A partridge	Títhara H
Cord, thin	Rhim
Thread	Mayo rhim (mayo=small)
Needle	Gyap
A bee	Túmbá
The human body	Mhá
The head	Tolong
The hair	Min
The face	Khén
The forehead	Jyél
The eye	Mik T
The nose	Gné Nyé
The mouth	Móthong
The chin	Kám-tyó
The ear	Nô T
The arm	Krût
The hand	Kûtpá
The leg	Dom
The foot	....
The belly	Túkh

* The separate twelve months and seven days have no names.
† Wá is the generic of birds of the fowl kind.
‡ No other grain named, but wheat and rice.

English.	*Chépáng.*	*English.*	*Chépáng.*
Bone	Rhús T	The senses	. . . .
Blood	Wí	Touching	Dinang
Blood-vessel	Só	Smelling	Gnamang
A house	Kyim T	Seeing	Yorsang
A door	Kharók	Hearing	Saisung
A stone	Báng	Pen	Ré syáng
A brick	. . . .	Ink	Hildang
A temple	Ding tháni	Sovereign	Rájah H
An idol	Simtá	Subject	Parja H
A boat	. . . .	Citizen	Béráng mo
Dinner	Amjia	Countryman, rustic	ó moy
A dish	Lô	Soldier	Gal moy
A plate	Mila	Villager	Désing moy
Flesh	Mai	Priest	Jhákri
Bread	Lang	Physician	Chimé
Vegetables	Kyáng	Druggist	Osa yilong
Honey	Túm	Master	Sing chopo
Wax	Main P	Servant	Mayo (small)
Milk	Gnútí	Slave	Gráng
Gheu	Gheu H	Cultivator	Kámin chara
Cloth	Nai	Cowherd	Góthála H
Clothes, apparel	Nai	Carpenter	Bing kami N
Bed clothes	Lou	Blacksmith	Kami N
Upper vest	Doura	Weaver	Naik yousa
Lower vest	Súmbá	Spinner	Rhim rhousa
Shoe	Panai P	Tailor	Rûpsa
Stocking	Dócha P	Basket- master	Gráng kióni
Wool, raw	Min	Currier	Pún rúpo
Cotton, ditto	Kapás H	Tanner	Pún lai
Hemp, ditto	Kyou	Cotton-dresser	Rhim rhowan
Bow	Lúï	Iron	Phalám P
Arrow	Láh T	Copper	Támba H
Axe	Wárhé	Lead	Sísa H
Spade, hoe	Taik	Gold	Liáng
Plough	You sing	Silver	Rúpá H
Loom	. . . .	Rain	Nyông wá
Knife	Phiá ghúl	Frost	Chépú
Brush, broom	Phék	Snow	Rápáng
Basket	Tokorong	Ice	Chépú
Rope, thick	Rá	Fog	Khású
Beer	Han	Lightning	Marang
Spirits	Rakshi P	Thunder	Maranh múra
A still	Kúti pông		

English.	*Chepáng.*	*English.*	*Chepáng.*
A storm	Marhú	Four	Phôï-zho
A road	Liam T	Five	Púma-zho
A path	Mayo liam	Six	Krúk-zho
A spring (water)	Tíshakwô	Seven	Chana-zho
Trade	Yinláng	Eight	Práp-zho
Capital	Rás	Nine	Takú-zho
Interest	Chô	Ten	Gyib-zho
Coin	Tanka H	Half	Bákh
Robbery	Latiláng	The whole	Yágúr
Theft	Ditto	Some, any	. . . .
Murder	Jénsatáng	Many	Jhô
Rape	Kútyáláng	None	Dômánalo
Cultivated field	Blú	Near	Lôkô
City or town	Béráng	Far	Dyángtó
Village	Dési	Blind	Mikchángna
Horn	Róng T	Lame	Domtonga
Ivory	Laik	Dumb	Nósa chúl
Stupid	Waiva chúl	Deaf	Nósa mal
Honest	Waba pina	Clean	Bhangto
Dishonest	Wada pilo	Dirty	Gálto
Great	Bronto	Strong	Jokto
Small	Maito, mayo	Weak	Joklo
Heavy	Lito	Good	Pito
Light, (levis)	. . . .	Bad	Pilo
Tasting	Youngsang*	Ugly	Pilo
Hunger	Rúng	Handsome	Dyángto
Thirst	Kiôp	Young	Dyáng mai
Disease	Róg H	Old	Burha H
Medicine	Osá N	Clever	Chimo
Fever	Aimang	To stand up	Chingsa
Dysentery	Boárláng	To sleep	Yémsa
Small-pox	Brôm	To wake	Tyoksa
Fear	Rai	To give	Búïsa T
Hope	Aphrô	To take	Lísa T
Love	Mharláng	To lend	Búïsa
Hate	Ghrim náng	To borrow	Lísa
Grief, sorrow	Manbharáng	To buy	Yingsa
Joy	Yang náng	Black	Gálto
One	Yá-zho**	White	Phámto
Two	Nhi-zho T	Green	Phelto
Three	Sùm-zho T	Blue	Gálto

* Sá I think is the infinitive sign, and áng, the participial ; and one or other should appear uniformly here. Query? Sá the sign of neuter verbs.

** [*Zho* is evidently the *sho* 'number,' of the Chinese. J.S.]

English.	*Tibetan.*	*Lhópa.*	*Chepáng.*
Red	Dúto	To shit	Yésa
Yellow	Yérpo	„ piss	Chúsa
Sweet	Nimto	„ ascend	Jyáksa*
Sour	Nimlo	„ descend	Púsa
Straight	Dhimto	„ cut	Stalchisa
Crooked	Dòngto	„ break	Tlésa
Hot	Dháto	„ join, unite	Chòsa
Cold	Yéscho	„ jump	Jyésa
Dark	....	„ sit down	Músa
Light, luminous	Takto	„ write	Résa
Great	Bronto	„ read	Bròsa
Greater	Mhák tálto	„ sing	Mansa
Greatest	Mhák tálto	„ dance	Syáksa
Small	Maito	„ lie down	Kontimúsa
Smaller	Cholam	„ get up	....
Smallest	Cholam	„ tell a falsehood	Hekaksa
To stand	Chimsa	„ see	Chésa, yorsa
To fall	Chônsa	„ sell	Yinlangalsa
To walk	Whása	„ exchange	Gyésa
To run	Kìsa	„ live	....
To climb	Jyáksa†	„ die	....
To question	Hótsa	„ reap	Rása
To answer	Dyengnuksa	„ sow	Warsa
To request	Bajhináng?*	„ thresh	Rhapsa
To refuse	Bainanglo?	„ winnow	Krapsa
„ fight	Kaichinsa	„ hear	Saisa‡
„ kiss	Chopchisa	„ taste	Lyémsa
„ laugh	Nhísa	„ smell	Namsa
„ cry	Rhiása	„ touch	Dimsa
„ eat	Jhisa	„ count	Théngsa
„ drink	Túmsa	„ measure	Krúsa
„ talk	Nhosa	„ remember	Mhardangsa
„ be silent	Ashimanga	„ forget	Mhoiyangsa

N.B.—T postfixed indicates a Tibetan etymon for the word, H Hindi origin, P Parrbattia or Khas, and N Néwár ditto. It was not in my power to do more than collect vocables. I could not ascertain the structure: but comparing all the words, I conceive the anomalies of the verbs may be set right by assuming *sá* to be the infinitival sign, and *áng*, varied to chang, yang, and rang, the participial one.—B. H. H.

———o———

* These should be Chésa and Saisa I apprehend ; and so of the rest.

† If, as I suppose, Sá be the infinitival sign, there must be error, and the rather that all the verbs should have one form. A'ng, I think, is the participial sign.

List of Chépáng Words derived from the Tibetan Language, and specially the Bhútanese Dialect of it.

English.	*Tibetan.*	*Lhopa.*	*Chepáng.*
Eye	Mig	....	Mik
Sun	Nyima	Nyim	Nyam
Sky	Namkháh	Nam	Nam
Ear	....	Nó	Navó
Mountain	Rí	Rong	Rías
Star	Kárma	Kam	Kar
Tree	Shing	Shing	Sing-tak
Wood	Shing	Shing	Sing
Leaf	Lò-ma	....	Ló
Salt	Thsá	Chha	Chhé
Road	Lam	Lam	Liam
House	Khyim	Khim	Kyim
Moon	zLava (pron. Da-va)	....	Lámé
Bone	Rúspa	....	Rhús
Fire	Mé	Mí	Mí
Arrow	Dáh	Dáh	Láh
Dog	Khyi	Khi	Kúï
Buffalo	Mahi S	Meshi	Mísha
Day	....	Nyim	Nyi
Earth	....	Sá	Sá
Fish	Nyá	Gná	Gná
Hog	Phag	Phag	Piak
Horn	Rá	Ròng	Róng
Two	Nyis	Nyi	Nhí-zho§
Three	Súm	Súm	Num-zho
Give	Búh	Bin	Búï
Take	Lan	Ling	Lí

§ Zho is an enumerative servile affix, like thampa in the decimal series of Tibetan.

5. A CURSORY NOTICE OF NAYAKOTE AND OF THE REMARKABLE TRIBES INHABITING IT.

Náyakóte, or the Hither Náyakóte, as it is often called, to distinguish it from Náyakóte of the Choubisi, is the name of a petty town and district lying W. N. W., seventeen miles from Káthmándú, by the high road to Górkha. The town (so to speak) is situated at the northern extremity of the district, upon a spur descending south-westerly from Mount Dhaibung, or Jibjibia, at about a mile distant from the River Trisool on the west, and the same from the River Tádi, or Surajmatti, on the south and east. The town consists of from sixty to a hundred pakka three-storied houses, in the Chinese style of Káthmándú, chiefly owned by the court and chiefs; of a durbar, called the upper, to distinguish it from the lower one on the banks of the Tádi; and of a temple to Bhairavi, all in the like style of architecture. The town forms only a single street, lying in an indentation on the crest of the ridge, and is consequently not visible from below on any side, though the durbar and temple, from being placed higher, are so partially. Náyakóte, up to the late war with the English, was the winter residence of the present dynasty of Népál; but as the situation of the town is bleak and uncomfortable at that season, the court and chiefs then usually resided in mansions still standing at the base of the hill towards the Tádi, but now a good deal dilapidated, like the town residences, owing to the court having been stationary at Káthmándú since 1813. The district, like the edifices of the great, bears marks of neglect, which are the more palpable, by reason of a considerable portion of it being devoted to gardens and orchards, the property in a great measure of the owners of those edifices. The elevation of the town above the level of the Trisool must be from 800 to 1,000 feet, and the effect of this elevation in concealing it is aided on the side towards the Tádi by a fine forest of saul-trees occupying the whole declivity. On other aspects, the saul-trees, inherent to the whole site, are reduced to scrubby brushwood, by perpetual injudicious cutting and defoliation, the leaves being used as plates to eat from, and being perpetually carried to Káthmándú for sale there. This ridge has a soil of a deep red clay and its general form is rounded, but broken by deep ruts and ravines in most directions. Towards the Trisool west, and towards the Tádi south and east, the declivity of the ridge of Náyakóte is precipitous; but towards the junction of the two streams, in a south westerly direction, the hill falls off more gently, and about 1½ mile below the town spreads into an undulating plain, which occupies almost the whole space between the rivers to their junction and the ridge on which the town stands. This tract may be represented as a nearly equilateral triangle, two of the sides of which are formed by the rivers, and the third by the ridge. This triangle is a plain, exclusive of the declining spur of the ridge, and is an *elevated* plain, exclusive of that north-easterly angle lying on either side the Tádi, towards and to its junction with the Sindhu at the base of Bhálu Danra. This north-east corner is on the level of the rivers; the other parts are variously from one to four hundred feet above that level; and together they constitute the chief part and body, as it were, of the valley of Náyakóte, the rest or legs (so to speak with some aptness) of the district being the glens of the Tádi and of the Sindhu as far upwards, respectively, as the confluence of the Likhu and the base of Burmándi. The mountain ridges enclosing the district of Náyakóte, as above defined, are, beginning with the Náyakóte ridge itself, and circling east back again to it—Maha Mandal Nerja (north of Tádi,) Kabilás (dividing the Tádi and the Likhu), Bhálu (dividing the Likhu and the Sindhu), Dang-mai or Burmándi Madanpore, and Ghoor (enclosing the glen of the Sindhu on the south), Bélkote (carrying on the same southern barrier down the Tádi to Dévi Ghát), Jhiltoong (below the Ghát but still on the south of the river), Thirkiab (opposite to Jhiltoong on the north of, and across, the river), and Gowri and Sámari-bhanjáng (running northerly up the Trisool to the Sánga, or bridge at Khinchát), where we complete the circuit by linking the last to the Náyakóte ridge, the two in that spot pressing close on either bank of a river. With regard to size, if we speak of this tract as a whole, it will not be easy to be at once precise and distinct; but we may observe in regard to the body of the district, inclusive of the north-east corner on the low level, that from Dévi Ghát direct up the Trisool to the Sanga at Khinchat the length is four miles, by the road five miles; from Dévi Ghát to the town of Náyakóte from four to five miles, through the middle of the elevated portion of the district; from Dévi Ghát up the Tádi to its junction with the Sindhu, four miles and the same from the latter point to Khinchát across the base of the triangle, from the Tádi to the Trisool; again, and inclusively of the legs of the district, from Dévi Ghát to Burmándi, up the glens of the Tádi and the Sindu, is six miles ; and from the same point up the Tádi to its junction with the Likhu, eight miles. The maximum breadth of the entire district is at the base of the triangle just adverted to, and here the distance by the road from Bhálu Dánrá to Khinchát is four miles. The mean maximum of breadth, however, is not above three miles; that of the plateau alone, between the principal river, two miles. But, in speaking of breadths especially, we should distinguish between those parts which have been called the legs and the body of the district, the legs being the subsidiary vales of the Sindhu and of the Tádi. The former of these, then, from the base of Burmándi to the apex of the Bhálu ridge, where this glen merges in the larger one of the Tádi, is only from two hundred to four hundred yards wide ; whilst the width of the vale of the Tádi in that portion of it which extends lengthwise from the apex of the Bhálu ridge to that of Kabilás at Choughora, is from half to three-quarters of a mile; and, if we distinguish (as well we may) the low tract lying on both banks of the Tádi, between the western extremity of the two last-named divisions, and the point where the Tádi gets compressed into a mere gully on the upper confines of Bélkote (forming the north-east corner just spoken of inclusively,) we have a third tract, which is some 1,200 yards in medium breadth. The length, again, of the first of the sub-divisions of Náyakóte is two miles; of the second, four miles; of the third, one mile. All these three are tracts of the same character, that is, they are hot, swampy rice beds on the level of the streams that water them, except in the instance of the glen of the Tádi, which, upon the right bank of the river, possesses a widish strip of land considerably raised above the stream, and running under the Maha Mandal and Náyakóte ridges (where the court and chiefs have houses) to where the latter spreads into the chief elevated plain of the district above spoken of. That plain cannot be watered from the Trisool or Tádi by reason of its elevation; and as the Náyakóte ridge, whence it is derived, yields no efficient springs of water, the plain is condemned to exclusive dependance on rain. Every such plain or plateau is, in the language of Népál, a Tár; whereas the lower and perpetually waterable tracts, above contra-distinguished, are, in the same language, called Biási. The first of the three is the Sindhu Biási, from the name of its streamlet, the Sindhu; the next the Tádi Biási, from its river; and the third, either Tádi Biási also, or Sangum Biási, from the conflueuce of the Sindhu and Tádi within it. The Tár, or chief tract, is numerously sub-appellated, as Pullo Tár, next Dévi Ghát; then Manjhi Tár; then Bur Tár, next the Náyakóte hill; with various others parallel to these and nearer the Trisool, towards which the plateau in general has a tendency to sink step-wise, though never nearer the deep narrow bed of that river than several feet, twenty or more. These Társ are rather more wholesome and habitable than the Biásis, and capable of more various culture, though chiefly of trees, since trees alone can flourish deprived of water, except from rain; and thus is, in part, explained the great predominance of mangoe and other groves over fields of agriculture in the Tár or Társ of Náyakóte, which, however lovely at all seasons, boast no winter or spring crops, despite of the high temperature of the place; the Társ are too dry, and the Biásis too wet for spring crops, though they be common in the much colder valley of Népál Proper. The difference of temperature between the valleys of Náyakóte and of Népál Proper is occasioned by the difference of elevation above the sea. This difference amounts to 2,250* feet; and the same cause affords us also the only apparent, but far from satisfactory, explanation of the fact, that whilst Náyakóte is pestilently malarious from March to November, Népál Proper is free from this scourge, all other circumstances being the same in each valley. The lowlands of Náyakóte, consequently, are but very thinly peopled, the only permanent dwellers therein being several singular and affined races of men, called Dahi or Dari, Kúmhá, Kuswar, Botia, Bhrámu, and Dénwár, of whom more hereafter, and some few Parbattias and Néwárs. The Néwárs build and dwell solely on the Társ. The Parbattias will not adventure even so far, but usually have their houses on the hills around, and never suffer themselves to

* See Dr. Campbell's excellent paper, *apud* J. H. and A. S.

sleep in any part of the lowlands for a single night between April and November. In the Biásis, then, are houses of Dénwárs and their compeers only: in the Társ, those of the above people, and of some few Parbattias and Néwárs also, but in neither do the clusters of cottages hardly ever reach the size of a village and the dwellings stand for the most part single and scanty. The whole district is said to contain 700 houses, but I doubt it, even allowing 100 or 150 houses to the town; and half the number in either case would probably be nearer the mark.

The soil of Náyakóte contains a juster proportion of clay to silex and calx than the soil of the greater valley of Népál Proper, which is derived principally from the débris of granitic formations; and hence we obtain an explanation of the reputed eminent fertility of the former, and, more surely, of its celebrated potteries. The heights around Náyakóte are of inferior size, consisting on the northern side especially, mostly of iron clay, of very deep red tint; and the superficial soil of the Társ is for the most part the same, the substratum being, however, usually gravel, whence the dryness of their soil is increased.

The soil of the Biásis also is clayey, but untinted luteous white, and where unmixed with silex or other ingredients, even more tenacious than the red clay. The pottery clays are exclusively of the former sort. Mica, so common in the great valley of Népál, is here never witnessed. The high temperature of Náyakôte admits of most of the trees, forest and fruit, as well as of the superior Cerealia, of North Behar and the Tarai, being cultivated with success, though they cannot be raised in the great valley. Náyakóte has, besides, distinguished products of its own, which are not found, or not found so good, in the plains of Behar—these are the orange and the pine-apple. The forest trees peculiar to the district, not found in the great valley, and identifying this of Náyakôte with the Tarai and plains, are the Saul (*Shorea robusta*), Burr and Pipal (*Ficus Indica et Religiosa*), Semal or Cottontree, Prás, Neem, and Mohwa. The *Pinus longifolia*, and other mountain-growths, are frequently found mixed with these on the declivities around.

The chief of the fruit-trees is the Mangoe of various sorts, many exotic and superior, though the celebrated Bombay mangoe is apt to lose its flavour by swelling into undue and dropsical dimensions; the tamarind, the abir, the jack-fruit or bél, the kathur, the bádhur, the pukri, the guava, the custard-apple or sharifa, and, in a word, all the ordinary fruit-trees of India, none of which, it should be added, flourish in the larger valley. To the above we must subjoin the following exotics grown in the gardens of Khinchát, belonging to the Government—naril or cocoanut, supari or betel, vine, pear, apple, apricot (native), and plums of many kinds. All but the two first of these, however, flourish as well, or better, in the greater valley, being European products.

The smaller horticultural products of Náyakóte are pine-apples (excellent), plantains of many kinds and good, jamans of four sorts, melons, but no grapes nor peaches. Pines, platains and jamans are denied to the greater valley, where however the orange—that boast of Náyakóte—flourishes. The better kinds of the Náyakôte oranges are equal to any in the world, so that our horticulturists in

India should endeavour to procure and propagate them. The agricultural products of Náyakóte resemble in general those of the greater valley of Népál Proper; and as the latter have been fully described in print,* I shall on the present occasion specify only the peculiarities of Náyakóte produce, resulting from its more tropical climate. It has already been observed, that whereas there are two crops per annum in the greater valley, there is only one in the lesser, because of the excess of moisture and want of drainage in the Biásis, and of the total absence of means of artificial irrigation in the Társ. The Biásis yield only rice, which is not planted nor reaped at the early periods prevalent in the greater valley, but at the later ones usual in the plains of Behar; and the like is true of the sugar-cane which is grown on the skirts of the Biásis. In the great valley every blade of rice has disappeared by the beginning of November, and half the crop by the middle of October; the untransplanted sorts of Ghaiya even sooner. In Náyakóte the rice-harvest lasts till the beginning of December, nay to the middle of that month, and there are then no means of desiccating the fields rapidly enough for a spring crop. The rices grown in the Biásis are different from those grown in the greater valley, with the exception of Malsi and Touli, and even of these two sorts there is but little. Munsera is the staple crop of Náyakóte, and of its several kinds, as Doodia, Gouria, &c. It is of a bright golden hue, straw and grain, and longer in the stalk than our rices, to the best of which it is equal in quality. Among the seventeen to twenty sorts of rice grown at Náyakóte, are the Mal-bhóg, Krishen-bhóg, and other fine descriptions, for which Pillibheet is so famous. None of these last can be raised in the greater valley. The following are the names of the Náyakóte rices:—

Málsi,	Krishen-bhóg,	Isegoon,
Touli,	Bairini,	Anandi,
Doodraj,	Chárinagari,	Roodra,
Mansera,	Jarasari,	Katónja,
Gouria,	Mal-bhóg,	Tharia,
Kála Gouria,	Jhágri,	&c., &c.

The Ook, or sugar-cane of Náyakóte, is incomparably superior to that of the greater valley, and indeed to that of most parts of India. There are five principal sorts, fonr of which are yellowish, and the fifth, dark red. I purpose to send specimens of these to Calcutta for examination, Ook is grown on the skirts of the Biásis, as well as on the declivities of the hills near them. On the Társ, or plateaux or upper levels, are grown, besides the ordinary rain's produce of similar sites in the greater valley, the superior sorts of Dall, such as Arher, and cotton of inferior quality, neither of which can be raised at all in the greater valley. Of the whole surface of the Társ of Náyakóte, a half probably is devoted to gardens and orchard; a quarter to fields of dry produce; an eighth to rice or wet produce; and the remaining eighth may be barren.

* See Dr. Campbell's excellent paper, *apud.* J. H. and A. S.

The genera of mammals and birds observed during a hurried visit, under disadvantageous circumstances, were Nemorhedus (Ghoral), Stylocerus (Ratwa), Martes (Flavigula), Sciuropterus (Magnificus), Sciurus (Locria), all common to the greater valley; Corvus, Pastor, Coracias, Alauda, Anthus, Motacilla, Budytes, Pyrgita, Phœnicura, Saxicola, Phœnicornis, Dicrurus, Muscicapa, Tichodroma (Muraria), Picus, Palœornis, Clorhynchus,* Totanus Tringa, Egretta, Anas, Querquedula, Carbo, Mergus, Turtur, Euplocomus, Gallus, (Jungle-cock, Bankiva,) Chœtopus, Perdix, Coturnix, Hemipodius. Of these, Gallus, Coracias, and Palœornis, unknown to the greater valley, proclaim the *quasi*-Indian climate of Náyakôte; as Carbo and Mergus, also unknown there, do its larger rivers. For the rest, the species, as well as genera, are those common to both districts. The wall-creeper of Europe, supposed to be confined thereto, is frequent in both.

The commerce and manufactures of Náyakóte are too inconsiderable to claim specific notice; but in the cold season, in this, as in all other smaller valleys of Népál, booths are erected on the riverside by traders and craftsmen from the great valley, who reside there for the four coldest and salubrious months (December to March inclusive), exchanging grain for rock salt with the Bhótias, both Cis and Trans-Himálayan, dyeing the home-spun cloths of the neighbouring hill tribes with the madder supplied by them and the indigo of Tirhoot, and tinkering and pedlaring, and huckstering, for the assembly collected at this petty sort of fair.

It has been already observed, that the inhabitants of Náyakóte consist of several peculiar races, besides the ordinary Parbattia tribes and the Néwár. Both the latter have been described elsewhere, I shall therefore confine myself in this place to a short notice of the former, or Dénwár, Dari, Kuswar, Botia, Bhrámu, and Kúmhá. These tribes are exceedingly ignorant, and moreover are disposed to use the little wit they have in cunning evasion of all enquiry into their origin and history, affecting to be hill-men, employing the Parbattia language, and pretending to have forgotten their father-land and speech. In their (comparatively with reference to the Tartaric type) dark-hued skins, slender forms, oval faces, elevated features, and peculiar dialect, barbarous *patois* as the last now is —may perhaps be traced the apparent signs of a Southern origin. These men certainly do not all, if any, belong to the ordinary or dominant Tartaric stock of the mountaineers of Népál, but either to the ordinary stock of the Indian population (Indo-Germanic) or to some of those fragmentous branches of it, which still here and there represent a preceding Turánian race or races, as the Hós, Múndas, Uráuns, Gónds, Bhils across the Ganges, and the Thárus and Bóksas of the Népálese Tarai. Between the last-mentioned and the Dénwárs in particular, a distinct affinity may be traced: but to verify and illustrate this affinity through Tháru helps, is as little feasible, as to do it through Dénwâr ones; and I shall only therefore venture to say at present, that whether the Thárus of the Tarai, and the Dénwárs and their compeer cultivators of Náyakóte, and of other simi-

* Ibidorhynchus. Gould.

lar low and malarious tracts within the hills (for in many others they are found), belong to the aboriginal or to the ordinary stock of Indian population, they are closely connected among themselves, separate from the dominant Tartar breeds of the mountains, and possibly emigants from the plains countless generations back.*

The Kúswár, Bótia, Kúmhá (not own name), Bhrámú, Dénwár, and Dari or Dahi inhabit with impunity the lowest and hottest valleys of Népál, just as the Thárus, etc. do the Tarai, and also, the Múndas and Uráons of Chota Nagpore, but as recent servants and settlers merely, in the case of the last two, who are chiefly mentioned here, because of their participating with the races now before us, in that singular immunity from malarious affection, which is not known to be the attribute of any other people whatever.

Wherever malaria rages from March to November, beyond the Saul forest and within the hills, there the Dénwárs, Daris, Bhrámús, Kúmhás, and Mánjhis§ dwell, and dwell exclusively, sometimes collected in small villages, more usually in scattered cottages, comfortably built of unhewn stone, or wattles laid over with plaister, and furnished with a pent and overhung roof of grass or rice straw, which is verandahed towards the east. They follow the avocations of agriculturists, potters, fishermen, and ferrymen, and at all these crafts, and more especially at the second, they are very expert; the Kúmhás of Náyakóte in particular being renowned for their workmanship even in the vicinity of the very able craftsmen in that kind, whom the great valley produces.

These races of men affect a distinctness *among themselves*, which is apt to make a stranger smile, though it may possibly indicate different periods of immigration and of settlement within the hills, or immigrations from different places. In general, the five tribes or races will not intermarry among themselves, nor with any of the races around them; and they allege that their languages (dialects) were, and customs are, distinct. But they all now commonly use the Khas language, and call themselves Hindus, though they neither believe in the sacred

* I have, since this was written (sixteen years back), obtained samples of the languages of most of the above named tribes, which I am thus enabled to class with the broken Turanian tribes of the Himálaya, inclusive of its Tarai. These tribes, by their complex languages and altered physical type, form most interesting links between the Himálayan normal or unbroken tribes, as well as their confrères beyond the snows, and the broken and unbroken tribes of the Turanian stock in Central and Southern India, *viz.* the Dravirians or Tamulians and the Múndas, Hós, and Sontals. I cannot subscribe to Müller's or Logan's doctrine of a separate Gangetic sub-family of Turanians, nor to that of a separate Lohitic sub-family. Very remotely divided *times* of Turanian immigration may be conceded, but not totally sundered *routes*, and still less such broad distinctions of race among the immigrants as seem to be contended for. The hundred gates of Himálaya were ever open to admit immigrants, and the population beyond the snows has been in all time one and the same, or Turanian with subordinate distinctions equally found beyond and within the Himálaya. It may be that the Ugric stock of the immigrants found their way into India by rounding the N.W. extremity of the Himálaya. But there are closely allied Turkic tribes in Central Himálaya, which certainly entered by the Himálayan Gháts, *e.g.* the Kúswár and Bótia. (not Bhótia).

§ This is a Khas term and includes with the tribes of which the proper and separate names are Kúswár and Bótia (not Bhotia or Tibetan).

scriptures of the Hindus, nor accept the sacerdotal offices of the Brahmans. With a general remembrance of manners and customs, they have some trivial diversities of usage, as follows.

*Mánjhis.**—Their priests are the old men of the tribe; in making burnt and other offerings to their deities, they use no sacred or other words or prayers. On account of births, they are impure for four days: they cut the navel on the day of birth, and four days afterwards make a feast. On account of deaths, the impurity lasts forteen days, but under stress of business, one day's observance will suffice at the moment, so that the other nine are observed afterwards.

Dénwárs.—They allege that they came from the Western hills; their priests are their daughters' husbands and sisters' sons.§ Impurity at births lasts for ten days, and the same at deaths: they will not eat pulse dressed by Brahmans, but rice, if it have ghee in it, they will. They sometimes enter into trade and service. *Dahi vel Dari, Kúmhá, Bhrámú,* have a general resemblance of manners and customs with the last; but they will not eat rice dressed by Brahmans, whether it have ghee in it or not, but will eat other things of Brahman's dressing. None of the five races has any written language or characters; but the investigation of their common connection, and of their affinity with other aboriginal races inhabiting other more or less secluded localities throughout the plains of India,† might still be meuaged, through their speech, their physical attributes, their manners and customs, if the Argus jealousy of the Népál Government could by any means be charmed into a more discriminating use of Chinese maxims of foreign policy.

RIVERS FALLING WITHIN THE ABOVE LIMITS.

1. The *Sindhu*|| rises from Sindubhanjung, an off-set from Mount Manichur, or the most eastern part of Sivapoor, the northern barrier of the greater valley. The Sindhu has a course of about fifteen miles almost due west behind, or to

* Divided in Kúswár and Bótia, which are the proper tribe names. Mánjhi refers only to their profession as fishermen, and is a name imposed by the Khas.

§ These purely arbitrary customs may serve hereafter as helps in tracing the affinity of these and other semi-barbarous races throughout the mountains and hills of the Indian Continent, the *disjecta membra* of its original population.

The Dadhi or Dahi, Kumha (not own name), Kuswar, Botia (not Bhotia), Denwar, Boksa, Tharu, have tongues which are now almost merged in Hindi, though still retain ing some structural traits of Turanian origin, .*g.*, the Kuswar with its conjunct pronoun suffixed to uonn and verb in the Turkic[a] way. The Bhramu (who are allied to the Dadhi) like the Háyu, the Chepang, and the Kusunda of the hills, have tongues of purely Turanian character still.

[a] Kuswár supra :—

Baba-im 'my father.'	*Thatha-im-ik-an* 'I strike.'
Baba-ir 'thy father.'	*Thatha-ir-ik-an* 'thou strike.'
Baba-ik 'his father.'	*Thatha—-ik-an* 'he strike.'

Ik, the transitive verb sign. It is the conjunct form of the third pronoun.

† See a paper on the Nilgirians, in a recent number of thé *Asiaiic Society's Journal.*

|| Sindhu, a petty feeder only of Upper Likhu, rises at a village of Sindhu, soon merged in Likhu. The Sindhúria is separate and rises from eastern end of Bhálu Dánrá, where it links on to Burmándi. Thárakhóla, from Káhúlia, joins at base of Burmándi, and both flow about four miles to the Tádi. The stream spoken of as No. 1 is therefore the Sindhúria as now defined. The Likhu and Sindhu are one in all the limits noted, or rather the Sindhu is nothing.

the north of, Sivapoor and Burmándi, through a narrow fertile glen, which is somewhat interrupted by the projection of the base of Burmándi, where the main road from Káthmándú runs. Above this point the glen often bears the name of Tansen; the river is a mere streamlet, drawing half its water moreover from the west aspect of Burmándi, below the Resident's Powah or bungalow. It falls into the Tádi at Narain, or Ghur Ghát, being divided from the Likhu by Bhálu Dánrá, or the Bear's Ridge.

2. The *Likhu*, a somewhat larger stream than the Sindhu, parallel to it on the north, and separated from it by Bhálu Dánrá. The Likhu rises from above the Kabilás ridge, which divides it from the Tádi on the north. The course of the Likhu, though in general parallel to that of the Sindhu, yet radiates towards the north, as the Tádi does still more. The Likhu is about double the size of Sindhu, and has a course of perhaps twenty miles; it falls into the Tádi at Chou-ghora, four miles above the lower Durbar of Náyakóte. Its glen is cultivated throughout, and has an average width of 300 yards in its lower part. It is not a third the size of the Tádi.

3. The *Tádi*, classically styled Suryávati, from its taking its rise at Súryakúnd, or the Sun's Fount which, in the most easterly of the twenty-two little lakes of Gosain-thán, is thrown off towards the east, as is the Trisool from the same point towards the west, by the loftiest of the snowy peaks in the region of Népál Proper, and which is consequently the point of divergency of the nearest seven Gan-daks on the one hand, and of the seven Cósis on the other. The Tádi, however, though at first put off in an easterly direction, is drawn round westerly to mingle with the seven Gandaks, instead of joining the proximate Milamchi and Indhani, or first feeders of the Sun Cósi, by a large ridge running south from Gosain-thán nearly to Sivapoor, and putting off laterally towards the west the inferior ridges of Kabilás and Nerja, which separate the rivers Likhu and Tádi in all their lower and parallel courses. The Tádi proceeding at first easterly is gradually bent to the west by the great ridge just mentioned. The whole course of the river to Dévi Ghát, where it merges in the Trisool, may be thirty miles, ten east and south, and the rest W. S. W. In its lower course, before reaching Náyakóte, it is bounded on the left bank by the narrow ridge of Kabilás, and on the right by that of Nerja. It receives the Likhu at Choughora, four miles above, or east of, the lower Durbar of Náyakôte, and the Sindhu, at Narain Ghát, opposite to that Durbar. In the rest of its course of about four miles W.S.W. to Dévi Ghát, it confines the great Tár or plateau of Náyakóte on the south, just as the Trisool does no the north. At Narain Ghát the Tádi in December is thirty to forty yards wide and two feet deep. It is but little wider or deeper at Dévi Ghát, and consequently is not a tenth of the size of the Trisool, which at the Sunga of Khinchát is thirty-six yards broad and twenty-two and a half feet deep. The glen of the Tádi is cultivated throughout nearly, and in its uppermost parts is said not to be malarious.

4. The *Trisool*, or most easterly of the seven Gandaks of Népál, rises from the principal of the twenty-two Kunds, or lakelets (pools) of Gosainthán. These

lakelets occupy a flat summit of considerable extent, that cannot be less than 16,000 feet high, and lies immediately below the unrivalled peak variously called Nilkant, Gosain-thán, and Dhawalagiri.* The lake, more especially called Gosain-thán, is probably a mile in circuit, and close behind it, from the perennial snow, issues by three principal clefts (hence the name Trisool†), the River Trisool, or Trisool Gandaki. Its course is at first due west almost for perhaps fifteen miles, but then turns S.S.W., running in that direction for twenty miles, and more, to Dévi Ghát. It is a deep blue, arrowy, beautiful stream, conducting not only the pilgrim to Gosain-thán, but the trader and traveller to Tibet; the road to Kérung in Tibet striking off from the river where it bends (as yon ascend) to the east, and the town itself of Kérung being visible from Gosain-thán in clear weather, at the distance of perhaps thirty miles. The Trisool, four miles above Náyakóte, receives the Betrávati at Dhaibung, from the N.E. It is a petty stream, not having a course above fifteen miles from one of the resilient angles or bosoms of Mount Dhaibung or Jibjibia, the continuation of which ridge towards the west, and across the Trisool, is called *Sálima Bhársia.* This latter ridge conducts another feeder into the Trisool from the N.W. called the Salankhu, of about the same size with the Betrávati. Considerably south of the Sálima ridge is the ridge called Samribhanjáng, whence flows a third and still smaller feeder of the Trisool, named the Samri Khola, which disembogues itself into the Trisool from the north-west, half a mile to a mile below the Sunga or suspension bridge of Khinchát. The valley of the Trisool is narrow, and without any Biási or plain on the level of its waters, which flow in a deep bed. The heights, however, on one or both sides, supply numerous rills for occasional cultivation, which is maintained as far up as ten miles above Dhaibung (Dayabhang), a considerable village, where the ordinary Parbattia population begins to yield to the race called Kachár-Bhótias, or Cis-Himálayan Bhótias. At Dévi Ghát the River Trisool is passed by a ferry most jealousy guarded; nor is the river thence to Dévi Ghát permitted to be used for any sort of transport, nor even for the floating of timber, though the rapids (there are no cataracts) may help the prohibition. A few miles below Dévi Ghát, the streamlets poured into the Trisool by the glen of Dhúnibyási, afford much better access to the great valley of Népál, by the route of Trisool, than that which follows that river to Náyakóte and thence leads over Burmándi. These latter routes issue into the great valley at Thankote and at Ichangu Narain.

---

* Nilkant and Gosain-thán may be called proper names of this great snow mass. Dhoulagiri is rather a descriptive epithet, equivalent to Mont Blanc and Lebanon, and its application to this peak is unadvisable, because it has now become the settled name of the next great peak to the west of Gosain-thán.

† The legend of the place states that Maha Déva went to the snow to cool his throat, which had been burnt by swallowing the kalkut poison, which appearing at the churning of the ocean, threatened to consume the world. Maha Déva is called "blue throat," from the injury he sustained. He produced the river by striking his Trisool into the snows.

5. ON THE TRIBES OF NORTHERN TIBET. (HORYEUL AND SOKYEUL) AND OF SIFAN.

I now submit my promised Sifán and Hórsók vocabularies, with such geographic illustrations as may tend to render them more easily and fully appreciable. I intended to have retained these vocabularies till I had completed my ending investigation of the grammar of the Gyárúng and Hórpa tongues. But the high interest attaching to the discovery of another surprising instance of the wide-spreading relations of these tongues, made in the course of that investigation, and which discovery is sufficiently verifiable even by the vocabularies, though by no means limited to their evidence, together with the bearings of these vocabularies upon my two last communications, induces me not to postpone the sending of them. I can follow them up, by and bye, by the proposed grammatical elucidations. In the meanwhile there is abundant matter for the present communication in such a statement as I now propose giving of the present discovery, in some general remarks on the characteristics of the vast group of tongues to which the vocabularies, now and priorly submitted, belong, and in some descriptions of the physical attributes of the almost unknown races more immediately now in question. Nor do I apprehend that the want of the grammatical details adverted to will materially impair the interest of the present communication, since I have anticipated so much on that head in the way of practical exposition by samples as to make the special discovery I announce perfectly appreciable without those details, which, moreover, speaking generally of this vast group of tongues, I have shown reasons for deeming less important than they are wont to be held both philologically and ethnologically.

This series of vocabularies is entirely my own work in a region equally interesting and untrodden. It consists of seven languages, *viz.*, the Tróchú, the Sókpa, the Gyámi, the Gyárúng, the Hôrpa, the Tákpa, and the Mányák; and so novel is a deal of the matter, that it will be necessary to explain at once what these terms mean, and to shew where the races of men are to be found speaking these tongues. Hórsók is a compound Tibetan word, by which the people of Tibet designate the nomades who occupy the whole northern part of their country, or that lying beyond the Nyénchhén-thánglá* range of mountains, and between it

* This important feature of the geography of Tibet is indicated by the Nian-tsin-tangla of Ritter's *Hoch Asien* and by the Tanla of Huc. I have, following native authority, used in a wide sense a name which those writers use in a contracted sense; and reasonably, because the extension, continuity, and height of the chain are indubitable. Nevertheless, Ritter and Guyon have no warrant for cutting off from Tibet the country beyond it up to the Kuenlún, nor are Katché and Khór, the names they give to the country beyond, admissible or recognized geographic terms. Khór, equal Kór, is purely ethnic. and Katché is a corruption of Khâchhén or Mahomedan, literally Big-mouth.

and the Kwanleun or Kuenlún chain. Hórsók designates the two distinct races of the Hór or Hórpa and the Sók or Sókpa, neither of whom, so far as I have means to learn, is led by the possession of a native name at once familiar and general, to eschew the Tibetan appellations as foreign; though it will soon be seen that they are really so, if our identifications fail not. The Hórpa occupy the western half of the region above defined, or Northern Tibet; and also a deal of Little Bukharia and of Songaria, where they are denominated Kao-tsé by the Chinese, and Ighúrs (as would seem) by themselves.

The Sókpa occupy the eastern half of Northern Tibet as above defined, and also the wide adjacent country usually called Khokhonúr and Tangút by Europeans but by the Tibetans, Sókyeul or Sók-land.

In Southern Tibet, or Tibet south of the Nyenchhen-thánglá chain, there are numerous scattered Hórpas and Sókpas, as there are many scattered Bódpas in Northern Tibet; but, in general, that great mountain chain, the worthy rival of the Himálaya and the Kuenlún, may be said to divide the nomadic Hórpas and Sókpas from the non-nomadic Bódpas or Tibetans proper. Though the major part be Buddhists, yet are there some followers of Islam among the Hórpas and Sókpas of Tibet; more beyond the Tibetan limits. They are all styled Kháchhén by the Tibetans, of which word I think the Chinese Kao-tsé is a mere corruption, despite Cunningham's ingenious interpretation of Kao-tsé.

The Islamites are also called Godkar, of which term again Klaproth's Thógar seems to be metamorphosis.

Between the Hórpa and Sókpa, in the central part of Northern Tibet, are the Drókpa* vel Brógpa whose vocables I have as yet failed to obtain; and also, numerous "Kazzâk" or mounted robber bands, styled by the Tibetans Chakpa vel Jagpa, who recruit their formidable association from any of the neighbouring races, but especially from the Bódpa (Tibetans proper), the Hórpa, the Sókpa, and the Drókpa.

The language of the Chakpa is the ordinary Tibetan, and therefore, and because also of their very mixed lineage, they are of little ethnic importance, though always cited by the Tibetans, with fear and trembling, as a separate element of their population. The predatory habits of the Chakpa often carry them beyond their own limits, and they and the erratic Drókpa are often seen in Nári, where Gerrard and Cunningham speak of them under the designations of Dzakpa and of Dókpa. I doubt the ethnic independence of both, and believe them to be mixed associations, composed of people of the above specified races, from among which the Hórpa or Turks contribute an element even to the Himálayan population of Kanáwer, as is proved by the infinitives in "mak" of the Taburskad tongue.

From Khokhonúr to Yúnnán, the conterminous frontier of China and Tibet, is successively and continuously occupied (going from north to south) by the Sókpa above spoken of; by the Amdóans, who for the most part now speak Tibetan; by

* Quite distinct from the Dúkpa vel Brúkpa of Bhútan. The 'vel' indicates the distinction of the written from the spoken word.

the Thóchú; by the Gyárúng; and by the Mányak, whose vocabularies are all subjoined; whilst returning back westward, along the "pente septentrionale" of the Himálaya, we have, after passing through the Kham districts of Chyárúng and Kwombo, the region of the Tákpas, or Tákyeul, styled† Dákpo by Ritter, who, however, places it east of Kwombo, whereas it lies west of that district, written Combo by him. The Brahmapútra or Yárú quits Tibet in the district of Kwombo, as he states.

Tákyeul, the Towáng Raj of the English, is a dependency of Lhása. Its civil administrator is the Chonajúng-peun; its ecclesiastic head, the Támba Láma, whence our Towáng.

The people of Sókyeul, of Amdo, of Thóchú, of Gyárúng, and of Mányak, who are under chiefs of their own, styled Gyábo or King, Sinicé Wang, bear among the Chinese the common designation of Sifán or Western aliens; and the Tibetans frequently denominate the whole of them Gyárúngbo, from the superior importance of the special tribe of Gyárúng, which reckons eighteen chiefs or banners, of power sufficient, in days of yore, often to have successfully resisted or assailed the Celestial Empire, though for some time past quietly submitting to a mere nominal dependency on China. The word Gyá, in the language of Tibet, is equivalent to that of Fan (*alienus*,* barbaros) in the language of China; and, as *rúng* means, in the former tongue, proper or special, Gyárúng signifies alien *par excellence*, a name of peculiar usefulness in designating the whole of these Eastern borderers, in order to discriminate them from the affined and approximate, but yet distinct, Bódpa of Kham. Others affirm that Gyárúng means wild, rude, primitive Gyáa, making *rúng* the same as *túng* in Myamma; and that the typical Gyás (Gyámi) are the Chinese, though the latter be usually designated specially black Gyás (Gyá-nak).

The Gyárúngs themselves have no general name for their country or people, a very common case. When I submit the interesting itinerary I possess of a journey from Káthmándú to Pekin, I shall more particularly notice the topography of Sifán. At present it will be sufficient to add that this country, which extends from the Blue Sea to Yúnnán, with a very unequal width, varying from several days' march to only two or three, forms a rugged mountainous declivity from the lofty plateau of Kham to the low plain of Szchuen, and which is assimilated by those who well know both, to the Indian declivity of the Himálaya, the mountains being for the most part free from snow, and the climate much more temperate than that of Tibet. *Within* this mountainous belt or barrier of Sifán are the Tákpa, who are consequently Tibetans: *without* it are the Gyámi, who are consequently Chinese, as will be seen by their respective vocabularies—vocabularies, not the less valuable for being dialects merely (if no more) of languages well known, because the dialectic differences of the Chinese and the

† I should add that Ritter's Gákpo and Gangpo, and Dákpo, are not three separate places, but merely various utterances of the single word Tákpa, and no more admissible therefore than his Katché and Khór before explained. This great geographer is rather too prone to give a "local habitation" to the airy nothings of the polyglottic region, as I have formerly had occasion to point out, though no one can more admire than I do his immense learning and the talent that guides and animates it.

Tibetan tongues are little understood,§ at the same time that they are very important for enabling us to test the alleged distinctness of the great groups of people nearest allied to these divisions.

For my part I apprehend that the true characteristics of the Chinese and Tibetan languages have been a good deal obscured by bookmen,* native and European; and, though it be somewhat premature to venture an opinion before I have completed my pending investigation of the Gyârúng and Hórpa tongues, I still must say that I suspect few competent judges will rise from the attentive study of this and my two prior series of vocabularies, with out feeling a conviction that the Indo-Chinese, the Chinese, the Tibetans, and the Altaians, have been too broadly contra-distinguished, and that they form in fact but one great ethnic family, which moreover includes what are usually called the Tamulian or Dravidian and the Kôl and Mūnda elements† of Indian population, as well as nearly every element of the population of Oceania.‡

* Hence Gyá philing, or Frankish stranger. European foreigner is the name for Europeans in Tibet. Philing = Frank, Indicé Feringi, *not* as interpreted by M. Huc.

§ Leyden reckoned ten Chinese tongues (As. Res., X. 266). Others hold that there is but one. Again, Rémusat (*Recherches sur les langues Tartares*) insisted that there must be several tongues in Tibet, whereas Csoma de Koros (Journal No. 4) considers that there is but one. This comes in part of the want of a standard of ethnic unity, whether lingual or physical, and in part of the mixture of distinct races by regarding them under a large geographic and political unity, thus the Hórsók belong undoubtedly to Tibet, but do not belong to the Bódpa race. I have given, I believe, all the languages of Tibet, that is, the languages of all the races now and long settled in Tibet. My Gyámi vocables exhibit a vast difference from the Kong one of Leyden, *ut supra*. But I do not rely on mine, nor have I means to test it.

* A deal of Csoma's abundant grammatical apparatus of the Tibetan tongue is positively repudiated by the people of Tibet, whilst the learned and sage Rémusat teaches us to question the over-strained aud unintelligible assertions about the monosyllabism of the Chinese tongue, as if there were no dissyllables, no adjuncts to the roots! and as if the roots of Sanskrit, Hebrew, and Arabic were *not* monosyllables. For some valuable remarks on monosyllabism, see *Recherches sur les langues Tartares*, i. 351-4, and compare what occurs in the sequel as to the monosyllabic polysyllabism (different aspects of the case) of Gyárúng and Tagala. Thus in Gyárúng the root *zo* becomes Masazangti by mere cumulation of particles, *ma*, *sa*, *za*, *ang*, and *ti*.

† For some proofs of the reality of this element, see a paper on the Nilgirians in a recent number of the *Asiatic Society*. Adverting to recent denials, it may be worth while to give here a Himálayan sample of Dravirianism from the Kiranti language :—

Wá popo, my *J'popo*, thy *A'popo*, his	uncle.	*Wá gû*, my *I'gu*, thy *A'gu*, his	hand.
Pog-ú, I *Pog-í*, thou *Pog-á*, he	beget.	*Teub-ú*, I *Teub-i*, thou *Teub-á*, he	strike.

Of that complex pronomenalization of the verb, for which the Hó and Sontál tongues are so remarkable, I shall shortly have to produce some still more perfect samples from the Central Himálaya. In the paper referred to, I have demonstrated the forthcomingness also of the Turkic, *viz.*, Kuswár tongue which has conjunct contracted pronoun suffixed to noun and to verb, and Mantchuric elements in the languages of Himálaya.

‡ The elder oceanic element, or Alforian, = our Tamulian and the analogous dispersed and subdued tribes of the Himálaya, Indo-China, and China: the younger oceanic element, or Malayo-polynesian, = the now dominant tribes of Indo-China, China, Tibet, and Himálaya. I must content myself, at present, with pointing to the special illustration of the *latter* part of this reunion of the continental and insular races in the sequel, though every proof of the wide common domain of the continentals is also an illustration, inferential, yet clear, of both parts of it.

My former vocabularies showed how intimately the Indo-Chinese tongues are allied with the Himálayan and Tibetan by identity of roots, of servile particles, and even of entire words, as the integral results of the combination of the two former, provided only that the comparison be drawn from a field large enough to exhibit the necessary range of admitted mutation, both in the primary and secondary parts of words, in use for ages among widely sundered, and often also extremely segregated, races. How large that range of admitted mutation is, I have illustrated by examples in the note appended to the present series of vocabularies, and I recommend those who would properly appreciate the great apparent deviations from a type of language, which is, as I suppose, one and the same, to take good heed of what is there instanced. In the meanwhile, without fatiguing the reader with more analyses at present, I proceed to remark that the analogies and affinities indicated by the last series of vocabularies between the Himálayan and Tibetan tongues on one hand, and the Indo-Chinese on the other, are carried on and confirmed by some of the present series, whilst others extend the links to the Altaic group of languages; the Gyárúng, Tákpa, and Mányak carrying the chain of connexion onwards from the south-east, and the Thóchú, Hórpa, and Sókpa, transmitting it over the Kwanleun to the north and west; the Gyárúng by its grammatical structure exhibiting also marvellous correspondencies with remoter regions; with Caucasus, as has been separately shown already, and with Oceania, as will appear in the sequel of this communication. How far precisely the other languages now submitted may participate these express and peculiar features of grammatical affinity, I am not yet prepared to say. But the whole of them certainly exhibit a great general resemblance in the broader traits of syntactic,* and yet a greater in those of etymological, construction. In a word, they are evidently members of that single and vast family of languages, the singleness and the vastness of which I conceive to be justly inferrible even from its vocables—1*st*, because of the similarity of the roots; 2*nd*. because of the similarity of serviles; 3*rd*. because of the similar principles governing the uses and the mutations of both, and the consequent composition and the character of the integral words, which exhibit an essential identity in numberless terms of prime necessity, after due allowance for synonymous changes in their roots and for euphonic and differential changes in their serviles within known limits and upon a demonstrably single plan.

I infer that the differences characterizing this vast family of languages, however striking at first sight, are subordinate, because when the languages are examined upon a broad enough scale, these differences are seen to pass away by insensible gradations. Such as they are, they arise from—1*st*, a greater or lesser use of the pre-fixed, in-fixed, and post-fixed particles, amounting to nearly con-

* I may instance the universal substitution of continuative gerunds and participles in lieu of conjunctions and of conjunctive (relative) pronouns, because this feature has been supposed to be specially characteristic of the Altaic group. It is no more so than the vocalic harmony of Turki, or than the inverted style and tonic system of the Indo-Chinese tongues. These appear to me to be blending differences of degree only, not absolute differences of kind, and to have been used to sever unduly the several groups.

stant employment of some or all of them in some tongues, and to nearly total disuse of some or all of them in others; [The disuse or non-use is often only apparent, for the surplus "silent" letters are really pre-fixes, with a blended, instead of a separate utterance. That this is so may be proved to demonstration by identity of *function* (differential) in the two; and yet the blended or separate utterance makes all the difference between monosyllabism and its opposite, besides causing other differences that are apt to conceal the essential identity of words.] *2nd*, from a preference by one tongue of the pre-fixes, of the in-fixes by another, and of the suffixes by a third; *3rd*, from that transposed position and function of the primary and secondary part of words (root and particle), which is a law of these languages eminently obscurative of identities in its partial operation; [compare 'overleap' and 'leap over;' what holds good chiefly as to our verbs, holds good equally as to the verbs and nouns of these tongues, wherein indeed the two classes of words are but faintly distinguishable, or not at all so. Abundant fresh evidence of the law may be found by comparing Leyden's Indo-Chinese with my Tibeto-Himálayan vocabularies: compare *mimma* and *sa-mi*, Burmese, with *mi*-sā, Newari, root *mi;* and *ma-nek*, Burmese, with *nyi-mā*, Tibetan, root *nyi*, 'Day, sun, and morning,' when compared, speak for themselves.] *4th*, from the substitution of a reiterated root, for a root and particle in the composition of words, when the various meanings of the root might otherwise transcend the differencing power of the particles, or, at all events, not satisfy the demand for an unusually broad distinction; [in Gyárúng, the root *pyé* 'bird,' is so near to the root *pé* 'father,' that they have been segregated by the application to one of the usual prefix, to the other of the iterative principle, or root repeated, whence *ta-pé* 'a father,' and *pyé pyé* 'a bird,' for *san et pé pé*. I might add, as a fifth cause of difference between these tongues, the different degrees in which each employs the tonic or accentual variant, which principle has been most erroneously supposed to be exclusively Chinese and Indo-Chinese, whereas it prevails far and wide, only more or less developed; most where the servile particles and so-called silent letters are least in use; least, where they are most in use; so that the differential and equivalent function of all three peculiarities, that is, of "empty words" (see Chinese Grammar), of "silent letters," and of tones, is placed in a clear light, such as Rémusat vainly strove to throw upon one of the three, viewing it separately.*] *5th*, from the disjunct or conjunct (elided vowel) method of using the pre-fixed serviles, whence results at once all the difference of soft polysyllabism or harsh monosyllabism.

The resulting disparities of the vocables are certainly often very marked, as in the Wa-tú and U-í instance of Gyárúng and Circassian, (so singularly confirmed by the Malay and Tagala *itú* 'that') [*I-tú*, *Wa-tú*, and *U-í* are easily explained, and show how congruous all these tongues are at bottom. Few of them

* See *Recherches sur les langues Tartares*, p. 355-7, vol. i. Csoma de Koros strangely enough says nothing about tones or servile particles, and hence his remarks on the silent letters want point and significance. The language of Népál Proper is remarkable for its numerous tones and its scanty serviles, whether literal or syllabic.

have any proper third pronoun, they use as equivalents the demonstratives, which are *í* and *á*, or *ú* or *w*, or *wa=u*. *Ta*, with or without the nasal ending, *ta*, *tan*, *tang*, is a synonym (*Ti*, *di* Tibetan, *Thi* Burmese, etc.) constantly added to the near or far demonstrative, and repeating its vowel thus, *i-thi* Burmese, *wa-thi* Hayu, *i-ta* and *u-ta* Khas, *wa-tu* Circassian, whose *u-i* is a mere combination of the two demonstratives, either of which is equal to the third personal. The *ta* is prefixed or suffixed, in the sense of Latin *ejus* to nouns, and thus we have *ta-yú* Lepcha and Tamil for a woman, *ta-gri* Lepcha and *tandri* Tamil, a man, and *tangkos* Uraon, a son, etc., as samples of its prefixed use. Müller is, I think, wrong in citing the crude *pa* and *ma* as normal samples to be opposed to the Arian *pa-ter* and *ma-ter*. Few Turanian tongues use the crude forms, and many use the identical root and servile.] The case is similar with those given at the end of the present series of vocabularies, so that it is no great wonder that the Mongolidan or Turanian tongues have been referred to many groups so trenchantly separated as virtually to fall under different families. And, if I incline so strongly to unitise the family, it is only because, as far as my investigations have gone, I have been able to discern nothing absolute and invariable in the distinctions—which though no doubt distinctions proper to the vocables only, and not effecting structural diagnostics (in the usual narrow sense, for composition of words *is* structure), are yet unusually, and as I conceive decisively important, owing to the extremely inartificial character which belongs to the grammar of these tongues, with some apparently borrowed exceptions, such as that of the Turkish verbs. Not that the grammatical or the physical evidence of this assumed family identity conflicts with that of the vocables—much the contrary, as we shall soon see—but that the latter has unusual relative value. [I may mention here an interesting sample of this identity, derived from the substantive verb. It is *da* in Myamma, *a-da* in Malay, *da* in Hórpa, *gdah* in Tibetan, *dan* in Uraon, etc. So also it is *mena* in Sontál and *mna* in Tibetan; and again, it is *dug* in Tibetan, *dong* in Bodo and Garo, *du* in Newari, *dong* and *kam dong* in Gyárúng.] And, would we speak plainly, we should say that grammar relates equally to the construction of words and to the construction of sentences, and that the former sort of putting together, or syntax, is always equally, and often more, important than the latter. Certainly, it is more so in the Mongolidan tongues, which are as much distinguished by their immensity of nicely discriminated terms, most of them *necessarily* compounds—and compounds of no unskilful contrivance—as by the scantiness and simplicity of the contrivances by which those terms are held together in sentences. [See vocab. voce 'give' and 'take.' A Tartar cannot endure that confusion of the precative, optative, and imperative, which our imperative mood exhibits. But he remedies the defect not by the multiplication of grammatical forms but by the use of distinct words or distinct modifications of the same word, thus *Davo* 'commands' and *Davong* 'solicits,' *et sic do cæteris*. Compare the disjunctive *we*, so common in these tongues. *Davo* means 'give him,' *Davong* 'give to me, by the annexed pronouns, and just so in Limbu *Pire* and *Pirang*, and in Vayu *Hato*, and Hasing, Lepcha, and Néwári, which eschew suffixed pronouns, have

Bo and *Bi*, *Byu* and *Ti*, for the respective senses, the former modifying the one root, the other using two distinct roots. Observe the identity of *byu*, *bo*, *bi* and *pi* (of *pi-re*, *pi-rang*.)] Nay, if we look carefully to what has been so well done in one's own day for the elucidation of our own language, we shall discern that the new lights have been principally etymological, borrowed from, as thrown upon, the construction and composition of words, not of sentences.

Perhaps it will be urged that, after all, the structural analogy I have established between the Gyárúng and Circassian tongues belongs rather to the etymological than to the syntactic department of languages. Let it be granted, and I would then ask whether the analogy be therefore less important? And is it not singular and a proof wherein resides the essential genius and character of these tongues, and where therefore we are to seek for their true and closest relations, that my scanty knowledge of the Himálayan and Tibetan group of them should enable me unhesitatingly to analyse the words of the Caucasian group, of which I know nothing, and to pronounce, for instance, *Di-di* to be a re-duplicate root, and *Dini* to be a root and servile prefix, with perfect confidence, and, as I doubt not, with equal accuracy? *That* will, at all events, be known by and bye, and should the result be such as I look for, the consequent affinity of the Caucasian and Mongolian tongues will take an unquestionable shape and stand on the unassailable basis of words similarly constructed in all their parts and similarly employed throughout.

I must, however, whilst thus insisting on the pre-eminent importance of Mongolidan vocables, freely admit that those of all my present series are by no means entitled to equal confidence,* my access to the individuals who furnished the Sôkpa and Gyámi words in particular having been deficient for such analytic dissection as I hold by, and the competence of my informants, moreover, not beyond question. I am likewise much in want of adequate original information respecting the Altaic group, and of the books that might supply it. Nevertheless, I think, I may safely affirm upon the strength of my vocabularies, that the Sókpo of the Tibetans are, as has been already assumed in this paper, no other than the Oelet and Kalmak of Rémusat and Klaproth,† whilst their *confrères*, the Hórpa, are almost as evidently Turkish, the Turkish affinity of the latter being inferred, not only from the vocables, but from the complex structure of Hórpa verbs and from the quasi-

* Unfair use has been made of this admission. The vocabularies, such as they are, are exceedingly valuable, though perhaps without analysis incapable of supporting such a towering superstructure of theory as has been raised on them by their impugners.

† I might now add, having just laid my hand on M. Huc's book, the synonym of Turgot to those of Kalmak and Oelet, but that Turgot, like Dúrbét, designates only a tribe of this race, and a tribe whose tribunal denomination, as well as its migration to the Volga and back to the Ili, had been already stated by Rémusat. M. Huc's amusing work, in fact, adds nothing to our stores of accurate ethnological knowledge, his mere assertion, for instance, that the Hiongnu were Huns throwing no fresh light upon a long debated point, and the nullity of the absolute identity of names in reference to the Sog, teaching us yet more to doubt vaguer identifications of this sort. Let me add that M. Huc's account of the habits, manners and characters of the several peoples is capital, and most evidently, accurately, as vividly, delineated.

Arian physiognomy of the samples I have seen of the Hórpa race.† And thus, quoad Sókpo, is dissipated the dream of twenty years, during all which time I have been in vain endeavouring to get access to the Sókpo, assured from the identity of names (Sók pronounced Sog), that in the much talked-of people of Eastern Tibet, I should discover that famous race which gave their appellations to the Sogdiana and Sogdorum regio (or the Indus) of the classics, and whose identity with the Sacæ of Indian and Grecian story, whose genuine Arianism and resplendent renown I never permitted myself to doubt. Reverting to what I have better assurance of, I shall next note a fact as extraordinary almost as that which formed the subject of my last communication to the Society, to wit, that some of Humboldt's characteristics of the Malayo-Polynesian tongues hold good as to the Gyárúng language even more strangely than Rosen's of the Circassian; so that we may have possibly, in the unsophisticated tongue of this primitive race of mountaineers, situated centrally between the Chinese, the Indo-Chinese, the Tibetans, and the Altaians, and protected from absorption, assimilation or conquest by their fastnesses, the main and middle link of that vast chain which unites the insular and continental nations of the East and the most dispersed scions of the immensely diffused family of the Mongolidæ* !! Those who are acquainted with the famous Kavi Sprache (known to me alas! only at second hand) will know what I mean, when I solicit their attention to the accompanying Gyárúng vocabulary, as bearing on the face of it evidence, that in the Gyárúng tongue almost all the words in their ordinary† state are dissyllables, whilst I can assert positively from my own knowledge of the language, that the two syllables may be resolved into a monosyllabic root and its affix, *or* into a repeated monosyllabic root. Now these features (which by the way are very noticeable even in the small samples accessible to me of the Circassian tongue) Humboldt has denoted as special characteristics of the Malayo-Polynesian languages; and they are certainly most conspicuous attributes of the Gyárúng tongue. Thus, in the first column of the Gyárúng vocables, there are thirty-five words, whereof not less than thirty-one are dissyllables and only four monosyllables, and the dissyllables are all re-

† Müller doubts, but the Tibetans cannot mistake, and with them Hór = Turk and Sók = Mongol. I have failed to get fresh access to these people, which I the more regret, inasmuch as the name Hór, even to the guttural *h* and to the omissible *r*, tallies exactly with the appellation given by themselves to the so-called Lerka tribe of Singbhúm. See Tickell's narrative and vocabulary. I have elsewhere pointed out the Turkic affinity of one Himálayan tribe (Kúswár) and the Mantchuric of another (Váyu or Háyu). See paper on the Nilgirians. *(J.A.S.B.)* Tibet has been absurdly isolated by philologers and geographers. The northern half of it actually belongs rather to the Altaic than to the Bódpa tribes, and hence is called by the latter Hóryeul and Sókyeul. I am indebted to the Múndas for the knowledge that Hó is pronounced Khó and Khór, just as it is to the North.

* It may reconcile some of my readers to this startling announcement to hear that there are historical or traditional grounds for supposing this very region to be the common nest and original seat of the Chinese and Tibetan races. See Klaproth's *Tabl. Histor. and Mêmoires relatifs à l'Asie,* and Rémusat's *Recherches sur les langues Tartares.*

†I say ordinary state, because, when all the apparatus of composition attaches, they become polysyllabic. See the sequel, and mark the consequence as to the monosyllabic test.

solvable into a monosyllabic root and its customary pre-fix (Ta, mutable into Ka,) save those (Pyépyé, Nyényé) that are formed by re-duplication of the radical.

That *Pyê* 'bird,' and *Nyé* 'cow,' are roots, any one may prove for himself by turning to their Tibetan and Chinese equivalents; and that in the Gyárúng tongue the root is in these instances repeated to constitute the current term or integral word is self-apparent. That, again, in Gyárúng, *Ta* is the common and almost indespensible prefix, and is mutable into *Ka*, both liable to euphonic changes of the vowel, to suit that of the radical, the vocabulary also demonstrates, testably to any extent by its predecessors of the allied tongues. And if it be urged, as in truth it may be, that the above constitution of the vocables belongs in essence to all the continental tongues, as Humboldt's sagacity divined it did to all the insular ones, the more frequent use of the prefix and consequent dissyllabism being all that is exclusively Gyárúng, I have still to produce another Gyárúng trait, which it shares with what has been deemed the most primitive Malayo-Polynesian type; and I shall do so by the following quotation from* Leyden:—"Few languages present a greater appearance of originality than the Ta-gala. Though a multitude of its terms agree precisely with those of the languages just enumerated (the Western Polynesian), yet the simple terms are so metamorphosed by a variety of the most simple contrivances, that it becomes impossible (difficult—*B.H.H.*) for a person who understands all the original words in a sentence to recognize them individually, or to comprehend the meaning of the whole. The artifices which it employs are chiefly the pre-fixing or post-fixing (or in-fixing—*B.H.H.*) to the simple vocables (roots) of certain particles (serviles) which are again combined with others; and the complete or partial repetition of terms in this re-duplication may be again combined with other particles." The above, as well as what follows (pp. 211-12) upon Ta-gala verbs, is in general remarkably coincident with Gyárúng, the differences being such only as, when compared with other allied tongues, to show that the characteristics, however pre-eminently, are by no means exclusively, Gyárúng among the continental tongues, any more than they are exclusively Ta-gala among the insular ones. [Here are some samples as significant as Leyden's illustrations of the Ta-gala verbs. From the root *Ching*, 'to go,' we have almost indifferently *Yaching*, *Kaching*, *Daching*, *Naching*, in a present sense, and *Yataching*, *Kataching*, *Dataching*, *Tataching*, *Nataching*, in a past sense, with some speciality of sense as to the *na* and *ta* pre-fix that need here be particularized. Next we have *Yatachinti*, *Katachinti*, *Datachinti*, *Tatachinti*, *Natachinti*, meaning, 'one who goes or went, or the goer,' if one's self; and, if any other, then the series becomes *Yatachisi*, *Katachisi*, etc. The negatives are *Matachinti vel Matachisi*, according to the person, the particle of negation displacing the first of the pre-fixes indifferently. So from *Máng*, 'to sleep,' *Carmáng*, *Marmáng*, *Tatarmáng*, *Matarmángti*, *Tatarméti*, *Matarmési*, 'I sleep, I sleep not, I slept, I who slept not, thou who sleepest, he who slept not,' or 'the sleepless,'

* *Researches*, *B.A.S.*, vol. x., p. 209.

(other than one's self). From *Zo*, 'eat,' *Tasazo* 'feed,' *Tasazangti*, 'I who feed,' *Tasazési*, 'he who feeds,' *Masazángti*, 'I who feed not.' Of these I give the analysis of the last as a sample, *Ma*, negative pre-fix—*Sa*, causative in-fix. *Záng*, 'I eat,' from the root *Zá* with suffixed pronoun. Ti mutable with Si, the participial attributive suffix. These are the simplest verbal forms and the most usual, whence the prevalent dissyllabic character of the verbs, as of the nouns, as seen in the vocabulary, consisting of a root and one pre-fix. But the vocabulary, whilst it demonstrates this, indicates also the more complex forms, put rather too prominently forward by Leyden in his Ta-gala samples. Thus, in our Gyárúng vocabulary, the words, cry, laugh, be silent, run, or four out of twenty-four verbs, instead of a single prefix, have a double and even a treble supply in the simple imperative form there used; as *Da-ka-krú* from the root *Krú: Kana-ré* from the root *Ré; Na-ka-chúm* from the root *Chúm; Da-na-ra-gyúk* from the root *Gyúk*. Hence compounding as before, we have from the last cited simple term, *Danarasagyúk*, 'cause to run'; *Mada narasagyúk*, 'do not cause to run'; *Danarasagyúngit* 'I who cause to run'; *Manarasagyáti* or *Madanara-ságyúti*, 'he who does not cause to run.' I believe also that the reiterative form *Matarmáng* is quite as usual as the substitutive form *Marmáng*, and *Matsazangti* for *Matasazángti*, as *Masazángti*, time and tense notwithstanding. Repetition and other changes above illustrated in the prefixes belong much less to the roots, infixes and suffixes, whether in verbs or nouns, and when the root is repeated, the suffix is commonly dropt, as has been explained as to substantive. But there are instances in the verbs of root repeated and yet pre-fix retained, though the vocabulary affords none such as its *Kalarlar*, 'round,' which is a root repeated yet retaining its pre-fix; whilst the adjectives of the vocabulary, unlike the substantives, also afford several instances of the doubly and trebly reiterated pre-fix, as Kamgnár, 'sweet,' Ka-ma-gnár from the root *gnár*, and Kavándro, 'cold,' Ka-va-na-dro from the root *dro*. The elided forms, however, and particularly Kamgnár, show that leaning towards dissyllabism, which has been dwelt on, perhaps, too strongly, though it assuredly be a most marked feature of this tongue, and one too which Leyden's mistake as to his own sample verb shows to be preëminently proper to Ta-gala; for "*tolog*, to sleep," is *not* the radical form of the word, as he assumes, but a compound of the root and its customary pre-fix, *ta*, with the vowel harmonised to that of the root. The prefixes are the great variants, and besides being so much repeated, they can be transposed and interchanged almost at pleasure, owing to their synonymous character, and these variations of the pre-fixes, with the elisions consequent on much reiteration of them, constitute the greatest part of that enigma which Leyden emphasizes; though it be in the actual use of the speech much less excessive (I still speak of Gyárúng) than his sample would lead any one to suppose. In the above samples of Gyárúng I have given the verbs alone, without the added pronouns of Leydens' Ta-galan instances—such additional complication being rather suited to create wonderment than to promote sound knowledge.] Humboldt considers that the Ta-gala (a specimen by the way of the inseparable pre-fix) preserves the primitive type of the whole group; and that that type is revealed in the Gyárúng I am inclined to assert, without however forgetting that my investigation is far from complete, and without insisting so much upon the primitiveness of this type as upon its much more interesting feature of a connecting bond between the so-called monosyllabic aptotic and the so-called polysyllabic* non-aptotic classes—classes which appear to me to have no very deep or solid foundation, much as they have been insisted on to the obscuration of the higher branches of philology and ethnology, rather than to their illustration (as I venture to think), and but for which obscuration our Leydens and our Joneses, our Bopps and our Humboldts, could never have been found at such extreme apparent diversity of opinion. I may add, with reference to the disputed primitiveness of Ta-gala, owing to its use of the "artifices" above cited, that throughout the Himálaya and Tibet it is precisely the rudest or most primitive tongues that are distinguished by useless intricacies, such as the interminable pronouns, and all the perplexity caused by conjugation by means of them, with their duals and plurals, and inclusive and exclusive forms of the first person of both. In this way, Kiranti,* for instance, has thirty-three personal forms for each tense; and, as many tense-forms as there are thus constituted, so many are there of the gerunds and of the participles—a Manchuric trait of great interest. The more advanced tribes, whether of the continent or of the islands, have, generally speaking, long since cast away all or most of these "artifices."

I have thus, in the present and two former communications, shown what a strange conformity in the essential components of their speech still unites the long and widely sundered races inhabiting now the Himálaya, Tibet, Indo-China, Sifán, Altaia, Caucasus and Oceanica; and, as a no less strange conformity of physical conformation, unites (with one alleged exception) these races, it cannot much longer be doubted that they all belong to one ethnic family, whose physical attributes it shall next be my business to help the illustration of by describing the heretofore unknown people, whose languages have been submitted to inspection and examination. Before, however, I turn to the physical characteristics, I must add that all the languages, whose vocables are herewith submitted to the Society, are, and always have been, devoid of letters and of literature; what

* Compare the monosyllabic roots and dissyllabic simple vocables of Gyárúng with the sesquipedalians just given. The comparison is pregnant with hints, especially as there are in the cognate tongues all grades of approximation. Thus, *Kanarê*, 'laugh,' in Gyárúng, with its double pre-fix, is *Yeré* in Limbu with one, and *Ré* in Magar without any; and thus *Taliáng*, 'air,' in Lepcha, with its pre-fix and suffix, is *Tali* in Gyárúng, with pre-fix only, and *Li* or *Lê* in Burmese, without either. Innumerable instances like this make me conclude that the Gyárúng differs only in degree, not in kind, notwithstanding that its verb, like that of the Ta-gala, certainly presents an extraordinary and seemingly unique spectacle in some aspects, but not in all; for, in the sentence *tizé-kazé papun*, 'he called them to feast,' though the root *za*, 'to eat,' be repeated, and each time with a differently vowelled servile attached; yet the combination is not grotesque, nor the root smothered.

* See a memoir on this tongue and another on the Háyu *vel* Váyu tongue in the forthcoming Nos. of the Bengal Asiatic Society's Journal. (Printed in 1857 very incorrectly. Corrected copies sent to Pott, Lassen, Schiefner, etc.)

writing there is among these races being confined to the Tibet-trained monks, whose religious ministry they all accept, and who (the monks) use the Tibetan system of writing applied solely to the Tibetan language, and never to that of their flocks, the several races now in question, or any of them.

I cannot learn that in Tibet the Sókpo or the Hórpa ever employ any system of writing of their own, though I need not add (assuming their identification to be just) that the Mongols and the Eastern Turks have each their own system quite distinct from the Tibetan. Having always considered the physical evidence* of race quite as important as the lingual, and the one as the true complement of the other, I have not failed to use the opportunity of access to the people whose vocables are now submitted in order to note their physical traits.

The following are ths chief results of that investigation:—

	Amdoan. I.	Horpa. II.	Gyarung. III.	Manyak. IV.
Height without shoes	5.8.½	5.7.½	5.3.0	5.4.0
Length of head, from crown to chin (with calipers)	0.8.½	0.8.½	0.9.0	0.9.½
Girth of head	1.10.0	1.9.¼	1.10.¾	1.10.¾
Length of head, fore and aft, or forehead to occiput	0.7.¾	0.7.¾	0.8.0	0.8.0
Width of head, between parietes	0.6.½	0.6.0	0.6.⅞	0.6.⅞
Crown of head to hip	2.4.⅓	2.4.0	2.3.½	2.3.0
Hip to heel	3.3.¾	3.3.½	2.11.½	3.1.0
Width between the shoulders ..	1.4.0	1.1.0	1.1.½	1.4.0
Girth of chest	3.1.0	2.9.0	2.11.¼	2.11.¾
Length of arm and hand	2.6.¾	2.6.0	2.4.¾	2.4.0
Ditto of arm	1.0.0	1.0.0	0.11.½	0.11.¼
Ditto of fore-arm	0.11.0	0.10.0	0.9.½	0.9.3
Ditto of hand	0.8.0	0.7.⅓	0.7.¾	0.7.¼
Ditto of thigh..............	1.8.0	1.7.0	1.6.½	1.7.0
Ditto of leg to ankle	1.4.½	1.5.0	1.3.0	1.5.0
Ditto of foot	0.11.0	0.10.0	0.9.½	0.9.¼
Width of hand	0.4.¾	0.4.⅜	0.4.0	0.4.0
Ditto of foot	0.4.¾	0.4.¼	0.4.⅛	0.4.0
Girth of thigh	1.9.0	1.4.¾	1.6.¾	1.7.½
Ditto of calf	1.3.½	1.1.¾	1.2.0	1.1.½
Ditto of fore-arm	0.11.0	0.9.¾	0.10.0	0.9.½

No. 1.—A native of Amdó, aged thirty-five years, a finely formed and very strong man, capable of carrying three *maunds* or 250 pounds over these mountains, which he has done several times, in order to turn a penny during his so-

* Some attempts have reeently been made (see last vol. of Brit. Assoc. and Journal of Roy. As. Soc.) to disparage the value of this evidence. But no one well acquainted with the Tartars in various remote locations could for a moment think of so doing. I refer with confidence to Dr. Buchanan's remarks on the subject in vol. V. As. Res.

journ here, though the lax state of his muscles shows that he is usually an idler, and not now in training for such work, nor much used to it.

A Gélúng or monk of the mendicant class, and of course a shaveling, so that his head has been examined with unusual advantage. Five feet eight and a half inches tall, and more than proportionably broad or bulky, with large bones and ample muscle, not however showing any bold development, the surface on the contrary being smooth and even, like the body of an idler; nor fat at all, but well fleshed. Colour of the skin, a very pale clear brown, of isabelline hue, like dry earth, or dirty linen, or unbleached paper; not yellow, nor ruddy at all. No trace of red on the cheeks, which are moderately full. Colour of eyes, dark brown; of hair, generally, black, but that of moustache, auburn. No hair on chest, nor on legs or arms. Moustache spare. No beard nor whisker. Hair of head, so far as traceable, abundant, strong and straight. Cranium not compressed nor depressed; not raised pyramidally, yet brachycephalic rather than dolichocephalic, and the occiput truncated or flush with the thick neck, but not flattened. Vertical view of the head, ovoid not oval, widest between the ears, and thence narrowing equally to the forehead and to the occiput. Facial angle good. Profile inconspicuous. Contour of the face (front view) rather ovoid than angular or lozenge-shaped, the cheek-bones having no conspicuous lateral saliency nor the forehead and chin any noticeable attenuation. Forehead sufficiently high and broad, and not appearing otherwise from any unusual projection of the orbitar periphery or of the zygomata. Eyes sufficiently large and not noticeably oblique, but remote from each other, and flush with the cheek and the upper lid, drooping and constricted to the inner canthus, which is large and tumid. Nose, good, straight; the bridge well raised between the eyes and the terminal part, nor spread nor thickened, though the nostrils be shorter and rounder than in Europeans, and the saliency of the whole organ less than in them. Ears large and standing out from the head, but occupying the usual relative position. Mouth good, but large, with fine vertical teeth, not showing the least symptom of prognathism in the jaws. Very full lips, but not gaping, nor at all Negro-like in their tumidity. Chin not retiring, nor yet roundly salient, but level with the gums, or in the same plane with the teeth, and square and strong, as well as the jaws, which afford ample room in front for an uncrowded set of beautiful teeth. Body well-proportioned, but somewhat long (as well as massive and square) in the trunk and in the arms, relatively to the legs. Hands and feet well made and large, but rather as to breadth than length. Head well set on the short thick neck, and shoulders high. Chest, splendid, wide and deep, and general form good. Expression Mongolian, (but not at all markedly so as to features,) and calm and placidly good-natured. Ears bored, but not distended; and tattooing or other disfigurement of the skin quite unknown to all these races, as I may say once for all.

No. II.—A Hôrpa of Tango, west of Gyârúng, towards Amdó, named Isaba. Age thirty-eight years. A man of good height (five feet seven and a half inches) and figure, but far less powerful than the Amdóan, and somewhat darker in colour. Spare of flesh, but not actually meagre. Colour, a pale brown, without yellow or red, like all the Himálayans and Tibetans, and the eye, of a dark clear brown, as usual with them. No trace of ruddiness on cheek. Hair of the head, moustache and whisker, pure black. Hair of head, long, straight, strong, abundant. Moustache small and feeble. Whisker rather ampler. No beard, nor a trace of hair on the chest, back or limbs. Head longer (fore and aft) than wide, but scarcely dolichocephalic, though not truncated occipitally, nor compressed, nor depressed, nor pyramidised. Vertical view, oval, the wider end being the posteal or occipital, and being wider there than between the ears. Facial angle, good. Contour of the face long and oval, without any trace of the lozence breadth and angularity. Forehead, narrow and rather low, but not retiring. Cheek bones not salient laterally nor the frontal sinuses or orbits prominent. Ears large and loose. Eyes of good size, remote, but not noticeably oblique, though the inner angle be tumid with the usual constriction thereto of the upper lid, which somewhat narrows the parting of the lids. Nose straight, not very salient, yet well raised between the eyes, and not dilated towards the tip, and the nares elliptic and long, but the bridge nevertheless broad and obtusely rounded. Mouth good, but large and prominent from the fulness of the lips,* which, however, are not gaping nor are the teeth at all prognathously inclined, well made and vertically set, but not sound. Chin not pointed, nor heavy, nor retiring, nor jaws unduly large and angular; whence, with the non-saliency of the zygomata, the face takes a good and Arian contour. Figure good, almost elegant, but the arms rather long, and the legs rather short in comparison of the European form. Hands and feet well made and well proportioned. Hair plaited into a tail, *á la Chinoise.* Ears bored, but not dilated, and furnished with small ear-rings. Expression pleasing, and cast of features but faintly Mongolian.

No. III.—A Gyárúng of Tazar, north of Tachindo, by name Máching, and by age thirty-three years. Height five feet three inches, or much shorter than either of the above. A well-made smallish man. Bony and muscular development moderate, especially the former. In moderate flesh, but thigh and calf very fine; arms much less so. Arms longish. Legs shortish. Colour of skin, a pale earthy brown or isabelline hue, without the least mixture of yellow or of red; like Chinese, but deeper toned. No ruddiness on the spare cheeks. Eye dark hazel. Colour of hair in all parts uniformly black; long, straight, abundant, strong, on head; spare on upper lip; none on chin, nor on body, nor on limbs. Cranium large, nor compressed, nor depressed, nor pyramidally raised towards the crown, though there be a semblance of that sort from the width of the zygomata (but this feature belongs to the face). Occiput not truncated posteally. Fronto-occipital axis the longer and vertical view oval with the wide end backwards,

* It is not so much the fullness of the lips as a certain thickening of the gums, particularly those of the upper incisive or front teeth common to Cis- and Trans-Himálayans.

the occiput being conspicuously wider than the frontal region, or than the parietal, and the maximum occipital breadth lessening regularly forwards to the forehead. Facial angle good, with a vertical, but inconspicuous profile. Contour of the face (front view) lozenge-shaped, widest between the cheek-bones, which project much laterally, and are flattened to the front, causing great breadth of face just below the eyes, whence there is a regular narrowing upwards and downwards. Forehead sufficiently high and not retiring, but narrowed apparently upwards, owing to the salient zygomata and molars. Frontal sinus not salient. Eye smallish and not well opened nor hollowed out from the cheek, and upper lid drooping and drawn to the inner, inclined and tumid canthus. Eyes wide apart and oblique. Nose long, straight, thick, with a broad base between the eyes, where, however, the bridge is not flat, but raised into a wide low arch. Width great there, and spreading into an expanded fleshy termination, with broad alæ and large round nostrils. Mouth large and salient, yet good. Lips moderate and closed, and teeth vertically set, and very fine in shape and colour. Chin pretty good, not retiring, nor yet projecting, flush with the teeth and somewhat squared, as also the large jaws. Ears long and loose. Figure good, with head well set on; neck sufficiently loug; chest deep and wide, and well made hands and feet. Hair worn plaited into a pig tail. Ears bored, but declaredly contrary to the custom of his country, and not distended. A very Chinese face and figure, and belonging to one who has, in his character, a deal of the shrewdness tending to knavery that marks the Chinaman.

No. IV.—The Mányaker is forty years old, and bears the euphonious name of I'drophúncho. He is a native of Rákho, six days south of Tachindo, and by profession a Gélúng or mendicant friar; and a cross made ugly fellow he is, as one could wish to see, with round shoulders and short neck, but stout and good-tempered exceedingly; and moreover, accomplished in reading, writing, drawing and carving, like most of the regular troops of Lámaism to which corps he belongs, though to the heterodox branch of it, or Bonpa sect, called by him Beunpo or Peunpo, and which he has enabled me to say is no other than Tántrika Búddhism, or what is commonly called Shamanism.* This very interesting and important discovery I therefore make no apology for inserting here,

* In saying that Shamanism is nothing but Tántrika Búddhism, I speak most advisedly, and fully aware of the opinions I oppose. That the Bonpa also are Buddhists, there can be no doubt, and my friend I'dro's statements and drawings show that his sect follow the Gyút or Tántras, which, though canonical, are in bad odour, and have been so since the Gelukpa reform. A Bonpa and a Moslem are alike odious to the orthodox in Tibet, though the Bonpas have many Vihárs of high name and date all over the country. Since this was written, I have found some interesting traces of the existence of the Bonpa sect in the Himálaya, where the Múrmi tribe for instance still call their exorcist Bonpa. The probable general solution is, that both the Brahmanists and the Buddhists, of all the various divisions of those creeds, adopted largely into their systems the prior superstitions of the country, whence in Java, in Népál, in Ava, as in India, Buddhist and Brahmanical remains exhibit so much of a common character, sometimes wearing the aspect of Vaishnaism, more commonly than of Saivaism. Compare my remarks on the subject (*apud* volume on the Búddhism of Népál) with Leyden's Fahian and Yule on the Remains of Pagan (*apud* A. S. J. B.) Yule describes exactly the Padmapani, Manjusri, etc., of Népál, and I have myself found them at Karnagurh on the Ganges.

though it be somewhat out of place; and as I am digressing, I may as well add that to confound the Lamas with the Gélúngs as Huc and Gabet invariably do, is a worse error than it would be to confound the Brahmans with the Pandits in India. To return to my friend Idro, whose shaven head has afforded me a second excellent opportunity for closely examining the cranial characters of these races, I proceed to note that he is a man of moderate height (five feet four inches), but strongly made, with large bones and plenty of muscle, but no fat. Colour, a pale whitey pure brown. No trace of red in the spare cheeks, winter though it be. Eye, dark rich brown, and hair throughout unmixed and pure black. Like the others, he has none of the Esau characteristic, but on the contrary is, as usual, scant of hair, having not a trace of it on the body or limbs, and not much on the face. No beard. No whisker. A very wretched lean moustache, and a spare straight eyebrow. Cranium brachycephalic and large. Vertical view of the head ovoid not oval, widest between the ears, as in the Amdóan. Thence regularly and equally narrowed to the frontal and occipital extremities. No compression, nor depression of the cranium, but on the contrary a distinct pyramidal ascension from a broad base, the point of crinal radiation being somewhat conically raised from the interaureal and widest part of the scull. Occiput truncate and flattened, that is, not projecting beyond the neck, nor rounded posteally, like most heads. Facial angle pretty good, but rather deficient in verticality of profile. Contour of the face lozenge shape owing to the large laterally salient cheek bones, though the forehead be not very noticeably narrowed (except with reference to its bulging base), nor the chin pointed. Forehead sufficiently good, high but some what compressed and retiring, and appearing more so by reason of the heavy frontal sinuses and zygomata, which project beyond the temples towards the sides and front. Ears big and salient. Eyes remote and oblique, with the inner angle down and tumid, and the upper lid drooping and drawn to the inner canthus. Nose rather short, straight, not level with the eyes, nor yet much raised to separate them, nor elsewhere. Not clubbed at the end, but the alæ spreading, and the nares large and round. Mouth large and forward, with very thick lips, but no prognathism, the teeth being vertical and the lips not gaping so as to expose them. Teeth well formed and well set in an obtusely convex large arch, those of the upper jaw, however, overhanging those of the lower. Chin rather retiring, or flat and square. The partial retirement of the chin and large frontal sinuses are what mar the verticality of the profile, which moreover shows little of nasal and much of oral projection. Figure bad, with thick goitrous neck, high forward shoulders, and somewhat bowed legs. Hands and feet well made. Muscular development of arms poor, of legs good. A thoroughly Mongolian face, but the ugliness in part redeemed by the good-natured, placid, yet somewhat dull, expression.

Note. — The orthography of the comparative vocabulary is in general that sanctioned by the Asiatic Society of Bengal, but there are a few deviations necessitated by the peculiar articulation of these races, whose gallic *j* and *u* are of incessant recurrence. I have represented the former sound by *zy* and the latter by *eu*. Both sounds are found in the French word *jeu*. The system of tones or accents, so important for discriminating the many otherwise-identical roots in these tongues, there is no practicable method of doing justice to. But I have marked the chief one, or abrupt final, y an underscored h, thus *h*. In Thóchú and in Hórpa, the *h*, *kh*, and *gh*, have often, nay generally, a harsh Arabic utterance. I use the short vague English *a* and *e*, as in *cat*, *get*, for their common equivalents in these tongues, but *u* has always the *oo* sound, whether short or long. It so occurs in English though rarely, as in *put*, *pudding*. The continental (European) and Eastern system of the vowels is that pursued, and the long sound of each is noted by accent superscribed. But there is a great evil attendant on this Jonesian use of accent as marking quantity; for the Tartar accent denotes the radical syllable or syllables, irrespective altogether of the long or short sound of vowels. I cannot, however, at present, remedy this evil, though hereafter I shall use the accent to denote roots putting it over the end of the radical syllable, whether ending in vowel or consonant. *Quoad* vowels, mine is the common vocalic system, the English being wholly beside the mark. *Y* is always a consonant. It blends with many others to give them a sliding sound as in the *zy*, above instanced. It gives *S* the sound of *Sh*, as in the Syán (Shan) tribe's name. It must never be made a vowel, *à l'anglaise*, for that makes monosyllables dissyllabic, and totally changes the proper sounds of words. The same as to *w*, which we English are however more familiar with. From *é*, I make the diphthong *ai;* from *á* that of *au;* from *ó* that of *ou*, sounded as in *aye aye*, *hawfinch*, *how;* which, with the gallic *eu* (*beurre*), are invariably diphthongs, each with a single blended sound. If two vowels come together and require separate utterance, the latter is superscribed with two dots, as *dáï*. I have marked off the prefixes (*tir-mí*, 'man,' see Gyárúng column) to facilitate access to the root and comparison on a large scale, such as that lately employed to illustrate ethnic affinities. This and the like marking off of the suffixes will be a great aid to those who wish to make such comparisons without knowledge of these languages. But the procedure is hardly correct, since the root and its prefix in particular are apt to be blended in utterance by transfer of the accent (*mí*, *tír-mi*), and since the sense also of the roots is occasionally as dependant (though in a different way) on that of their prefixes, as it is in regard to the prepositions of the Arian tongues (*tir-mi*, 'man': *ti-mi*, 'fire'). Nevertheless these important particles are liable to a large range of mutations, synonymous as well as differential, merely euphonic, as well as essential, whilst some of the tongues use them very amply, and others very rarely. Add to these features the infixes and the suffixes, with the occasional change of place and function between all these, and you have before you the causes of the differences of these languages, which are often so operative as to merge their essential affinity and make it indiscernible, except by those who, knowing the roots, can pursue them *and* the servile portions of the vocables through their various metamorphoses and transpositions.*

* Compare in Tibeto-Himálayan and Indo-Chinese series, as follows:—
Day.—*Nyi'-ma, Ma-ni', Nye'-n-ti, Nhi'-ti-ma, Sak-ni'*. Root *Nyi*.
Eye.—*A-mik, Mi-do, Mi-kha, Ta-i-myek, Myé-t-si*. Root *Mig*.
Dog.—*Khi-cha, Ko-chu, Chói-ma, Khwé, Ta-kwi, Ka-zeu*. Root *Khyi*.
Ripe.—*Kas-sman, Mhai-ti, Mhin, Min-bo*. Root *sMin*.
Sour.—*Kúch-chúr, Kyúr-bo, Da-chu'*. Root *sKyúr*.
Hear.—*Khep-ché, Nap-syé, Ta-ché-n*. Root *Shê*.

These are extreme cases, perhaps, of mutation; but they are therefore all the better adapted to illustrate my meaning; and links enough will be found in the vocabularies to bind them surely together.

6. ON THE COLONIZATION OF THE HIMALAYA BY EUROPEANS.*

As the interesting subject of the fitness of the Himálaya for European colonization is beginning to excite the attention of individuals and of the Government, it may be worth while to state distinctly may own conviction on the subject, together with the chief grounds of that conviction, because I have resided some thirty years in the Central and Eastern parts of the range, and have also served awhile in the Western, and all that time my attention has been directed to studies calculated to make my observation and experience more effective.

I say, then, unhesitatingly, that the Himálaya generally is very well calculated for the settlement of Europeans, and I feel more and more convinced that the encouragement of colonization therein is one of the highest and most important duties of the Government.

In the long, and throughout the globe quite unparalleled, gradation of heights, from the plains to the snows, every variety of climate is found with correspondent capabilities for the successful culture of various products suited to the wants of Europeans, for their own consumption or for profitable sale; and in this extraordinary gradation of heights, the high and the low are *juxtaposed* in a manner alike favourable to the labours of the healthful and to the relief of the ailing.

A healthy cultivator of our race could have his dwelling at four to six thousand feet, and his farms, both there and at various higher and lower elevations, yet still close to his abode; so that quasi-tropical and quasi-European products might be raised by him with the greatest facility; and in defect of health and strength, the colonist, like the visitor, would enjoy the vast advantage of entirely changing his climate without cost or fatigue of journey, besides having the additional resource of easy access to medicinal waters of universal diffusion and of proved efficacy in many kinds of ailments. The greatest variety of climate has of course relation to the transverse section of the Himálaya, or that from plains to snows; but the longitudinal section, or the S. E. and N. W. one, likewise presents as much and the same variety of climate as is proper to the plains in Bengal, Benáres, and the north-western provinces; and it is quite a mistake to allege of the South-East Himálayas, or of Bengal, that their climate differs *only* for the *worse* from the drier climate of the hills or plains further west and north.

* Written in 1856.

Undoubtedly, the South-East Himálaya has much less sun and much more moisture* than the North-West Himálaya. But those Europeans, who have experienced the effects of the climate of both, frequently prefer that of the former, and it is quite certain that, in the past twenty years, the South-East Himálaya has suffered much less from epidemics, and has also enjoyed a complete exemption from those severe dysenteries and fevers which have afflicted the denizens of the North-West Himálaya. It is as certain that the obscured sun of the South-East Himálaya is the cause of the difference,§ and that, though our clouds and mists may hurt our popular reputation with strangers, they are welcome to ourselves from their experienced and admitted beneficialness. Cloudy and misty as is our climate for five to six months, rheumatism and pulmonary affections are unknown. That the Himálaya, generally speaking is a region eminently healthful, can be doubted by no competent judge, and is demonstrable at once, and readily, by pointing to the finely developed muscles,pure skins, cheerful countenances, and universally well-formed strong-boned bodies of the native inhabitants, whose health and strength, and capacity of enduring toil and carrying heavy burdens, are as notorious, as are their exemption from bodily malformations and from most of the direst diseases to which flesh is heir, as well in the tropics as in the high latitudes of Europe—results owing to the preëminent equability and temperateness of the climate,† added to the simple active habits of the people.

The fearful epidemics of the plains seldom penetrate the Himálayas, which, moreover, seem to have a positive exemption from endemic diseases, or those proper to any given country. For forty years cholera has ravaged the plains continually almost. But in all that period Népál has been visited only twice and Darjeeling scarcely at all. In the same forty years at Káthmándú, only two deaths (Mr. Stuart and Lieutenant Young) have occurred among Europeans, and both those were occasioned by diseases wholly apart from local influences; and in the escort of the Resident, the salubrity in my time was so great, that promotion came hardly to be calculated on at all, and a Sepahee would be a Sepahee still, after fifteen to twenty years' service.‡

The Civil medical statistics of Népál, as of Darjeeling, have always told the same story; and if the Military statistics of the latter place have been, till lately,

* The fall of rain is no accurate test of mean moisture, but the following facts have their value :—Mean annual fall of rain at Darjeeling 130 inches ; at Káthmándú, in the Valley of Népál, 60 ; at Simla 70 ; at Cherrapúnji 500. It must always be remembered, that the amount of rain and moisture at any given spot in the Himálaya depends greatly on the number of covering ridges intervening between such spot and the course of the great column of vapour borne by the monsoon from the ocean. The fact, that the fall of rain in the Concan is five-fold what it is in the Deccan, owing to the intervention of the Ghát range, will make this more intelligible.

§ Very imperfect sanitary arrangements to the north-west, where large multitudes are assembled yearly such as are unknown to the Sanitaria of the south-east, must be added in explanation of the dysenteries and fevers noted.

† In my sitting-room, which is freely ventilated, the thermometer ranges only from 60 to 65, day and night, between the end of June and the end of September. In December, January, and February, the range is about the same, or but slightly greater.

‡ The Escort or Honorary Guard formerly consisted of 200 men ; it now consists of 100.

less favourable, the reasons of this had nothing to do with the hill climate, bnt resulted wholly from the senseless selection of cases sent up; the absurd neglect of seasons in sending up and taking down of the invalids; and lastly, the shameful abandonment of all care and supervision of the men on the way up and down.

The appearance of the European children at Darjeeling might alone suffice to prove the suitableness of the climate of the Himálaya at six to eight thousand feet for European colonization, confirmed, as such evidence is, by that of the aspect and health of such adult Europeans as came here with uninjured constitutions, and have led an active life since their arrival. Finer specimens of manly vigour the world could not show; ‡ and though none of the individuals I allude to have lately toiled all day in the open air at agricnltural labours, yet I am credibly informed that some of them did for several years after their arrival here, and with perfect impunity; their agricultural pursuits having been abandoned for reasons quite apart from either injured health or inability to support themselves and families comfortably by such labours.

That Europeans would sustain injury from exposure during agricultural labours at any period § of the year, seems therefore refuted by fact; and when it is remembered that such persons would be working here, as at home, amid an indigenous arboreal vegetation of oaks, hollies, chesnuts, sycamores, elms, horn-beams, birches, alders, elders, willows, and, more westerly, pines and firs, such* a fact derives from such an analogy double strength; and the attempted inference from both is further justified by the healthful growth in the Himálaya of such of our own cereals and vegetables and fruits as we have thus far tried to introduce, with the sole exception of *delicate* and *soft* pulped fruits, not of an early or spring maturing kind, such as peaches, grapes, and the like. These rot, instead of ripening in the central region of the Himálaya, owing to the tropical rains and rarity of sun-shine at the ripening season.

But such soft fruits as become mature before the rains set in, as strawberries, come to perfection, as do all hard fruits, such as apples. There is, in fact, no end no end of the mineral and vegetable wealth of the Himalaya, and if the absence of flat ground, with the severity of the tropical monsoon or rainy season, present considerable drawback to agricultural success, on the other hand the endless inequalities of surface offer a variety of temperature and of exposure, together with signal modification even of the element of moisture and rain, all highly conducive to the advantageous cultivation of numerous and diverse products proper to the soil or imported from elsewhere.

Temperature changes regularly in the ratio of 3° diminution of heat for every thousand feet of height gained; and every large ridge crossing ths course of the

‡ We may now add that the children and, in a few instances, grandchildren born at Darjeeling of the Europeans in qnestion, and the children generally of the gentlemen resident there, are as healthy and vigorous as any children in Europe.

§ Agriculture does not require much exposure at the hottest season, when the crops are growing.

* The beech is the only European tree not found in the Himálaya. The rest are very common.

monsoon modifies almost as remarkably the amount of rain in the several tracts covered by such ridges. The ratio of decrease of heat with elevation, which has just been stated, must however be remembered to be an average and to have reference to the shade, not to the sun, for it has been found that the direct rays of the sun are as powerful at Darjeeling as in the plains, owing probably to the clearness of our atmosphere; and this is the reason why our clouds are so welcome and beneficial during the hottest months of the year. In other words, the constant cloudiness of that season is beneficial to the European. It is otherwise, however, as regards his crops, which being ripening at that period, would be benefited by a clearer sky; and thus it is that a certain degree of oppugnancy exists between the sites most congenial to the European and to his crops; for, whilst a height of six to seven thousand, perhaps, might be most congenial to him, one of four to five thousand would certainly suit them better, not so much for the average higher temperature, as for the larger supply of sun-shine. But the oppugnancy is only one of degree, and whilst four thousand is a very endurable climate for the European, there is no reason why he should not have his abode, as is the frequent custom of the country, at a somewhat higher level than that of his fields, should he find such an arrangement advantageous upon the whole.

The fertility of the soil is demonstrated by the luxuriance of the arboreal and shrub vegetation, a luxuriance as great in degree as universal in prevalence. True this luxuriance has its evils* and, in its present unpruned state, may be one great cause why the feeding of flocks and herds is scantily pursued by the people, and without much success, speaking generally; for there are exceptions even now, and European energy would soon multiply these exceptions, besides grappling successfully with the presumed source of the evil, or too much and too rank vegetation, not to add, that, in the districts next the snows and Tibet, that hyper-luxuriance ceases, and herds and flocks abound, and the latter yield fleeces admirable for either fineness or length of fibre.† The soil consists of a deep bed of very rich vegetable mould from one to three feet deep, to preserve which from being carried away by the tropical rains after the removal of its natural cover of forest and under-growth, by terracing and other known expedients, must be the colonist's first care, for the underlying earth is almost always a hungry red clay but happily one whose tenacity and poverty are much qualified by better ingredients derived from the debris of the gneisses and schists that constitute the almost sole

* The paucity of gramineæ is, I believe, a feature of the Himálayan Botany, and every observant person must notice the absence of meadows and grazing land and hay fields throughout the hills. But this is to be accounted for and explained by the uncommon strength and abundance of the indigenous vegetation; for, whenever a tract of land is kept clear, grass springs up; and the European grasses that have been imported, including clovers and lucern, flourish exceedingly, the moist climate being very favourable to them. Such, however, are the richness and high flavour of the native vegetation, that large and small cattle, even when provided with the finest European pasture, are apt to desert it in order to graze at large amid the forests and copses. I here speak of the *central region* of the Himálaya, wherein leeches are the great enemies of the cattle, and a peculiar disease of the hoofs to which they are subject.

† The samples I sent to Europe of the wool of the sheep and goats of the Northern region of the Himálaya and of Tibet were valued at seven to nine pence per pound.

rocks. The argillaceous constituents of the soil are perhaps in good proportion; the siliceous, perhaps, rather too abundant; the calcareous, deficient. Heretofore, the superficial mould has been the sole stay of the agriculturist and floriculturist. How far that would continue to be the case under abler culture, I know not. But, so long as it did continue, the caution above given would demand the most vigilant and incessant attention.

The common European cereals, or wheat, barley, rye, aud oats, are little heeded in the Himálaya, where I never saw crops equal those grown in various parts of the plains. But this, though no doubt attributable in some measure to a deal of the Himálayan population being located at heights above those where, in the present forest encumbered state of the country, a sufficiency of snmmer sun for such crops can be safely calculated upon, is likewise attributable in part to the preference for rices, maizes, sorghums, panicums or millets, buck-wheat, and amaranth, on the part of the people, whose cultivation of wheat is most careless, without manure, even in double-cropped and old lands, and the plant is allowed to be over-run, whilst growing, by wild hemp or artemesia, or other social weed of most frequent occurrence in the Himálaya. Observe, too, that the system of double cropping now occasions the sacrifice of the despised wheat crop which is a spring one to the cherished autumal crop which is a rice one; and that were the former allowed due consideration and treated with reference to its furnishing a main article of food, instead of being regarded merely with reference to the still, as is now very generally the case among the native population, we might reasonably expect to see fine crops of wheat as high at least as five thousand feet and more, especially so when the clearance of the land, conducted judiciously, was enabled to produce its due and experienced effects in augmenting the sun-shine and diminishing the rain and mist in such properly cleared tracts. Heretofore, skill and energy have done absolutely nothing, in these or other respects, for Himálayan agriculture, and yet there is no country on earth where more advantage might be derived from skill and energy applied to the culture of agricultural products. As already said, the infinite variety of elevation and of exposure (both as to heat and moisture), together with the indefinite richness of the soil, as proved by the indigenous tree and shrub and other vegetation, are premises one can hardly fail to rest soundly upon in prognosticating the high success of European culture of the Himálayan slopes, notwithstanding the drawbacks I have enumerated. There need hardly be any end to the variety of the products, and good success must attend the cultivation of many of them, after a little experience shall have taught the specialities of the soil and climate, so that the subject should be incessantly agitated till the Government and the public are made fully aware of its merits. How much iteration is needed, may be illustrated by the simple mention of the fact, that the fitness of the Himálayas for tea-growing was fully ascertained twenty-five years ago in the valley of Népál, a normal characteristic region, as well in regard to position* as to elevation. Tea seeds and plants were procured from

* It is equi-distant from snows and plains, and has a mean elevation of 4,500 feet.

China through the medium of the Cashmere merchants then located at Káthmándú. They were sown and planted in the Residency garden, where they flourished greatly, flowering and seeding as usual, and moreover, grafts *ad libitum* were multiplied by means of the nearly allied Eurya (Camellia) kisi, which, in the valley of Népál, as elsewhere, throughout the Himálaya, is an indigenous and most abundant species. These favourable results were duly announced at the time to Dr. Abel, Physician to the Governor General, an accomplished person, with special qualifications, for their just appreciation. And yet, in spite of all this, twenty years were suffered to elapse before any effective notice of so important an experiment could be obtained.

I trust, therefore, that the general subject of the high capabilities of the climate and soil of the Himálayas, and their eminent fitness for European colonization having once been taken up, will never be dropped till colonization is a "*fait accompli*" and that the accomplishment of this greatest, surest, soundest, and simplest of all political measures for the stabilitation of the British power in India, may adorn the annals of the present Viceroy's administration.

But observe, I do not mean wholesale and instantaneous colonization, for any such I regard as simply impossible; nor, were it possible, would I advocate it. The distance and unpopularity of India, however, would preclude all rational anticipation of any such colonization, whatever might be the wish to effect it. What I mean is, looking to these very obstacles and drawbacks, seeming and real, that some systematic means should be used to reduce their apparent and real dimensions, to make familiarly and generally known the cheapest methods and actual cost of reaching India; to afford discriminating aid in some cases towards reaching it and settling in it; and to shew that, in regard to the Himálaya, the vulgar dread of Indian diseases is wholly baseless — to show also, that its infinite variety of *juxtaposed* elevations, with correspondent differences of climate, both as to heat and moisture, and the unbounded richness of its soil at all elevations, offer peculiar and almost unique advantages (not a fiftieth part of the surface being now occupied) to the colonist, as well on the score of health as on that of opportunity, to cultivate a wonderful variety of products ranging from the tropical nearly to the European.

A word as to the native population, in relation to the measure under contemplation. In the first place, the vast extent of unoccupied land would free the Government from the necessity of providing against wrongful displacement; and, in the second place, the erect spirit and freedom from disqualifying prejudices, proper to the Himálayan population, would at once make their protection from European oppression easy, and would render them readily subservient under the direction of European energy and skill to the more effectual drawing forth of the natural resources of the region. Located himself at an elevation he might find most conducive to his health, the colonist might, on the very verge of the lower region (see *Essay on Physical Geography of Himálaya*, in another part of this work), effectually command the great resources for traffic in timber, drugs, dyes, hides,* horns, ghee, and textile materials, not excluding silk, which that region affords; whilst, if he chose to locate himself further from the plains and devote himself to agriculture and sheep-breeding, he might make his election among endless sites in the central and higher regions (see paper above referred to) of the Himálaya, of a place where these or those sorts of cereal flourished best, and where cattle and sheep could be reared, under circumstances of surface, vegetation, and temperature as various as the imagination can depict, but all more or less propitious ; the steep slopes and abundant vegetation, rank but nutritious, of the central region, giving place, in the higher region, to a drier air, a more level surface, and a scanter and highly aromatic vegetation, peculiarly suited to sheep and goats, whose fleeces in that region would well repay the cost of transport to the most distant markets.

Not that I would in general hold out to the colonist the prospect of growing rich by the utmost use of the above indicated resources for the accumulation of wealth—to which might, and certainly in due course would, be added those of the Trans-Himálayan commerce||—but would rather fix his attention, primarily, at least upon the *certain prospect* of *comfort*, of a full belly, a warm back, and a decent domicile, or, in other words, of food, clothes, and shelter for himself, his wife, and children, unfailing with the most ordinary prudence and toil, and such, as to quality and quantity, as would be a perfect god-send to the starving peasantry of Ireland and of the Scotch Highlands. *These* are the settlers I would, but without discouraging the others, primarily encourage by free grants for the first five years, and by a very light rent upon long and fixed leases thereafter, looking to compensation in the general prestige‡ of their known forthcomingness on the spot, and assured that, with the actual backing upon occasions of political stress and difficulty of some fifty to one hundred thousand loyal hearts and stalwart bodies of Saxon mould, our empire in India might safely defy the world in arms against it.

* Countless herds of cattle are driven for pasturage annually, during the hot months, from the open plains into the Tarai and Bháver, and of the thousands that die there, the hides and horns are left to rot, for want of systematic purchase, and this whilst the demand is so urgent, that cattle-killing has become a trade in order to meet it.

|| In 1832 I furnished to Government a statement of the amount of this commerce, as conducted through Népál proper. the exports and imports then reached thirty *lakhs*, and this under circumstances as little encouraging to commercial enterprise as can well be imagined, for monopolies were the order of the day, and those in power were often the holders of such monopolies, as I believe is still the case in Népál and also in Cashmere.

In the paper adverted to, I also pointed out, by comparative statements, how successfully Britain could compete with Russia in regard to this commerce.

‡ We are, it should never be forgotten, '*rari nantes in gurgite vasto*,' occupying a position quite analogous to that of the Romans, when one of their ablest statesmen exclaimed '*quantum nobis periculum si servi nostri numerare nos cœpiscent.*' We cannot, for financial reasons of an enduring kind, create an adequate guard against the perils of such a position, nor materially alter it for the better *quoad* physical security, *save* by having such a body of our countrymen as above contemplated *within call.*

To ward off Russian power and influence, we are just now entering on a war (in Persia) as immediately and immensely costly, as full of perplexities and difficulties even in any of its better issues. Were one-tenth, nay, one-fiftieth, of the money which that war, if it last, will cost, bestowed on the encouragement of European settlements in the Himálaya, we might thus provide a far more durable, safe and cheap barrier against Russian aggression, and should soon reduce her land-borne commerce with Eastern Asia to *nil*. (A.D. 1856.)

7. ON THE COMMERCE OF NEPAL.

[The following papers, which are of special interest just now, were addressed to the Political Secretary at Calcutta in 1831, and were published in a volume of "Selections from the Records of the Government of Bengal, No. XXVII," in 1857. —Ed.]

No. I.—A precise practical account of the commercial route to Káthmándú, and thence to the marts on the Bhote or Tibetan frontier; with the manner and expense of conveying goods, the amount and nature of the duties levied thereon by the Nêpál Government, and the places where they are levied.

No. II.—Lists of imports and exports, with remarks.

It is scarcely necessary for me to remark, that a connexion with this country was originally sought by us purely for commercial purposes, which purposes the government, up to the beginning of this century, directly and strenuously exerted itself, by arms and by diplomacy, to promote. Now, though I would by no means advise a recurrence to that mode of fostering the commerce in question, but, on the contrary, entirely adhere to the opinions expressed by me in my public despatch of the 8th of March, 1830, yet I think it is possible we may fall into the opposite error of entire forgetfulness and neglect of the matter. I conceive, therefore, that a few remarks tending to reveal the actual and possible extent and value of the trade in question will, at the present moment, be well timed and useful, in which hope I shall now proceed to make some such remarks, and to point out, in the course of them, the specific object for which each of the two accompanying documents was framed. Why that great commerce, which naturally ought to, and formerly did,* subsist between the vast Cis- and Trans-Himálayan regions, should seek the channel of Nêpál rather than that of Bhûtan on the one hand, or of Kumaon on the other, I have already explained at large, in my despatch above alluded to, and to which I beg to refer you, should the subject seem worthy of any present consultation or consideration. But I shall probably be met at the threshold of the discussion with the reasonable questions—what has been the effect of sixteen years'

* I recommend a reference to the old records (inaccessible to me)of the commercial Residency of Patna and of its out-post Bettia. In 1842, an official reference was made to me, too immediately before my departure from Népál to be answered, the object of which was to ascertain whp the imports from Tibet through Népál, and particularly that of gold, had fallen off so much

peace and alliance with Nêpâl?—what is now the positive amount of this commerce? —what its extent as compared with any like preceding period. If the mustard-seed be indeed, to attain its promised dimensions, there ought to be now some distinct symptoms of its great power of increase.

To meet in some sort, and prospectively, these reasonble enquiries, I have drawn up the paper No. II. I have myself searched in vain tnrough my record for any — the vaguest—data, by which I might judge of the amount of this commerce at the times of Kirkpatrick's‡ and Knox's missions to Káthmándú, or, at the period of Mr. Gardner's arrival here (1816), and the vexation I have experienced at finding none such, has led me thus to place on record the best attainable data for the present time. Fifteen years hence these data will furnish a scale of comparison by which to measure the justness of the views now entertained respecting the power of increase inherent in the trade of Nêpâl. It will readily be anticipated this government neither makes nor keeps any express record of the annual amount of exports and imports, and that it is no easy thing for one in my situation to get possession of the indirect, yet facile, measure of this amount furnished by the sum-total of the duties annually realized upon it. So far as attainable, I have used this measure. I have also, sought and obtained other measures. I have secretly and carefully applied to some of the oldest and most respectable merchants of Kâthmándû, and the other chief towns of the Valley, for conjectural estimates of the total annual amount of imports and exports, and of the number and capital of the chief commercial firms of the Valley. These estimates are given in Number II. In the absence of statistical documents, these are the only accessible data, and when it is considered that I have been many years at this place, it may reasonably be presumed, that I have the means of so applying to the merchants in question as to procure from them sincere statements to the best of their knowledge.

It appears then that at this present time there are, in the great towns of the Valley of Nêpál, fifty-two native and thirty-four Indian merchants engaged in foreign commerce, both with the South and the North, and that the trading capital of the former is considered to be not less than 50,18,000, nor that of the latter less than 23,05,000.* A third of such of these merchants as are natives of the plains have come up subsequently to the establishment of the Residency in 1816, since which period, as is thought by the oldest merchants of Káthmándû, the trade has been tripled.

Turning again to No. II., Part I., we have, for the annual prime cost value of the imports in Sicca rupees 16,11,000, and Part II. of No. II. affords, for the annual value, at Káthmándû, of the exports, 12,77,800 of Nêpálese rupees, equivalent to Kuldars 10,64,833-5-4, thus making the total of imports and exports 26,75,833-5-4 of Kuldar rupees. But, from particular circumstances, the imports of 1830-31 were above what can be considered an average specimen, and should be reduced by one

‡1792 and 1801, respectively.

* Before I left Népâl, I had some reason to suppose these estimates to be too high by a third.

lakh, in the articles of precious stones, English fowling pieces, horses, velvets, and kimkhabs, owing to the extraordinary purchases of the Durbar in that year. After this deduction, there will remain a total of annual imports and exports, according to the lists of No. II., of something short of twenty-six lakhs, which sum agrees sufficiently well with the twenty-five lakhs yielded by the subsequent calculation upon the amounts of duties and of exemptions from duty. I am aware that, after the deduction from the imports adverted to, there will still remain an excess of imports over exports, amounting to four and a half lakhs of rupees,† which may seem to want explanation, if considered as a permanent relation. But I think it will be felt, on reflection, that to attempt to reduce these estimates to rigorous precision, or to raise on them a nice speculation would be to forget that they are necessarily mere approximations. In other respects, I hope and believe both parts of No. II. likely to be very useful; but in regard to the precise accuracy of its sum-totals of annual transactions, I have no wish to deceive myself or others.

In respect to the annual amount of duties realized by this government upon this trade, I cannot ascertain it upon the northern branch of the trade, but upon the southern branch, or imports and exports from and to India, (which is farmed and more easily discoverable,) it reached last year (1830) the sum of one lakh and sixty-thousand three hundred and sixty-four Népálese rupees. Now, if we take (as there are good grounds for doing) the duty, upon an average, of 6 per cent. *ad valorem*, the above amount of duty will give a total annual value of imports and exports, with the plains of India alone, of 26,72,733⅓ Nêpálese Paisa rupees, equivalent to Siccas 17,81,821-10-8. But to this sum must be added the whole amount of imports and exports passing duty free, and which cannot be rated at less than seven lakhs of Kuldars per annum. There are exemptions, from principle, of a general nature, such as those affecting the export of gold, pice, and Nêpálese rupees; and which articles alone amounted for 1830-31, to fully five lakhs of Siccas, as per list of Part II. No. II. There are also exemptions from favoritism, which, by the usage of the Nêpál government, are largely extended to its more respectable functionaries, civil and military—all of whom, if they have a penny to turn, or expense to meet abroad, at once dabble in trade, and procure for themselves freedom of export and import for the nonce. The goods so exported and imported must be rated at a lakh per annum, nor can the Durbar's own purchases or imports be set down at less. We must add, therefore, seven lakhs of exempted goods to the nearly eighteen lakhs pointed out by the duties, and we shall have, in this way, little short of twenty-five lakhs of Kuldars for the total amount value of the exports and imports, to and from the plains, as indicated by the amount of duties and of exemptions. Such, according to data, of some worth at least, is the present extent

† The deficiency of exports is made up, and more by the agricultural produce of the lowlands, especially grain, six lakhs of which are annually sent to Patna, etc., where it is paid for in money wholly. The means of export afforded to Népál by her Tarai agriculture escaped me in drawing up the tables of commerce.— *B.H.H.*, 1834.

The total of exports and imports must, therefore, be set down at upwards of thirty lakhs.—*B.H.H.*, 1857.

of the trade of Nêpâl. If we would reasonably conjecture to what a height that trade might easily grow, we may do so by turning to the statistical documents touching the amount and nature of the Russian commerce with China *viâ* Kiachta; and then, comparing the facilities and difficulties of such a commerce with those which present themselves to a commerce with the same country *viâ* Káthmándû and Lhása. From St. Petersburg to Peking, by any feasible commercial route, cannot be less than 5,500 miles;* and though there is water carriage for a great part of the way, yet such is the savage sterility of the country, and such the rigor of the climate, that the water passage takes three years, and the land route one entire year, to accomplish it. The Russian government levies high duties on this trade, not less than 20 to 25 per cent., save on Russian products, which are scant, compared with the foreign. There are some monopolies, and many prohibitions, especially those mischievous ones affecting the export of either coin or precious metals.

I have mentioned the interval separating St. Petersburg and Peking. It is further necessary to advert to the yet more distant seats, both of production and of consumption, in reference to the more valuable articles constituting the Russian trade. The Russians export to China peltry, woollen and cotton cloths, glass-ware, hardware, hides, and prepared leather. Of these, not more than half of the first is produced in Siberia, the other half is obtained from North America, either *viâ* England, or by way of Kamtschatka and the Aleutian Isles. Of the cotton and woollen cloths, the coarse only are Russian made, the fine come chiefly from England; and the like is true of the glass-ware and hard-ware. The hides are, mainly, of home production. Russia imports from China musk, borax, rhubarb, tea, raw and wrought silk, ditto ditto cotton, porcelain, japan ware, water colours, etc. But the best musk, borax, and rhubarb by far are those of Tibet, and especially of Sífán, the north-eastern province of Tibet; and no tea is better or more abundant than that of Szchuen, which province is only eighty-seven days' journey from Káthmándû; whilst, of course, the musk, borax and rhubarb regions (as above indicated) are yet nearer to us, yet more inaccessible to the Russians, than Szchuen.

What more I have to say on these products will fall more naturally under my remarks on the line of communication with these countries through Nêpál; and to that topic I now address myself. From Calcutta to Peking is 2,880 miles. Of this, the interval between Calcutta and Káthmándû fills 540 miles, two-thirds of the way being navigable commodiously by means of the Ganges and Gandák. The mountains of Nêpál and of Tibet are steep and high; but they are, excepting the glaciers of the Himálaya, throughout chequered with cultivation and population, as well as possessed of a temperate climate. It is only necessary to observe the due season for passing the Himálaya, and there is no physical obstacle to apprehend; so that the journey from Káthmándû to Peking may be surely accomplished in five months, allowing for fifteen days of halts. But wherefore speak of Peking? At the eighty-seventh stage only, from Káthmándû, the merchant enters

* Mr. Brun gives 4,196 miles for what I take to be the direct, or nearly direct, way. Coxe. in one place, gives 5,363, in another place 4,701 miles. Bell's Itinerary yields 6,342. These are obviously the distances by various routes, or, by a more or less straight course, I take nearly the mean of them.

that rich and actively commercial province of China Proper, called Szchuen,* whence by means of the Yang-tsz-kiang, and of the Hwangho, he may transport his wares, as readily as cheaply, throughout the whole central and northern parts of China, if he can be supposed to have any adequate motive for going beyond the capital of Szchuen, where he may sell his European and Indian products, and purchase tea or silk or other products of China. The mountains of Sífán and of Tibet, which yield the finest borax, musk and rhubarb in the world, lie in his way both to and fro; and, in a word, without deviating from his immediate course, or proceeding above ninety days' journey from Káthmándû, he may procure where they grow, or are wrought, all those valuable articles of commerce which Russia must seek indirectly and at a much greater cost. But England and China, and not Calcutta and China, it may be argued, must be the sites of the production and consumption of the truly valuable articles of this commerce, of which the Nêpálese and Indians would have little more than the carrying trade; and England is afar off! It is so, indeed; but, with reference to the cheapness and facility of ship freight, of how little importance to commerce is the distance of England from Calcutta—not to mention that, as I have oberved in reference to the Russian commerce, we must not suppose the Russian has no further to seek than St. Petersburg, but remember that England and Canada supply him with half he needs. From Canada Russia seeks through England our peltry, to convey it to the Chinese across the endless savage wastes of Siberia. What should hinder our Indian subjects and the Népálese from procuring these same furs at Calcutta and conveying them through Népál and Tibet to these same Chinese. At less than ninety stages from Káthmándû, they would arrive at the banks of the Hwangho in Sífán, or those of the Yang-tsz-kiang in Szchuen; and then the merchants might be said to have reached their goal. What, again, should hinder the same merchants from under-selling the Russian, in the articles of English woollens, hard-ware and glass-ware, by conveying them to Sztchuen from Calcutta, by the same route? Nothing, it may safely be said, but want of sufficient information upon the general course and prospect of commerce throughout the world; and that information we might easily communicate the practical substance of to them. There are no political bars or hindrances to be removed for the Nêpálese have used the Chinese commerce *viá* Tibet for ages, and our Indian subjects might deal in concert with Nêpálese by joint firms at Káthmándû. Nay, by the same means, or now, or shortly, Europeans might essay this line of commercial adventure. But of them it is not my present purpose to speak†. Let

* The route from Lhása to the central and western provinces of China is far more easy than that from Lhása to Pekin.

† Lord Elgin is now proceeding to China, in order to determine the footing upon which the civilized world, and especially England, shall hereafter have commercial intercourse with the Celestial Empire.

It may be worth while to remind His Excellency of the vast extent of conterminous frontier and trading necessity in this quarter, between Gilgit and Brahmakund. We might stipulate for a Commercial Agent or Consul to be located at Lhása, or for a trading frontier post, like Kiachta; and, at all events, it would add to the weight and prestige of our Ambassador, to show himself familiar

the native merchants of Calcutta and of Nêpàl, separately or in concert, take up this commerce, and whilst we, though not the immediate movers, shall yet reap the great advantage of it, as consisting in an exchange of European articles for others chiefly wanted in Europe, we shall have a better chance of its growing to a vigorous maturity than if Europeans were to conduct it through its infancy. I have only further to add, in the way of continued contrast between the Russian commerce and that here sketched, that whilst the former is loaded with duties to the extent of 25 per cent., the latter would, in Nêpál, be subject only to 8 per cent.* duty; in Tibet, to no duty at all; and in our provinces only, I fancy, to a very moderate one, which might perhaps be advantageously abolished. Having thus, in the best manner I was able, without numerous books to refer to, none of which are to be had here, given a rapid view of the grounds upon which I conceive a very flourishing commerce might be driven in European and Indian articles, between the great Cis- and Trans-Himálayan plains, by means (at least in the first instance) of our Indian subjects and those of Nepál, I need only add, that the document No. I. is designed to arouse and direct the attention of the native merchants of Calcutta; that I have given it a popular form with an eye to its publication for general information in the *Gleanings in Science;* that No. 2. might be similarly published with advantage, and lastly, that nothing further is necessary, in order to give this publication all the effect which could be wished, than simply to enjoin the Editor of that work to refer any native making enquiries on the subject to the Resident at Káthmándû, who, without openly aiding or interfering, might smooth the merchant's-way to Káthmándû, and assist him with counsel and information. To prove that I have laid no undue stress on this matter, I only desire that a reference be had to the circumstances and extent of the Russian commerce at Kiachta, as lately (*i.e.*, in 1829) laid before Parliament; and even if this parallel between the two trades be objected to in its *present extent*, (and I have run it the whole length of China on one side, partly from a persuasion of the soundness of the notion, partly to provoke enquiry,) let us limit our own views to Tibet and maintain the parallel so modified. It may instruct, as well as stimulate us. Tibet, in the large sense, is an immense country, tolerably well peopled, possessed of a temperate climate, rich in natural productions, and inhabited by no rude nomades, but by a settled, peaceful, lettered, and commercially disposed race, to whom our broad cloths are the one thing needful; since, whilst all ranks and ages, and both sexes, wear woollen cloths, the native manufactures are most wretched, and China has none of a superior sort and moderate price wherewith to supply

with his whole case, or with the landward, as well as the sea-board relations of Britainand China.—*Note of* 1857.

* That is, the 6 per cent. before spoken of and 2 per cent. more levied between Káthmándú and the Bhote Frontier; but the latter duty can hardly be rated so high; at all events, 8 per cent. will amply cover all Custom House charges within the Népálese dominions. In our territories, the duties appear to reach 7 per cent. See general remarks to Part 1.

the Tibetans. With her musk, her rhubarb, her borax, her splendid wools, her mineral and animal wealth, her universal need of good woollens, and her incapacity to provide herself, or to obtain supplies from any of her neighours, Tibet may well be believed capable of maintaining a large and valuable exchange of commodities with Great Britain, through the medium of our Indian subjects and the people of Nepál, to which latter the *aditus*, closed to all others by China, is freely open. Nor is it now needful to use another argument, in proof of the extension of which this commerce is capable, than simply to point to the recorded extent of the existing Russian commerce with China across Siberia.

——o——

P.S., 1857.—A costly road has been constructed recently over the Western Himálaya; but, adverting to proximity and accessibility to the various centres of supply and demand, I apprehend that a brisk trade between the Cis- and Trans-Himálayan countries would inevitably seek the route of the central or eastern part of the chain. To Delhi, Benáres, Patna, Dacca and Calcutta, on the one hand, to all the rich and populous parts of Tibet, extending from Digarchee to Sífán, on the other hand, either of the latter routes is far nearer and much more accessible. By the unanimous testimony of all natives and of written native authorities Western Tibet is very much the poorest, most rugged, and least populous part of that country. Utsáng, Khám, Sífán, and the proximate parts of China furnish* all the materials, save shawl-wool, for a trade with us, as well as all the effective demand for our commodities. All this points to Káthmándû Darjeêling or Takyeul (above Gowhatti in Asam) as the most expedient line of transit of the Himálaya.‡

I.

THE TRADE OF NEPAL.

When we consider how much intelligent activity the native inhabitants of Calcutta have, of late years, been manifesting, we cannot help wondering that none of the mercantile class among them should have yet turned their attention to the commerce of Népál. Do they not know that the Néwárs, or aborigines of the great Valley of Népál, have, from the earliest times, maintained an extensive commercial intercourse between the plains of India on the one hand and those of Tibet on the

By the terms of the Treaty of 1792, the duties leviable on both sides are limited to 2½ per cent. *ad valorem* of the *invoice*. The actual charges to which the trader is put far exceed the customs duties *eo nomine*, since tolls are levied by every Jageerdar on the transit of goods through the lowlands.

* See Cooper, *Bengal As. Soc. Journal* for May 1869.

‡ Since this was written the successful growth and manufacture of tea in the British Himálaya are accomplished facts adding greatly to the means of establishing without doubt or difficulty a flourishing commerce with Tibet and the countries immediately north and east of it. In Kumaon, Sikim, Asam, are found great and thriving tea growing establishments. Nothing is more craved for or less procureable, in Tibet and up to the Russian frontier, than good tea; and if we cannot open up the Takyeul route from Asam we can and have that through Sikim by the Chola pass. The recent treaty has given us a right of way and of road construction, and this pass is not liable to be closed by the snow nor is the access to Sikim from the south rendered dangerous by malaria. The southern half of Sikim is our own: the northern half belongs to our dependant ally to whom we restored it in 1816, and for whom we have preserved it ever since, from the grasp of Népál.

other; that Népál is now subject to a wise and orderly Native Government; that owing to the firm peace and alliance between that Government and the Honorable Company's, the Indian merchant has full and free access to Népál; that the confidence inspired by the high character of the native administration, and by the presence of a British Resident at the Court, has led the native merchants of Benáres to establish several flourishing kothees at Káthmándú, that the Cashmerians of Patna have had kothees there for ages past; that so entirely is the mind of the inhabitants of our territories now disabused of the old idle dread of a journey to Népál, that lakhs of the natives of Oude, Behar, and North East Bengal, of all ranks and conditions, annually resort to Káthmándu, to keep the great vernal festival at Pasupati Kshétra. Are the shrewd native merchants of Calcutta incapable of imitating the example of their brethren of Benáres, who have now no less than ten kothees at Káthmándu; and will it not shame them to hear, that whilst not one of them has essayed a visit to Káthmándu, to make enquiry and observation on the spot, very many Népálese have found their way to Calcutta, and realized, on their return, cent. per cent. on their speculations in European articles? The native merchants of Calcutta have, whilst there, a hard struggle to maintain with their European rivals in trade, but at Káthmándu, they would have no such formidable rivalry to contend with, because Europeans not attached to the Residency, have no access to the country and without such access, they probably could not, and certainly have not, attempted to conduct any branch of the trade in question. But every native of the plains of India is free to enter Népál at his pleasure, nor would he find any difficulty in procuring from the Government of the country permission to sojourn by himself or his agent at Káthmándu, for purposes of trade. With a view to arouse, as well as to direct, the attention of our native brethren of the City of Palaces, in regard to the trade of Népál, we subjoin some of the principle details respecting the route, the manner and the cost of carriage, and the nature and amount of the duties levied by the Népál Government. It cannot be necessary to dwell upon that portion of the way which lies within the heart of our own provinces—suffice it to say that, by the Ganges and Gandák, there is commodious water carriage at all seasons, from Calcutta to Govindgunge or Kesriah, situated on the Gandák river, in the Zillah of Sarun, and no great way from the boundary of the Népálese territories. Kesriah or Govindgunge, then, must be the merchant's place of debarkation for himself and his goods, and there he must provide himself with bullocks for the conveyance of his wares, as far as the base of the greater mountains of Népál, where again, he will have to send back the bullocks and hire men to complete the transfer of his merchandise to Káthmándú; and here we may notice a precaution of some importance, which is, that the merchant's wares should be made up at Calcutta into secure packages adapted for carriage on a man's back of the full weight of two Calcutta bazar maunds each; because, if the wares be so made up, a single mountaineer will carry that surprising weight over the huge mountains of Népál, whereas two men not being able to unite their strength with effect in the conveyance of goods, packages heavier than two maunds are, of necessity, taken to pieces on the road at great hazard and inconvenience, or the merchant must submit to have very light weights carried for him, in consideration of his awkwardness or inexperience in regard to the mode of adjusting loads. Besides the system of duties proceeds in some sort upon a presumption of such loads as those prescribed; and lastly, two such loads form exactly a bullock freight; and upon bullocks it is necessary, or at least highly expedient to convey wares from Kesriah to the foot of the mountains. Let every merchant, therefore, make up his goods into parcels of two full bazar maunds each, and let him have with him apparatus for fixing two of such parcels across a bullock's saddle. He will thus save much money and trouble. Kesriah and Govindgunge are both flourishing villages at which plenty of good bullocks can be had by the merchant, for the carriage of his wares, as well as a good tattoo for his own riding to the foot of the hills, whence he himself must either walk, or provide himself (as he easily can at Hitounda) with a dooly, for the journey through the mountains to Kathmandu, the hire of a bullock from Kesriah to Hitounda, at the foot of the mountains, is three Sicca rupees: besides which sum, there is an expense of six annas per bullock to tokdars or watch-men on this route, *viz.*, two an nas at Moorliah, two at Bichiako, and two at Hitounda. The total expenses, therefore per bullock, from Kesriah to Hitounda, are Sicca rupees 3-6-0. The load of each bullock is four pukka maunds. The stages are nine, as follows:—Kesriah to Bhopatpoor, 5 cos; to Lohiá, 7 cos; to Segoulee, 5 cos; to Amodahi, 5 cos; to Pursoni, 6 cos; to Bisouliah or Simrabasa, 4 cos; to Bichiako, 5 cos; to Chooriah Ghauti, 3 cos; and to Hitounda, 4 cos; being 44 cos in all. Hitounda, as already frequently observed, is at the foot of the great mountains, which, for want of roads, no beast of burden can traverse laden. Men, therefore, are employed, but so athletic and careful and trustworthy are the hill porters, that this sort of carriage is far less expensive or inconvenient than might be imagined. The precautions in respect to packages before prescribed having been attended to by the trader he will find the four maunds of goods, which constituted the one bullock's load as far as Hitounda readily taken up by two hill-porters, who will convey them most carefully in six days to Káthmándú. It is an established rule, that four maunds, properly packed, make two bakkoos, or men's loads, which are conveyed to Káthmándú at the fixed rate of two rupees of the country per bakkoo or load. The stages and distances are as follows:—Hitounda to Bhainsa Dobháng, 3½ cos; to Bhimphédy, 4 cos; to Tambakhâni, 3 cos; to Chitlong, 3 cos; to Thankot, 3 cos; to Káthmándú, 3 cos—Total, 19½ cos. At Hitounda, there is a Custom House Chokey, where packages are counted merely, not opened, nor is any duty levied there. At Chisapâni Fort, which is half way between Bhimphédy and Tambakhâni, is another Custom Chokey, and there the merchandise is weighed, and a Government duty is levied of one anna per dhârni of three seers, being two Paisa rupees per bakkoo: also, a Zemindary duty at Chitlong of two annas per bakkoo or load of 32 dhârni, in other words of 96 ordinary seers. At Thankot, the last stage but one, a further Zemindary duty is levied of four annas per bakkoo.

SUMMARY RECAPITULATION OF THE EXPENSES FOR CARRIAGE AND DUTY BETWEEN HITOUNDA AND KATHMANDU.

		Nepalese Rs.	Siccas.
Hire of Porters		4 0 0	3 4 0½
Duties Paisa rupees	4 12 0	3 12 8	3 0 9¼
Per bullock load ..		7 12 8	6 4 9¾

To which, if we add the 3-6-0 Sicca for bullock hire and watch-men, between Kesriah and Hitounda, we shall have a total of Sicca rupees 9-10-9¾ for the expense, for duty and carriage, of conveying four pukka bazar maunds and upwards, (64 dhârni or 192 ordinary seers exactly,) from the Ghaut of the Gandák to Káthmándú, where finally the goods are subject to an *ad válorem* duty of rupees 3-8-0 of the country or 2-13-6 Sicca, and where the merchant may get cent. per cent. upon Calcutta prices for his European articles, if they have been well selected.

The duties upon imports from the plains, leviable at Káthmándú, are farmed by the Government, instead of being collected directly. The farm is called Bhansâr —the farmer, Bhansâri. On the arrival of a merchant with goods from the plains, the Bhansâri, or his deputy, waits upon the merchant and seals up his bales, if it be not convenient to him to have them at once examined. When the bales are opened and the goods inspected, an *ad valorem* duty (for the most part) of 3½ per cent. is levied on them by the Bhansâri, thus :—

	Nepalese Rs.	Siccas.
For Kinăra or Kinăra, per cent.	2 0 0	1 10 0
For Nirikhi, per cent.	1 8 0	1 3 6
	3 8 0	2 13 6

The value of the goods, upon which depends the amount of duty, is settled by inspection of the merchant's invoice and by appraisement of a regular officer, thence called the Nirikhman. If th merchant continue to dispute the appraiser's valuation, and the consequent amount, of duty, and will not listen to reason, it is usual for the Government, in the last resort, to require the merchant to dispose of his wares to it at his own alleged valuation. Let no one therefore think to abate the duty by under-valuing his goods, for if he do, he may find himself taken at his word, when he least expected it. For the rest, if he be fair and reasonable and exhibit his invoice, he has nothing to fear from the Bhansâri, who is not a man of eminent place or power, and if he were, would not be suffered, under the present able administration, to oppress the merchant. In respect to the duties levied on the way up, (at Chisapâni and Thankot,) as already explained, they are called Sayer and Bakwaoon. If the merchant please, he may avoid paying them on the road, and settle for them at Káthmándu, in which case the Collector of Chisapâni takes a memorandum of the weight of the goods and forwards it to the Bhansári and to the Government Collector at Káthmándú, giving the merchant, at the same time, a note of hand to pass him on.

We have stated that the duty on Imports from the plains is, in general, an *ad valorem* one of 3-8-0 of the country currency; but as, there is a different rate in respect to some of the articles, and, as the enumeration of the chief Imports will serve as a sort of guide to the Calcutta trader, who may be disposed to adventure a speculation to Káthmándú. we shall give a list of these Imports with the duty assigned to each.

		Duty in Nepal Rupees	& Siccas
European broad cloths and other woollens of all sorts	per cent.	3 8 0	2 13 6
European chintzes and other cotton of all sorts ..	„	3 8 0	2 13 6
European silks of all sorts	„	3 8 0	2 13 6
European linens of all sorts	„	3 8 0	2 13 6
Amritsur and Cashmere shawls, good	„	3 8 0	2 13 6
Dacca muslins and Jamdanees, sahans, &c.	„	3 8 0	2 13 6
Malda and Bhaugulpoor silk and mixed silk and cotton stuffs	„	3 8 0	2 13 6
Benáres kimkhabs, toftas, mushroos. shamlas, dopattahs, &c.	„	3 8 0	2 13 6
Mirzapoor and Calpee kharwas and garhas ..	„	3 8 0	2 13 6
Mowsahans, andarsahs, &c. „ .. ..,	„	3 8 0	2 13 6
Behar, pagrees, khasas, &c.	„	3 8 0	2 13 6
Bareilly, Lucknow and Tanha chintzes	„	3 8 0	2 13 6
Enropean cutlery, as knives, scissors, &c.	„	3 8 0	2 13 6
European glass-ware chandeliers, wall-shades, &c. ..	„	3 8 0	2 13 6
European mirrors, window glass, &c.	„	3 8 0	2 13 6
Indian karanas, or groceries, drugs, dyes, and spicery of all sorts	„	5 0 0	4 1 0
Peltry of Europe and India, as Dacca, other skins, goat ditto, &c.	„	5 0 0	4 1 0
Quicksilver, vermilion, red and white lead, brimstone, jasta, ranga, camphor	„	5 0 0	4 1 0
Indigo pays *in kind*	„	10 0 0	8 2 0
Precious stones, as diamond emerald, pearl, coral	„	1 8 0	1 3 6
Indian laces, as Kalabuttu, Gotah, &c.	„	5 0 0	4 1 0

Whoever has sold his wares at Káthmándú will next look to purchasing a "Return Cargo" with the proceeds of such sale. We therefore now proceed to notice the manner and amount of the *Export* duties levied by the Népál Government upon goods exported to the plains. There is no difference between goods the produce of Népál and such as are the produce of Bhote (Tibet) or China, all paying on exportation to India at the same rate.

The Exports, like the Imports, are farmed, and it is therefore with the Bhansári that the merchant will have again to treat with.

The Export duty is an *ad valorem* one, and amounts, for the most part, to 4-11-1 per cent., which is levied thus:—

	Ostensibly.			Really.		
As Bakkooána	1	0	0	1	2	2
As Nirikhi	1	4	0	1	4	0
As Kinara	2	0	0	2	4	3
	4	4	0	4	11	1

These sums are Népálese currency. Their equivalents in Sicca rupees are 3-7-3 and 3-13-9. There are no further duties levied on the road, and the merchant, upon payment of the above *ad valorem* duty at Káthmándú, receives from the Bhansári a pass, or Dhoka Nikási, which will carry him, free beyond the limits of Nêpál.

The merchant's goods, on his return, should be made up, as on his approach, into bakkoos or men's loads of thirty-two dhárnis of three seers per dhárni, and he should have bullocks waiting his arrival at Hitounda, by previous arrangement. The following is a list of some of the principal exports, with their respective duties:—

Articles.		Duties in Nepal Rupees			& Siccas.		
Chours	per cent.	4	11	1	3	13	9
Tibetan, Himálayan and Chinese woollens, as Maleeda, Toos, Namda, Chourpat, Rahry, Bhot etc	„	4	11	1	3	13	9
Chinese damasked and brocaded satins & silks	„	4	11	1	3	13	9
Sohága or borax	„	4	11	1	3	13	9
Nêpálese, Bhotea and Chinese drugs — rhubarb, mihargiyah, zaharmohara, momira, jatamangsee, hurtal, &c.	per cent.	4	11	1	3	13	9
Bhotea and Nêpálese paper	„	4	11	1	3	13	9
Musk pods, per seer of 32 Sa. Wt.		1	4	0	1	0	3
Gold	Duty free.						
Silver	Prohibited.						
Rupees of the plains	Ditto.						
Rupees of Nêpál and copper pice of ditto.	Free.						
Bhote poneys or tanghans, each		7	0	0	5	11	0
Hard-ware, as iron phowrahs &c.	per cent.	4	11	1	3	13	9

Though we would not advise the native merchant of Calcutta to meddle, in the *first* instance, directly himself, with the trade of Bhote, whether in exports or imports, yet as that country causes the great demand for European woollens in particular, and is, on many accounts, of more consideration in a commercial point of view than Népál, we shall give some details relative to the trade with it, through Nêpál, analogous to those we have already furnished respecting the trade with Nêpál itself.

The duties upon the Bhote trade are levied by government through its own officers, not farmed, like the duties on the trade with the plains. Goods of the plains (whether the produce of Europe or India,) exported through Népál to Bhote, are made up into packages or *bàkkoos*, of sixteen *dhárnis*, or forty-eight *seers* only, owing to the extreme difficulties of the road, which will not permit a man to carry more than that weight upon his back; and there are no other means whatever of conveyance, until the Himálaya has been passed. Upon these *bakkoos* or loads the duty is levied, and amounts to Paisa rupees 1-0-1 per *bakkoo*, for all articles alike. The duty is levied at the Taksâr or Mint, and the collector is familiarly called Taksâri in consequence. The details of duty of the 1-0-1 are these:—

Taksár	0	6	0
Nikási	0	10	0
Bahidár.......	0	0	1
Paisa Rupees	1	0	1 = Siccas 0 10 10

Upon payment of this sum to the Taksári, that officer furnishes the merchant with a passport, which will pass his goods, free, to the frontier of Bhot or Tibet.

The chief exports to Bhote are:—European broad cloths (crimson, green, orange, liver, and brown- coloured), cutlery, pearls, coral, diamonds, emeralds, indigo and opium. Goods imported into Nêpál from Bhote (no duty levied there) payto the Taksár at Káthmándú as follows:—

Musk pods, per seer (in kind) .. 1½ tolahs.
Gold, per tolah .... 1 anna.

Silver is all necessarily sold to the Taksár and is received at the Sicca weight, paid for at the Nêpálese or Mohári weight, difference three annas.

Articles.		Duty.
Chours, white	..per dhárni	4 annas.
Ditto, black	,,	3 ,,
Chinese and Bhotea velvets, woollens, satins, silk thread, and raw silk	per cent.	4 rupees
Peltry of Mongolia and Bhote, samoor, kakoon, chuah-khal, garbsooth, &c.	,,	4 ,,
Borax	,,	4 ,,
Chinese and Bhotea tea	,,	4 ,,
Drugs	,,	4 ,,

From Káthmándú to Bhote frontier, or rather, to the frontier *marts* of Kooti and of Keroong, there are two roads, one of which is called the Keroong and the other the Kooti way, after the marts in question, which are respectable Botea towns.

The following are the stages and expenses:— Káthmándú to Kooti, eight stages, sixteen dhárnis, or forty-eight seers, a man's load. His hire, 2 rupees of Nêpál—or Siccas 1-10-0 for the trip.

The stages are Sankhoo, 3½ cos; to Sipa, 7½ cos; to Choutra, 5 cos; to Maggargaon or Dharapani, 3 or 4 cos; to Listi, 5 cos; to Khasa, 4 cos; to Chê-sang, 5 cos; to Kooti, 3¾ cos.

From Káthmándú to Keroong, the eight stages are:—To Jaiphal-kepowah, 4 cos; to Nayakot, 5 cos; to Taptap, 4 cos; to Prebeo, 4 cos; to Dhom-chap, 5 cos; to Maidan Pootah, 3 cos; to Risoo (frontier), 4 cos; to Maima, 4 cos; to Keroong, 4 cos.

The load is the same as on the Kooti road and the hire of the carrier the same. The Himálaya once passed, you come to a tolerably plain country, along which beasts of burden can travel laden. The usual carriage is on ponies and mules, which carry two *bakkoos* of sixteen *dhárnis* each, and can be hired for the trip, from Lhása to the Népál frontier, for twenty rupees of Bhote currency. They perform the journey in about a month, allowing for three or four days' halts.

P.S.—The Nêpálese *dhárni* is equal to three seers. The Nêpálese rupee is worth thirteen annas. It is called, after an ancient dynasty, Mahêndra Mally, or shortly and commonly Môhari. It is almost a mere nominal coin, from its scarceness, the common currency consisting of half rupees or Môhars. The Bhote rupee is called Kala Môhari. It ought to be equal to the Nêpálese, but is rendered five gundas less valuable by undue adulteration.

No. II.

TRADE OF NEPAL.

Import of Goods from the Plains in 1830-31.

Articles.	At what price purchased in Calcutta, or where manufactured or grown.	Estimated prime cost amount of Imports in Co.'s Rupees.	At what price sold in Nepal, in Nepalese Rupees	Amount of goods transported to Bhote.	Amount of goods consumed in all Nepal.
European crimson velvet[a]	From 5 to 16 Rs. per yard.	13,000	26,000	2,000	24,000
" blue and green ditto.	" 8 " 12 " "	8,000	16,000	4,000	12,000
" black ditto	" 5 " 9 " "	4,000	8,000	None	8,000
" brown, liver, and abiree ditto	" 5 " 10 " "	5,000	10,000	2,000	8,000
Broad cloth of scarlet colour[b] ..	" 2½ " 10 " "	30,000	60,000	10,000	50,000
" green ditto	" 2 " 12 " "	20,000	40,000	25,000	15,000
" brown ditto	" 2½ " 6 " "	25,000	50,000	20,000	30,000
" black ditto.	" 3 " 9 " "	8,000	16,000	None	16,000
" blue ditto	" 2 " 11 " "	21,000	42,000	2,000	40,000
" liver, yellow, and abiree ditto.	" 2½ " 4 " "	16,000	32,000	3,000	29,000
Europe shawls[c]	" 6 " 25 " per piece.	2,000	4,000	None	4,000
" rumal	" ½ " 2 " "	1,000	2,000	"	2,000
" satin	" 1½ " 2 " "	2,500	5,000	"	5,000
" chintzes,[d] 1¼ by 28 yds.	" 4 " 14 " "	22,000	44,000	2,000	42,000
French ditto, 1 by 14 to 24 yds.	" 14 " 30 " "	8,000	16,000	1,000	15,000
Country ditto, from Tanda, Bareilly, Furruckabad, &c., 1¼ to 2 by 6 to 9 yards . . .	" 1-6 " 4 " "	2,00,000	3,00,000	5,000	2,95,000

Remarks.

[a] The Tibetans care not for our velvets, but they are much admired by the Nêpálese, both males and females. Ladies wear velvet bodies; gentlemen velvet caps and jackets. Scabbards and saddles and cushions are covered with velvets.

[b] Our broad cloth is equally prized in Népál aud Tibet, but the Tibetans are superstitious about its colour. The colours set down are the only ones which the Tibetans will wear, and even of these the blue and black, though agreeable to the Nêpálese, are not so to the Tibetans. The quality most saleable is that which fetches 8 to 10 rupees per yard at Calcutta.

In Tibet all ranks and both sexes wear woollens throughout the year, and almost exclusively of other apparel; yet the only

ARTICLES.	At what price purchased in Calcutta, or where manufactured or grown.	Estimated prime cost amount of Imports in Co.'s Rupees.	At what price sold in Nepal, in Nepalese Rupees	Amount of goods transported to Bhote.	Amount of goods consumed in all Nepal.	REMARKS.
Europe cambrics,[e] 1 by 12 yards	„ 5 „ 12 „ „	17,000	34,000	1,200	32,800	native manufactures of woollen
„ long cloth, 1 by 24 yds.	„ 6 „ 10 „ „	6,000	12,000	2,000	10,000	cloths are of the most inferior
„ leno cloth, 1 by 10 yds.	„ 2½ „ 9 „ „	16,000	32,000	2,000	30,000	quality and insufferable weight. Nor is the total deficiency of
Dacca jamdanee, 1 by 10 yards.	„ 10 „ 20 „ „	5,000	10,000	None	10,000	good broad cloths capable of being supplied by the Chinese, who
Tanda jamdanee	From 5 to 10 Rs. per piece.	5,000	7,500	500	7,000	use not, nor make, woollens.
Dacca muslin, 1 by 18 yards . .	„ 10 „ 15 „ „	8,000	16,000	None	16,000	Tibet only receives a small quantity of Toos and Maleeda
Muslin of Behar, Santipoor, Chandrakona, and Kopamow, 1 by 16 yards.	„ 2 „ 10 „ „	40,000	60,000	500	59,500	from China; and those cloths,
Dacca Sahan, 1 by 18 yards . .	„ 8 „ 13 „ „	40,000	80,000	None	80,000	though made of fine wool, are of poor, loose-textured, flannel-like manufacture.
Sahan cloth of Santipoor, Chandrakona, Jahana, Kopamow, and Tanda, etc., 1½ & 2 cubits by 12 and 18 yards	„ 1½ „ 8 „ „	100,000	150,000	5,000	145,000	How large, then, ought to be the demand for our broad cloths in Tibet, if sufficiently intelligent and active industry was employed in putting them within the
Kora Gora cloth of Dhurbanga, Landa, Jahana, and Majhduwa, etc., 1½ & 2 cubits by 14 and 20 yards	„ 1½ „ 6 „ „	33,000	50,000	500	49,000	reach of the Tibetans. [c] Our imitations of Indian hand-
Dacca Turban, 1½ span by 12 and 30 yards	„ 5 „ 15 „ „	10,000	20,000	300 (bracketed with the next row)	19,000	kerchiefs and Cashmere shawls are becoming very popular among
Turban of Kopamow and Jahana 1½ span by 20 and 40 yards .	„ 1¼ „ 10 „ „	30,000	45,000			the middle and lower orders in Nêpál.
Gimtee of Bykantpore and Jahana, etc., 1 cubit by 6 & 8 yds.	„ 1½ „ 4 „ „	10,000	15,000	None	15,000	[d] English and Indian Chintzes are not worn at all in Tibet.
Indian and English Doriah cloth 1 by 16 yards	„ 2 „ 8 „ „	7,000	10,500	„	10,500	Both are much worn in Nêpál by the middle and lower orders.
Maldahee cloth, 1 by 9 or 10 yds	„ 2½ „ 5 „ „	7,000	14,000			

Moorshedabad cloth stuffs of sorts, 1 by 6 or 9 yards . . .	„ 3 „ 10 „ „	12,000	24,000		
Boolbool Chasma Kummerbund of Malda, 1 cubit by 6 yards	„ 3 „ 6 „ „	1,000	2,000		
Radhanagaree and Bhaugulpooree, Soosy and Loonga cloth, 1 by 6 yards	„ 1-12 „ 3 „ „	1,600	3,200		
Benáres Mushroo and Goolbadan 1¾ cubits by 5 yards	„ 3 „ 12 „ „	7,000	14,000		
Benáres Dhootees,[f] of sorts, 2½ cubits by 5 yards, with brocaded edges	„ 2 „ 4 „ p. ea. pr.	3,000	6,000	None	6,000
Gya Dhootees, ditto ditto . . .	„ 1½ „ 3 „ „	2,000	4,000	„	4,000
Benáres Kinkhabs, 1 by 5 or 6 yards	„ 50 „ 400 „ per piece.	28,000	56,000	11,000	45,000
Benáres Baphtas, 1 by 5 yards	„ 20 „ 500 „ „	4,000	8,000	5,000	3,000
Benáres Dopattahs, with brocaded edges, 1½ or 2 by 3 or 4½ yards	„ 5 „ 40 „ each.	1,200	2,400	None	2,400
Tash of Benáres, Lucknow, and Patna, 1¼ by 6 yards	„ 75 „ 300 „ per piece.	8,000	16,000	4,000	12,000
Golden Kallabuttu	„ 2 „ per tolah.	12,000	24,000	None	24,000
Silver ditto	„ ½ „ 1-2 „ „	5,000	10,000	„	10,000
English lace and Indian Gotah and Kinara of gold	„ 2 „ 2¼ „ „	22,000	44,000	„	44,000
Ditto ditto of silver	„ 12 as „ 1 „ „	6,000	12,000	„	12,000
Gold and silver Salmasitara .	„ 2½ „ 3½ „ „	9,000	18,000	„	18,000
Cashmeer and Amritsar Shawls,[g] 2 by 3½ or 4½ yards	„ 80 „ 1000 „ per piece.	45,000	90,000	„	90,000
Ditto ditto Rumals, 1½ & 2½ yds	„ 20 „ 150 „ „	6,000	12,000	„	12,000
Kalpi and Mirzapoor Kharwa cloth, 1¾ cubits by 6 or 7 yds.	„ 14 as „ 1-12 „ „	6,000	12,000	4,500	7,500
Mirzapoor and Patna Carpets, of sorts	„ 2½ „ 50 „ „	4,000	8,000	None	8,000
Patna and Govindgunge Settringhees	„ 1 „ 8 „ „	1,200	2,400	„	2,400

Women make gowns of them, the men jackets and linings to jackets—to the latter use they are occasionally applied in Bhote.

[e]No kind of cotton stuffs, fine or coarse, are used as apparel in Tibet, and the little the Tibetans purchase, is wanted merely for covering the pictures with which the walls of sacred edifices are literally covered in Tibet. On the contrary, all classes in Nêpál wear cotton wholly or partially throughout the year; and as Nêpál has no native manufacture of any but very coarse sorts, there is a large demand for the finer fabrics, both of Europe and India. Chintzes and Sahans are the kinds most used. The whole of the middle and upper classes are clad in foreign cottons. The poor manufacture, each family, their own supply of coarse goods.

The manufacture is purely and entirely domestic and peculiar to the females.

[f]English lace, the beautiful silk and muslin fabrics of Benáres embroidered all round with gold or silver, and the splendid brocades of the latter place are not much known or esteemed in Bhote. They are all well-known and much esteemed in Népál, but only to a very moderate extent within the

ARTICLES.	At what price purchased in Calcutta, or where manufactured or grown.	Estimated prime cost amount of Imports in Co.'s Rupees.	At what price sold in Nepal, in Nepalese Rupees	Amount of goods transported to Bhote.	Amount of goods consumed in all Nepal.	REMARKS.
English and Bengal silk thread, per seer of 80 Sa. Wt. . . .	From 4 to 16 Rs. per seer.	12,000	24,000	None	24,000	means of the people. Every man,
Cotton thread, Indian mostly .	„ 22 „ 30 „ per md.	4,000	8,000	1,200	6,800	however, who can afford it, will
Cotton, raw, from Mirzapore . .	„ 12 „ 16 „ „	32,000	64,000	None	64,000	have his cap banded with lace of
Pearls[i] (large)	„ 25 „500 „ „	40,000	80,000	30,000	50,000	silver or gold, and the Court an-
Ditto (middle-sized)	„ 10 „ 20 „ per corge	22,000	44,000	30,000	14,000	nually consumes a considerable
Ditto (small)	„ 1 „ 10 „ „	13,000	26,000	20,000	6,000	supply of brocaded goods of the
Coral (large) 1½ to 2½ tolahs per bead	„ 8 „ 12 „ per tolah	25,000	50,000	15,000	35,000	finest qualities. [g] Bhote receives its supply of
Ditto (middle) 5 to 10 beads per tolah	„ 3 „ 7 „ „	40,000	88,000	50,000	38,000	Shawls by way of Ladakh, to Népál they come *viá* the plains. All
Ditto (small) 12 to 20 do. do.	„ 1½ „ 2½ „ „	26,000	52,000	15,000	37,000	who can afford it wear them.
Diamonds, Emeralds, Rubies, and Sapphires		50,000	100,000	None	100,000	[h] Kharwas are used for lining the dresses of the commonalty of
Neshapoori Feroza stone . . .	„ 12as „ 1½ „ per corge	16,000	32,000	15,000	17,000	Népál and for their bed furni-
Cornelian	„ 1 „ 10 „ „	1,000	2,000	500	1,500	ture. The Népálese manufac-
Kahrobah or red amber. . . .	„ 2 „ 8 „ per tolah	6,000	12,000	7,000	5,000	ture cotton stuffs as fine as Khar-
Ditto yellow		3,000	4,000	None	4,000	wa, but they cannot give them a
Ditto Kuppori Poosah		600	1,200	1,000	200	permanent colour, such as Khar-
Pôth or glass beads, of different colours	„ 4as „ 1 „ „	2,000	4,000	None	4,000	wa receives in the plains. [i] Bhote has long been a capital
Chank	„ 50 „100 „ „	2,500	5,000	3,500	1,500	market for pearls and coral. All
Large Cowries	„ 2 „ 4 as. each	100	200	None	200	the women of rank wear the for-
Kuldar gold Mohur[j]	„ 18 Rs. 2 as. each	20,000	25,000	„	25,000	mer and all the priests of rank
Golochan	„ 30 Rs. per seer	3,000	6,000	3,000	3,000	carry rosaries of coral beads.
Indigo[k]	„ 80 „120 Rs. per md.	21,000	63,000	12,000	51,000	The market, however, is now
Otter skins	„ 8 as. each	2,000	4,000	2,500	1,500	glutted.
Green skins, of prepared leather for native shoes, etc.	„ 3 to 6 as. each	1,000	2,000	600	1,400	[j] Gold Mohurs fetch at Káthmándú 25 rupees of Népál each,

Saltpetre[l] …………………	From 4 to 5½ Rs. per maund.	40,000	80,000	1,200	78,800
Brimstone …………………	„ „ 13 „ „	6,000	12,000	500	11,500
Quicksilver …………………	„ „ 80 „ „	3,000	6,000	1,600	4,400
Singraf or vermillion cinnabar	„ „ 80 „ „	3,000	6,000	None	6,000
China red lead …………………	„ „ 60 „ „	2,500	5,000	„	5,000
Country ditto …………………	„ „ 8 „ „	1,000	2,000	None	2,000
Ruskapoor …………………	„ 8 „ 12 „ per seer.	800	1,600	„	1,600
Camphire …………………	„ „ 32 „ per maund.	7,000	14,000	„	14,000
White sandal …………………	„ 9 „ 11 „ „	4,000	8,000	1,000	7,000
Red ditto …………………	„ 4½ „ 6 „ „	2,600	5,200	600	4,600
Zangar or verdigris …………	„ „ 5 „ „	5,000	10,000	1,200	8,800
White lead …………………	„ „ 2 „ „	3,000	6,000	1,500	4,500
Sankhia, Hartal, & Tabkiah do.	„ „ 4 „ per seer.	1,400	2,800	None	2,800
Opium[m] …………………	„ 3½ „ 4 „ „	4,000	8,000	2,000	6,000
Sajikhar and Jowakhar ……		500	1,000	None	1,000
Ranga, Tin[n] …………………	„ „ 12 „ per maund.	16,000	32,000	1,500	30,500
Jasta, Pewter …………………	„ „ 4 „ „	19,000	38,000	None	38,000
Lead …………………	„ „ 4 „ „	10,000	20,000	„	20,000
Steel …………………	„ 12 „ 16 „ „	4,000	8,000	„	8,000
Gun flints,[o] European ………	„ 1 „ 1½ „ per 100.	6,000	12,000	None	12,000
Europe-made guns, of sorts….	„ 250 „ 600 „ each.	11,000	22,000	„	22,000
Monghyr-made ditto ………	„ 10 „ 35 „ „	2,000	4,000	„	4,000
English knives and scissors….	„ 4as. „ 2 „ „	1,200	2,400	300	2,100

the rupee in question being worth 13 annas and 4 pie.

[k] Népál and Tibet are always very inadequately supplied with good indigo.

There is a very great demand for it just now in the latter country.

[l] Saltpetre is produced in the Népálese Tarai, and brimstone in the hills, but neither in quantity equal to the demand of the government, who therefore prohibit all export to Tibet of that which comes from the plains. The transports or exports to Tibet are therefore clandestine.

The Népálese want skill to work their Sulphur mines profitably.

[m] Opium is in great demand in Tibet just now. A very insufficient quantity gets there *viâ* Népál at present. It is procured by the Népálese furtively, in the Tarai, from our ryots between the Narayani and Bagmatti.

[n] Népál is full of copper and iron and the people have great skill in working them. Tin, lead and zinc they get from below, and a variety of mixed metals. Népál produces plenty of zinc, but no skill to work the mines.

[o] The whole Goorkha army, which is armed with muskets, is

Articles.	At what price purchased in Calcutta, or where manufactured or grown.	Estimated prime cost amount of Imports in Co.'s Rupees.	At what price sold in Nepal, in Nepalese Rupees	Amount of goods transported to Bhote.	Amount of goods consumed in all Nepal.	Remarks.
English needles, etc.	From 11as to $1\frac{1}{2}$ Rs. per 100.	1,500	3,000	None	3,000	supplied with flints from below, chiefly from Calcutta. For English fowling-pieces and hard-ware, there is always a considerable demand in Népál; guns priced between 250 to 500 rupees sell very well, and so do good plain scissors and needles. Our scissors are also in demand in Bhote, and if we would condescend to work upon the models prescribed by the usage of these countries, we might obtain a large vent in them for our hard-ware, of most sorts. [p] Both Népál and Tibet constantly require and consume a large quantity of tropical spicery, drugs and dried fruits, and both countries repay India with large returns of drugs and samples peculiar to either region. Of the number and value of the medicinal substances furnished to commerce by these countries, European medical men are strangely uninformed.
Ditto iron and brass-ware		600	1,200	,,	1,200	
Atter of Ghazeepore, Jaunpore,	,, 1 ,, 15 ,, per tolah.	6,000	12,000	None	12,000	
Choha oil ditto	,, 12as ,, 1 ,, per seer.	1,200	2,400	,,	2,400	
Keorah Water	,, 8as ,, $1\frac{1}{2}$,, ,,	1,000	2,000	,,	2,000	
Abir	,, 4 ,, 7 ,, per maund	5,000	10,000	,,	10,000	
Saffron of Cashmeer	,, 6 ,, 8 as per tolah	1,500	3,000	,,	3,000	
Lobhan	,, 20 ,, 24 Rs. per maund	3,500	7,000	,,	7,000	
Paun of Behar and Bengal	,, 1 ,, 4 as per dholi	2,500	7,000	,,	7,000	
Cocoanut Hookah bottoms	,, 3 ,, 10 Rs. per 100	600	1,200	,,	1,200	
Almonds[p]	,, ,, 6 ,, per maund	3,000	6,000	,,	6,000	
Dates	,, ,, 5 ,, ,,	1,800	3,600	,,	3,600	
Cocoanuts	,, ,, 12 ,, ,,	3,000	6,000	,,	6,000	
Dakh raisins.	,, ,, 16 ,, ,,	500	1,000	,,	1,000	
Makhanah	,, ,, 7 ,, ,,	400	800	,,	800	
Desee betel nut	,, 4 ,, $5\frac{1}{2}$,, ,,	5,000	10,000	500	9,500	
Ditto ditto Dakheeni pepper, (Black)	,, 9 ,, 11 ,, ,,	10,000	20,000	2,000	18,000	
Cloves	,, ,, 25 ,, ,,	14,000	28,000	2,000	26,000	
Nutmegs	,, ,, $1\frac{1}{2}$,, per 100	4,000	8,000	1,000	7,000	
Pepal.	,, 8 ,, 10 ,, per maund	800	1,600	200	1,400	
Assafœtida	,, 16 ,, 30 ,, ,,	5,500	11,000	1,000	10,000	
Mace	,, ,, 155 ,, ,,	6,000	12,000	None	12,000	
Cinnamon	,, 13 ,, 14 ,, ,,	2,500	5,000	,,	5,000	
Zeera.	,, 6 ,, 9 ,, ,,	2,000	4,000	,,	4,000	
Small Cardamums	,, 3 ,, $6\frac{1}{4}$,, per seer	4,000	8,000	,,	8,000	
Jetmudh or Moolhati.		500	1,000	,,	1,000	

Gugal	„ 6 „ 7 Rs. per maund.	600	1,200	None	1,200
Nasadar	„ 7 „ 9 „ „	1,200	2,400	„	2,400
Dry ginger	„ 3 „ 4 „ „	1,000	2,000	„	2,000
Mangrela or Kalonjee	„ 2½ „ 3 „ „	300	600	„	600
Akarkora	„ 18 „ 22 „ „	300	600	„	600
Bakkam wood	„ 9 „ 11 „ „	2,500	5,000	„	5,000
Kusum flower	„ 8 „ 13 „ „	6,000	12,000	„	12,000
Tund ditto		500	1,000	„	1,000
Tutea		900	1,800	„	1,800
Surma	„ ½ „ 1 a. per tolah.	300	600	„	600
Moordasankh	„ 5 „ 6 Rs. per maund.	200	400	„	400
Mahawur		250	500	„	500
Golali	„ „ 15 „ „	150	3,000	„	3,000
Tokhumbalanga and Ispoghol		600	1,200	„	1,200
Sendhu salt	„ 8 „ 10 „ „	500	1,000	„	1,000
Kattah	„ „ 3 „ „	400	800	„	800
Soap, soft, Indian	„ „ 8 „ „	1,200	2,400	„	2,400
Dry leaf, tobacco.	„ 5 „ 7 „ „	10,000	20,000	8,000	12,000
Chooree and Tikoolee, from Benáres and Ghazeepore	„ 4 as „ 2 „ per pair.	1,500	3,000	None	3,000
Peacocks' feathers		400	800	„	800
Chapra Lakh	„ 11 „ 18 „ per maund.	500	1,000	„	1,000

[q] Within the last fifteen years the gentry of Népál have become universally horsemen.

The Court makes large and regular purchases, and pays usually, not in money, but in elephants, the produce of its Tarai territories. This sort of barter answers equally well to both parties in the transaction.

[r] Small mirrors are in equal demand for Népál and for Tibet.

ARTICLES.	At what price purchased in Calcutta, or where manufactured or grown.	Estimated prime cost amount of Imports in Co.'s Rupees.	At what price sold in Nepal, in Nepalese Rupees	Amount of goods transported to Bhote.	Amount of goods consumed in all Nepal.	REMARKS.
Hookah snakes	From 1 a. to 1 Re. each.	250	500	None	500	[s] Not merely wall-shades and chandeliers, but tumblers, wine-glasses, and lanterns sell well at Káthmándú. The people are beginning to use our crockery and glass-ware at their tables. The Tibetans never had any scruples about using our plates, dishes, and glasses. China however supplies them with crockery. For fine glass-ware they cause some demand.
Tazee horses[q]	„ 300 „ 1200 „ „	30,000	60,000	„	60,000	
Indian horse apparatus		1,000	2,000	„	2,000	
Buffaloes	„ 3 „ 12 „ „	41,000	1,23,000	„	1,23,000	
Goats, etc.	„ 6as „ $1\frac{1}{2}$ „ „	11,500	34,500	„	34,500	
Arhur doll	„ „ 1 „ per maund.	500	1,000	„	1,000	
Sugar	„ 5 „ 7 „ „	5,600	11,200	1,000	10,200	
Sugar candy	„ 15 „ 18 „ „	3,000	6,000	None	6,000	
English looking-glasses[r] . . .	„ 1 a. „ 15 „ each.	3,000	6,000	400	5,600	
English glass-ware[s]		5,000	10,000	2,000	8,000	
	Grand Total Rs. . . .	16,11,000	30,60,500	3,56,900	27,03,600	

GENERAL OBSERVATION.

Of the seven headings under which the information continued in this paper is arranged, the first and last speak for themselves, the others may be helped by illustration. The purpose of the second heading is, chiefly, to afford a useful hint as to the usual *quality* of the goods required for this market. A comparison of headings 3 and 4 will give the average profits upon the trade realized at Káthmándú, at this time, after allowance has been made for the difference of Kuldar and Népâlese rupees $\frac{100 \text{ to } 135}{(13\frac{1}{2} \text{ to } 16)}$ and for the expenses of duties and carriage on account of both, which latter may be seen in No. I., *so far as the Népâlese Territories are concerned;* and for the rest of the way, through our own territories, the means of accurate information are in everybody's hands. I may as well, however, observe, that according to the statements of Népâlese merchants of credit, the costs on account of Custom House charges, in our territories,*

* By Treaty they ought to be $2\frac{1}{2}$ only.

reach 7 per cent. of the value of the goods, half of the sum being paid at Calcutta, and the other half to various of the Customs on the way up. Every Custom House examiner on the river, they affirm, must be propitiated, and no doubt there is some truth in this assertion. The more's the pity! These two headings (3 and 4) may likewise, by showing the *proportionate* amounts of the several sorts of goods required, and by indicating, though less accurately of course, the total amounts of annual transactions, serve to regulate commercial adventure in the first instance, and afterwards to point out to the merchant and to the government, by collation with the known wants and means of their regions, the enlargement of which this commerce is probably capable. And by a reference to the same circumstances, in conjunction with the matter of headings 5 and 6, which show how much of the imports is consumed in Népál, and how much transported to the North, the intelligent merchant may gather further hints for the guidance of his conduct both present and prospective, and Government further indications relative to the probable future importance of the commerce in question. Where there is no such thing as statistical records, or, at least, none accessible to the stranger, we must be contented with the best opinions and estimates within our reach; and those which I have given have been gathered with care from native merchants of high respectability. So much for the form of the paper. In regard to its substance, I know not that much can be added to the marginal remarks. It appears (columns 3 and 4), that whilst traffic is languishing all over the world, almost for want of compensating profits, the merchants of Népál are procuring from 50 to 100 per cent. advance upon Calcutta prices at Káthmándú upon their speculations in European woollens, chintzes, hardware and glassware, as well as upon those in Indian cotton stuffs of all sorts, spicery, drugs, dyes, and pelagic produce, such as pearls and coral, and, in a word, upon the great mass of their commercial transactions. This fact, when taken in connection with the rapidity of the returns upon these speculations, is demonstrative of itself, that the trade in question is capable of an easy, immediate, and considerable extension. It must be remembered, however, that the 100 per cent. of profit in question is not actually such, but only 200 Népálese for 100 Sicca rupees, which proportion of 2 to 1, when reduced to one denomination of coin, is only * 27 to 16. It must be further remembered that this profit, or rather advance on Calcutta prices, is exclusive of the costs in duties and carriage. When, however, due allowance has been made on both these scores, there will still remain a net or real profit of 30 per cent., which is quite sufficient to support the inference based on it. There is one general remark in reference to the trade of Bhote that I cannot forbear making. It is this, that whereas, at present, of the whole goods imported from the plains, the greater portion is consumed in Népál, and a small portion only transported to the North, the reverse of this ought naturally to be the case with reference to the relative means and wants of the two countries. But Népál has already sensibly felt the benign influence of a pacific intercourse with us; Tibet has yet to receive it. Its indirect extension beyond the snows, however, though slower, must be equally certain with its prevalence on this side of them. Népál being within the reach of those whom it concerns to note *particulars* relative to this commerce, I shall pass on to such as belong to the Tibetan branch of it, and observe that the great staples exported through Népál to that country have heretofore been English broad cloth, pearls, and coral. Of the two latter, there is said to be just now a superfluity in Tibet, but the former is in good demand. The best *qualities* are those from 8 to 15 rupees per yard; the best *colours* liver, imperial purple, scarlet, yellow, lively green, and clean brown. Three or four times the quantity now sent there might at once be profitably transmitted by one who knew well the markets of Calcutta and Káthmándú, and had an establishment at the latter place, and such an one might also realise a large profit upon the transmission of select hardware, glassware, and peltry.

Opium and indigo are now in great demand in Tibet, and there is good reason to believe that a large quantity of both articles might be annually sent there with great advantage to the transmitter. Not merely the Tibetans, but the neighbouring Mongolians and Chinese, would eagerly purchase opium if they could get it at Lhása or Digarchee. How far the Company's regulations interfere with the export of this drug in this direction, I cannot say; but those whom it concerns can easily ascertain. The present Calcutta prices are four times as great as those at which the Népálese merchants now purchase, the small quantity they can procure clandestinely, and which is all that is transmitted to Tibet.

* Popular standard—Calcutta assay gives the value of Siccas to Moháris thus:—100 of former to $135\frac{3}{8}$ of latter.

TRADE OF NÉPÁL. No. II.—EXPORTS.

Export (of Tibetan and Népálese Goods) to the Plains in 1830–31).

Articles.	Prices at Káthmándú in Népálese Rupees.	Estimated amount of Imports from the North in Népálese Rupees.	Estimated total amount of Exports of Tibetan, Kachar, and Népálese Goods, in Népálese Rupees.	Estimated amount of Goods consumed in Népál in Népálese Rupees.	Produce of what place.	Remarks.
Gold—						
1st sort or Kundun	24 Rupees per tolah	20,000	20,000	None	Tibet	Bringing Gold from Bhot to Népál answers well as a commercial speculation, chiefly owing to the intervention of the adulterated coin of Bhot. Transporting to the plains is not profitable, but Silver and Indian Rupees being prohibited, the merchants of Népál, who have now more to buy than to sell below, carry down Gold to meet the difference, since they can dispose of it without loss, and look to gain from the returns. The Gold brought from Bhot is usually very impure. The Népálese refine it and gain by the craft. Much is used for ornaments in Népál, little for coin.
2d sort	22 do. do.	90,000	80,000	10,000		
3d sort	20 do. do.	60,000	40,000	20,000		
4th sort or Bukkee	18 do. do.	45,000	30,000	15,000		
5th sort (white or bad)	From 13 to 17 Rs. per tolah	15,000	5,000	10,000		
Silver	Paid by Government Monopoly 15 and $\frac{1}{100}$ per old Kuldar weight	10,000	None	10,000	Ditto	The chief coinage of Népál is of Silver, and much less than is required comes from the North, hence its transport to the Plains is prohibited, and all that is brought here is necessarily made over to the Government, which, with a short-sighted policy, perpetuating the short supply,

						pays the importer less than a just price. The Silver is weighed against old Kuldars, and paid for in Népál Rupees at 113 per 100 Kuldars. The Bazar rate is 120, the Calcutta assay $135\frac{3}{4}$ nearly.
Musk Pods—						
1st sort or Kagazee	From 64 to 70 Rs. per seer of 32 Sa. Wt.	75,000	35,000	5,000	Ditto	Bhot or Tibet produces the finest Musk, Rhubarb, and Borax in the world. Borax is, I think, an exclusive product of Tibet. A deal of Rhubarb is exported from Bhot for Europe *viâ* Siberia and Russia, and also *viâ* China and the Ocean. It would answer well to get it *viâ* Népál, but Europeans have never sought it hence, and the natives know not its use. The European demand for Musk is very slack now—has it lost its repute as a medicine?
2d ditto or Ganouta	60 Rs. per seer.		8,000			
3d ditto or Kacharee	From 45 to 50 Rs. per seer.		12,000			
4th ditto or Nekhana, or $2\frac{1}{2}$ Musk free of the Pod	Do. $4\frac{1}{2}$ to $5\frac{1}{2}$ Rs. per tolah.		15,000			
Revand Chinee	8 Rupees per seer.	Caret	Caret.	Caret	Ditto and Sífán	
Borax	From 4 to 5 as. per seer.	16,000	13,000	3,000	Tibet and Digarchee	
Hurtal (Mejensoor)	Do. 12 as. to 1 Re. do.	13,000	12,000	1,000	Ditto 1 Re. per Shoenee	
Charas	Do. 1 to 2 Rs. do.	6,500	3,500	3,000	Népálese Kachar.	
Wax	Do. 12 as. to 1 Re. do.	13,000	7,000	6,000	Ditto	
Honey	5 annas per seer.	1,400	400	1,000	Ditto	
Paper	Do. 5 to 6 as. do.	8,500	5,000	3,500	Ditto	The Paper of Népál is an admirable article, which ought to be substituted for the flimsy paper of the Plains, for the records of every Office in Bengal.
Zahr Mohara	Do. $1\frac{1}{2}$ to 4 Rs. do.	1,200	1,000	200	Kathai or Northern China proper	In the Pharmacy of India, Mahomedan as well as Hindoo, the drugs of the Kachar and of Tibet enjoy the highest repute, many selling far above their weight in gold, and in great quantities. European medical men seem almost utterly ignorant of them, though they are to be found in most of the druggists' shops of the great cities of India.
Silajeet, white and black	Do. 5 to 8 as. do.	4,000	2,500	1,500	Kachar	

No. II.—(*Continued.*)

Articles.	Prices at Káthmándú in Népálese Rupees.	Estimated amount of Imports from the North in Népálese Rupees.	Estimated total amount of Exports of Tibetan, Kachar, and Népálese Goods, in Népálese Rupees.	Estimated amount of Goods consumed in Népál in Népálese Rupees.	Produce of what place.	Remarks.
Bikhma	From 6 to 8 as. per seer	2,500	2,000	500	Kachar	
Nirbisi	1 Re. do. do.	800	500	300	Kathai	
Padmchal	From 3 to 4 as. do. do.	1,000	700	300	Kachar	
Jatamangsi	Do. 8 to 12 do. do. do.	10,000	8,000	2,000	Ditto and Kooti	
Kutki	3 annas do. do.	2,500	1,800	700	Helma, Listy, Kachar, and Lekh	
Mehr Gya	From 130 to 150 Rs. p. tolah	5,000	5,000	None	China	Comes through Lhása, and sells chiefly in Lucknow, Nagpore, and Hydrabad. It is, I believe, the famous Ginsing. Its price seems absurdly high. Be it what it will, it comes through Lhása.
Tanzoo	....	Caret	Caret	Caret	Kathai	
Chiraitah	6 as. per seer	2,500	2,000	500	Kachar	
Mamira	From 2 to 4 as. per tolah	700	300	400	Kathai	
Tea	Do. 2 to 3 Rs. per seer	4,500	2,500	2,000	Tibet	The Tea is chiefly produced in Setchuen, and is commonly compacted into cakes, like Chocolate, or Cocoa. The Bhoteas universally consume the substance of the Tea triturated and made up for convenience of carriage as above noted, and not like us and the Chinese; the mere infusion. Tea, however, such as we purchase in China, may be had at Lhása in any quantity that is called for, and some of it is annually brought to Népál.
Walnut	1 Re. per hundred	1,400	400	1,000	Ditto and Kachar	
Halva Bedud	From 1½ to 2 Rs. per seer	1,600	600	1,000	Tibet	
Dakh Raisins	2 Rs. do. do.	1,000	None	1,000	Ditto	
Bhotea Bair	Do. 1 to ½ Rs. do. do.	400	200	200	Ditto	
Salt (Rock)	3 annas do. do.	1,50,000	None	1,50,000	Ditto	All Népál proper and its adjacencies is supplied with Salt from Tibet. It is Rock Salt, and very good. The Government levies no duty on it. As there cannot be less than 100,000 families in the limits contemplated, nor the families con-
Cow-tail (white)	From 1½ to 2 Rs. do do.	27,000	26,000	1,000	Ditto	
Ditto (black)	Do. 1 to 1¼ do. do. do.	25,000	24,000	1,000	Ditto	
Shawls	Do. 60 to 100 Rs. pr. pair.	2,000	None	2,000	Ladak	

						sume, on an average, less than half a pice worth per day, I should take the import to reach $2\frac{1}{4}$ lakhs; my informants, however, say not.
Toos, $\frac{1}{2}$ by 30 yards.	Do. 24 to 45 Rs. pr. piece.	5,000	2,500	2,500	Pechin	Toos and Malidah are loosely woven, flannel-like, very narrow cloths, made of excellent wool—wretched substitutes for good Broad Cloth, but the best within the ordinary reach of the Bhoteas, who, though they all wear woollens throughout the year, have not a single good manufacture of their own, nor can China furnish them with anything better than Toos and Malidah. Half the nation would be clothed by Britain if such commercial relations as time and prudent activity may establish prevailed. The Toos and Malidah sell at a good profit in the Plains, Sircars, and upper servants liking them for chudders. Namdah and the rest to Namboo are blankets of good wool, but very coarse and heavy, save the Chourput, which is light and finer. It is made of the inner coat of the chowree ox or yâk: the others of sheeps' wool. The Kachar blankets sell well in the plains to the poor, being warmer and more durable than Indian blankets. Some of the former are felted, not woven.
Ditto ditto..........	Do. 15 to 21 do. do....	3,500	1,500	2,000	Siling	
Malidah, $\frac{1}{2}$ by 12 yards..............	Do. 14 to 24 do. do....	4,000	2,500	1,500	Lhása	
Namdah, $1\frac{1}{2}$ by 2 yards..............	Do. 8 to 12 do. do....	2,000	1,500	500	Ditto	
Pankhi, 1 cubit by 6 to 9 yards	Do. $1\frac{1}{2}$ to 3 do. do....	2,500	1,000	1,500	Kachar	
Bhot Blanket, 1 by 15 cubits	Do. 2 to 3 do. do....	5,000	2,000	3,000	Kooti, Keroong, and Kachar	
Rarhi Blanket	Do. 2 to $4\frac{1}{2}$ do. do....	7,000	3,000	4,000	Ditto ditto	
Chourput	Do. 1 to $1\frac{1}{2}$ do. do....	800	200	600	Ditto ditto	
Namboo, 1 span by 22 yards..........	Do. $4\frac{1}{2}$ to 16 do. do....	8,000	None	8,000	Ditto ditto	
Tur or Sambur	Do. 60 to 80 do. do....	4,500	1,500	3,000	Lhása	
Cakum (a kind of fur)...............	Do. 80 to 150 Rs. each.... Bakkoo or long coat ready made........................	2,800	1,300	1,500	Ditto	
Choowa-Khal.........	From 40 to 80 Rs. each..	2,500	1,200	1,300	Ditto	
Garbsoot (Peltry)....	Do. 12 to 23 do. do.	700	200	500	Ditto	
Cochin or Satin.......	Do. $2\frac{1}{8}$ to 6 do. pr. yard	15,000	9,000	6,000	China	
Bhooleea and Sansu (Silks), $1\frac{1}{2}$ by $2\frac{1}{2}$ yards..............	Do. $4\frac{1}{2}$ to 6 do. pr. piece	2,000	None	2,000	Ditto	Bhoolea and Sansu are much used by the Népálese women, but the Satins, Silks, and Velvets of China are in Népál generally giving way before our Broad Cloths and Velvets.
Velvet of Pechin.....	Do. 8 to 16 do. do.	3,000	do.	3,000	Ditto	
Kummerbund (Silk), $1\frac{1}{2}$ by 9 cubits...	Do. 9 to 13 do. do.	500	do.	500	Lhása	
China Sewing Silk...	8 annas per tolah....	1,200	do.	1,200	China	Comes through Lhása.
Sphatic or Crystal...	From 1-5 to 2 Rs. pr. seer	2,500	1,500	1,000	Helmoo and Kachar	

No. II.—(*Continued.*)

Articles.	Prices at Káthmándu in Népálese Rupees.	Estimated amount of Imports from the North in Népálese Rupees.	Estimated total amount of exports of Tibetan, Kachar, and Népálese Goods, in Népálese Rupees.	Estimated amount of Goods consumed in Népál in Népálese Rupees.	Produce of what place.	Remarks.
Yû or Oriental jade.	....	700	200	500	Lhása	Vases, Sword-handles, Snuff Boxes, &c., come ready made of Yû. The unwrought stone is not brought. It is called oriental jade by its analyser.
Sang-yesham	....	1,200	200	1,000	Ditto	
Hawk (female)	From 25 to 40 Rs. each	12,000	10,000	2,000	Lhása and Ke-roong	The bird in question, known by the name of Baaz below, and by the Tibetans called Sâyi, is a large tractable species of the short-winged kind of true Hawk. It is by far the best known to Indian Falconry, and is sent all over the Deccan, as well as Hindostan. It is the Gos-hawk of Europeans. Female called Bâz; male, Júrra.
Hawk (male).........	Do. 15 to 26 do. do.........	6,000	5,000	1,000	Ditto ditto	
Tangan (Pony).......	Do. 40 to 120 do. do.	10,000	7,000	3,000	Ladakh, Mosting, Pûmi, and Jumla	
Sheep and Goat, &c. (Changra)	Sheep 1½ to 3 Rs. do.... Changra 1 to 1½ do.... Khasee 2½ to 5 do....	8,500	None	8,500	Népálese Kachar	These mountains are not favourable to Sheep or Goats, and as the people consume a deal of their flesh, large numbers are annually brought to Népál proper, both from the juxta Himálayan tracts where they flourish and from the Plains. None are exported.
Copper Pots, &c......	From 1 Re 10 as. to Rs. p. sr.	....	4,5000	....	Népál	
Copper Pice	20 gundas per Rupee	....	250,000	....	Ditto	

Mahindramalee Rupees	13 as. 4 p. Sicca each	....	2,20,000	....	Ditto	Mahindramalee Rupees are the Rupees of Népál. They are exported only, because the exportation of Siccas or of Silver is prohibited. At Káthmándú the exchange in the bazar against Siccas is fixed by authority at 120 to 100. It ought to be by Calcutta assay 135 to 100, and in the Tarai the Népál merchants are glad to get 100 Siccas for 130 Mahindramalees. Népál is full of fine copper, and supplies copper currency to the whole tract of the plains between the Ganges and the Hills, Monghyr and Pillibheet. Iron too abounds in Népál, and much *coarse* Hardware is exported to the same tract of the plains.
Iron Phowrahs, Koodalies, &c.	From 4 to 5 seers pr. Rupee	....	5,000	....	Ditto	
Khookree or Népálese short Sword	Do. 8 as. to 10 Rs. each	....	1,000	....	Ditto	
Brass Pots, &c.	1 Re. 10 as. per seer	....	1,000	....	Ditto	
Elephants	From 100 to 700 Rupees per elephant	....	67,200	....	Népálese Tarai in common parlance, but really Forest and lesser Hills	
Ivory	Do. 1 Re. 4 as. to 2 Rs. p. seer	....	400	....		
Bootwal, Horses	Do. 40 to 150 Rs. each	....	45,000	....		
Wax	Do. 12 as. to 1 do. pr. seer.	....	7,000	....		There are Lead, Sulphur, and Zinc mines in Népál, but no skill to work them profitably. A deal of each is imported from the plains, and also of Tin, with which last, and with the Zinc got from us, the Népálese mix their own Copper, and make a great variety of mixed metals in a superior style.
Honey	Do. 5 to 8 as. do.	....	3,000	....		
Ganja	Caret	....	3,500	....		
Bháng	From 5 to 7 as. per maund	....	2,000	....	Népálese Tarai in common parlance, but really Forest and lesser Hills	
Harra, Bahera, Aonla	Do. 8 to 12 as. do.	....	3,000	....		
Small Harra	Do. 3 to 4½ Rs. do.	....	5,000	....		From the article Copper Pots to the end, the whole are the produce of Népál, except the silver of the Népál Rupees. Of course, therefore, the column indicating the quantities imported from Tibet is blank, and as the entire almost of the articles produced in the Népálese Tarai is exported to the plains, and no part of them consumed in Népál, the column indicating local consumption is likewise a blank. It was not thought worth while to alter the form of the paper for the sake of these articles, which have no interest for the great foreign merchant.
Bootwal Chasas	Do. 3 to 3½ do. do.	....	4,500	....		
Large Cardamoms	Do. 5 to 6½ do. do.	....	6,000	....	Hills	
Charaitah	Caret	....	2,000	....		
Resin of the Saul-tree	4 Rupees per maund	....	12,000	....	Népál Tarai	
Taj	From 1 to 2 Rs. do.	....	2,500	....		
Tezpaut	Do. do.	....	4,000	....		
Pipala	From 5 to 6 Rs. do.	....	3,000	....		
Kuth (Terrajapon)	Caret	....	4,500	....		
Moosli, Chiriakand, &c.	1 Rupee per bullock-load	....	1,200	....		
Stag Horn	From 1 to 2 as. per horn	....	1,200	....		

No. II.—(*Continued.*)

Articles.	Prices at Káthmándú in Népálese Rupees.	Estimated amount of Imports from the North in Népálese Rupees.	Estimated total amount of Exports of Tibetan, Kachar and Népálese Goods, in Népálese Rupees.	Estimated amount of Goods consumed in Népál in Népálese Rupees.	Produce of what place.	Remarks.
Deer Skins	Caret	......	1,000	......	Népál Tarai.	The Népálese Saul Forest is an inexhaustible mine of timber. Saul and Sissoo are the most valuable kinds of produce. The open low lands of Népál have been wonderfully resuscitated by the continued peace and alliance with our Government, and the energy of the Népálese administration since 1816. No regular troops are maintained there by the Government, and the Civil establishment is on a very moderate scale, nor do any of the mountaineers holding lands reside there. The whole net produce of the land, consequently, is exported to Patna, &c., and chiefly on Government account. It is paid for in money therefore, and these low lands not only supply the Government of Népál with bullion for its currency, but enable it to furnish itself with the luxuries of the plains, and to maintain the balance of a trade which, so far as the hill produce is concerned, is always apt to be against Népál.
Timber	Caret	......	3,00,000	......		
Grain	Caret	......	6,00,000	......		
Ox and Buffalo Skins	Caret	......	1,00,000			
Do. Horns	Caret	......		......		
	Népál Rupees	7,27,400	19,77,800	3,12,700		
	Sicca Rupees	6,06,16,610	10,64,833	5,42,60,583	54	

Népal Residency,
The 1st December 1831.

(Signed) B. H. HODGSON,
Officiating Resident.

No. II.—(*Concluded.*)

GENERAL OBSERVATION.

In regard to the trade of Népál, or rather to the trade through Népál, the principal objects of interest for the foreign merchant of every country, and especially for the European merchant, must be the production and consumption of the great *Trans-Himálayan* countries. On this account I have arranged the information I possess relative to the exports under heads calculated to show where the articles exported through Népál are produced; what quantity of them is, at present, consumed in Népál; and what quantity is transmitted to the plains. But I have adhered to the Káthmándú prices, because I know not the prices of Tibet, and because, for the immediate benefit of those for whom I write, the Káthmándú prices will suffice. It appears by Part I., that the prime cost value of Imports from the South per annum is 16,11,000 Sicca Rupees; by Part II., that the value of the Annual Exports to the South at Káthmándú is 12,77,800 of Népálese Rupees, which last sum is equivalent to 10,64,833-5-4 of Sicca Rupees. And as this disparity, if not supposed to be counterpoised, as I pretend not to say it is, by a "favourable balance" upon the Northern branch of the Trade, may excite sceptical remark, I must observe first, that the Imports of 1830–31 from the South were raised one and a half lakhs above the ordinary standard, with reference to the Trade in general, by an extraordinary amount of purchases on the part of the Government. With reference to these unusual purchases, I would suggest the following reductions to those who wish to form as exact an idea as possible of the present average imports:—

Velvets	5,000
Kimkhabs	12,000
Pearls	15,000
Coral	25,000
Diamonds and Rubies	25,000
Saltpetre	20,000
Horses	20,000
	1,22,000 Sicca Rupees.

Let us say, then, that the Imports are reduced from 16,11,000 to fifteen lakhs of Siccas, and that the Exports remain at 10,64,833 Siccas as before, the difference is about four and a half lakhs; and upon this I confess I have no accurate information to supply. Statistical documents alone could supply, with precise accuracy, the amount of Imports and Exports; none such have I the means of referring to. Let, then, the *rest* of the information contained in both parts of No. II. be taken as trustworthy, and let no further reliance be placed upon the *totality of the transactions* than its conjectural character may seem to warrant. It should be observed, however, that though the total amounts of Imports and Exports may err materially, the *proportional* amount of the several articles of them may still be nearly accurate, and such I verily believe they are; and being such, it is imagined they must be of material practical utility. For the rest, the marginal notes contain all the detailed information I have to supply upon the Exports of Népál.

(Signed) B. H. HODGSON,
Officiating Resident.

NÉPÁL RESIDENCY,
The 1*st December* 1831.

(True Copies)
B. H. HODGSON.

INDEX.